Accredited Standards Committee C2-2002

National Electrical Safety Code®

Secretariat

Institute of Electrical and Electronics Engineers, Inc.

Approved 5 February 2001

Institute of Electrical and Electronics Engineers, Inc.

Approved 14 June 2001

American National Standards Institute

2002 Edition

Abstract: This standard covers basic provisions for safeguarding of persons from hazards arising from the installation, operation, or maintenance of 1) conductors and equipment in electric supply stations, and 2) overhead and underground electric supply and communication lines. It also includes work rules for the construction, maintenance, and operation of electric supply and communication lines and equipment.

The standard is applicable to the systems and equipment operated by utilities, or similar systems and equipment, of an industrial establishment or complex under the control of qualified persons.

This standard consists of the introduction, definitions, grounding rules, list of referenced and bibliographic documents, and Parts 1, 2, 3, and 4 of the 2002 Edition of the National Electrical Safety Code.

Keywords: communications industry safety; construction of communication lines; construction of electric supply lines; electrical safety; electric supply stations; electric utility stations; high-voltage safety; operation of communications systems; operation of electric supply systems; power station equipment; power station safety; public utility safety; safety work rules; underground communication line safety; underground electric line safety

The Institute of Electrical and Electronics Engineers, Inc.
3 Park Avenue, New York, NY 10016-5997, USA

ISBN 0-7381-2778-7

1 August 2001 *SH94911*

Recognized as an
American National Standard

An American National Standard implies a consensus of those substantially concerned with its scope and provisions. An American National Standard is intended as a guide to aid the manufacturer, the consumer, and the general public. The existence of an American National Standard does not in any respect preclude anyone, whether he has approved the standard or not, from manufacturing, marketing, purchasing, or using products, processes, or procedures not conforming to the standard. American National Standards are subject to periodic review and users are cautioned to obtain the latest editions.

Foreword

(This foreword is not a part of Accredited Standards Committee C2-2002, National Electrical Safety Code®.)

This publication consists of the parts of the National Electrical Safety Code® (NESC®) currently in effect. The former practice of designating parts by editions has not been practical for some time. In the 1977 Edition, Parts 1 and 4 were 6th Editions; Part 2 was a 7th Edition; Part 3 was a revision of the 6th Edition; Part 2, Section 29, did not cover the same subject matter as the 5th Edition; and Part 3 was withdrawn in 1970. In the 1987 Edition, revisions were made in all parts, and revisions to all parts have been made in subsequent editions. It is therefore recommended that reference to the NESC be made solely by the year of the published volume and desired part number. Separate copies of the individual parts are not available.

Work on the NESC started in 1913 at the National Bureau of Standards (NBS), resulting in the publication of NBS Circular 49. The last complete edition of the Code (the 5th Edition, NBS Handbook H30) was issued in 1948, although separate portions had been available at various times starting in 1938. Part 2—Definitions, and the Grounding Rules, 6th Edition, was issued as NBS Handbook H81, ANSI C2.2-1960, in November 1961, but work on other parts was not actively in process again until 1970.

In 1970 the C2 Committee decided to delete the Rules for the Installation and Maintenance of Electric Utilization Equipment (Part 3 of the 5th Edition), now largely covered by the National Electrical Code (ANSI/ NFPA 70), and the Rules for Radio Installations (Part 5 of the 5th Edition) from future editions. The Discussion of the NESC, issued as NBS Handbook H4 (1928 Edition) for the 4th Edition of the NESC and as NBS Handbook H39 for Part 2 of the Grounding Rules of the 5th Edition, was not published for the 6th Edition.

The 1981 Edition included major changes in Parts 1, 2, and 3, minor changes in Part 4, and the incorporation of the rules common to all parts into Section 1. The 1984 Edition was revised to update all references and to list those references in a new Section 3. Rounded metric values, for information only, were added. Gender-related terminology was deleted. Section 1—Introduction, Section 2—Definitions, Section 3—References, and Section 9— Grounding Methods, were made applicable to each of the Parts 1, 2, 3, and 4.

The 1987 Edition was revised extensively. Definitions were changed or added. Requirements affecting grounding methods, electric supply stations, overhead line clearances and loading, underground lines, and work rules were revised.

The 1990 Edition included several major changes. General rules were revised. A significant change to the method for specifying overhead line clearances was made and the rationale added as Appendix A. Requirements for clearances of overhead lines from grain bins and an alternate method for determining the strength requirements for wood structures was added. Rules covering grounding methods, electric supply stations, underground lines, and work rules were changed.

In the 1993 Edition, changes were made in the rules applicable to emergency and temporary installations. In Section 9 and Parts 1, 2, and 3, rules were extended or clarified to include HVDC systems. The requirements for random separation of direct-buried supply and communication systems were modified for consistency and clarity, as was the rule in Part 4 on tagging electric supply circuits.

In the 1997 edition, the most notable general change that took place is that numerical values in the metric (SI) system are shown in the preferred position, with customary inch-foot-pound values (inside parentheses) following. A bibliography, Appendix B, which consists of a list of resources identified in notes or recommendations, was added. Changes were made to rules affecting grounding, electric supply stations, and overhead lines, particularly with regard to clearance rules applicable to emergency and temporary installations. Strength requirements contained in Sections 24, 25, and 26 were revised completely. Underground line requirements for random separation for underground lines of direct-buried cables were modified. The requirement for cable identification marking by means of sequentially placed logos was introduced. Work rules added a requirement that warning signs and tags comply with applicable ANSI standards, tagging requirements were clarified with regard to SCADA, and extensive requirements for fall protection were added.

In the 2002 Edition, several changes were made that affected all or several parts of the Code. Particularly, this edition clarifies interfaces between the NEC and NESC with regard to Code jurisdiction in the area of street lights and area lights. Also included is clarification for situations between utility workers and their authorized contractors and installations on industrial complexes.

In Section 2, definitions for communication lines, de-energized, fiber-optic cable—communications, fiber-optic cable—supply, readily climbable supporting structures, and vault were either modified or added.

No changes were made in the 2002 Edition regarding use of earth or sea as a continuous conductor in HVDC systems. Subcommittee 1 reviewed a change proposal with regard to using earth or sea as a continuous conductor and did not see an immediate need to revise the NESC for this type of operation for HVDC systems in the United States. Although several systems in Europe and one in the Philippines currently use sea as a conductor, there are concerns that should be addressed before revising the Code. The subcommittee recognized the comments submitted by members of the IEEE Power Engineering Society Substations Committee and Transmission and Distribution Committee, the work accomplished within joint working groups in IEEE and NACE International, and further endorsements received from recognized HVDC experts. If there is desire by some to design and install a monopolar system, it is permissible under Rule 013A2 to do so as an experimental system. The subcommittee also recommended that any such system be limited to using sea, not land, as a sole conductor for any part of the circuit.

In Section 9—Grounding Methods, a new rule was added to cover grounding requirements for a communication entity that wishes to use its own separate ground electrode. This requires that the communication ground be bonded with the grounding electrode conductor of the supply system to assure equal potential between the two systems.

In Part 1—Electric Supply Stations, the rules were clarified to be applicable to both ac and dc circuits. DC station clearances were added to the appropriate table. Requirements were revised to allow exceptions from prohibiting storage of equipment within an electrical supply station. The exception allows storage of material essential for maintenance and temporary storage of material related to work in progress in the station or a nearby transmission line. Restrictions are required to maintain safe clearances and working conditions. A Tentative Interim Amendment (TIA) was approved to remove the implication that grounding may be omitted when working on conductors normally operating at 25 kV or less where a visible open switch exists. The statement is not applicable to Part 1, and it may be a leftover practice that is no longer valid.

In Part 2—Overhead Lines, additional wording was added to address the situation of supply workers working in close proximity to communication antennas. New wording was also added to cover clearance of line conductors from communication antennas, and to clarify clearances of conductors from grain bins, boat launching areas, service drops, support arms, and lighting luminaires. Clarification was added to relationship of approach distances for non-utility workers from overhead lines and associated equipment. A major revision to Sections 25 and 26 on Strength and Loadings has begun and will span at least 3 Code cycles by the time it is completed. This is due to more sophisticated structural calculations that will be required, e.g., Load and Resistance Factor Design (LRFD), updated and improved input data from new wind and ice maps, and new pole materials that will supplement wood. Some change proposals were incorporated into the 2002 edition; others will be deferred until a later edition when the industry has become more familiar with the nature of these changes. For this edition, proposals were accepted to 1) treat wood, steel, and concrete poles with consistent Overhead Factors, and 2) include a new Extreme Wind Map that uses a new 3-s gust criteria, replacing the existing fastest mile wind. In addition, expanded calculations are required due to the use of gust response factor, importance factor, and velocity pressure coefficient.

In Part 3—Underground Lines, the requirements for random separation changed to prohibit both communication lines and power from being any closer than 12 in from fuel or steam lines. Other new rules address location of pad-mounted equipment and pedestals and cable protection by means of climbing devices required to be installed within the manhole. Also included are revised rules on burial depth for streetlight cables and security of above ground equipment.

In Part 4—Work Rules, the maximum use voltage rating of rubber protective equipment was added as a requirement.

Substantive changes in the 2002 edition are identified by a bar in the left-hand margin. In several cases, rules have been relocated without substantive changes in the wording. In these cases, only the rule numbers have been indicated as having been changed.

The Institute of Electrical and Electronics Engineers, Inc. was designated as the administrative secretariat for C2 in January 1973, assuming the functions formerly performed by the National Bureau of Standards.

Comments on the rules and suggestions for their improvement are invited, especially from those who have experience in their practical application. In future editions every effort will be made to improve the rules, both in the adequacy of coverage and in the clarification of requirements. Comments should be addressed to:

Secretary
National Electrical Safety Code Committee
Institute of Electrical and Electronics Engineers, Inc.
445 Hoes Lane
P.O. Box 1331
Piscataway, NJ 08855-1331

A representative Committee on Interpretations has been established to prepare replies to requests for interpretation of the rules contained in the Code. Requests for interpretation should state the rule in question, as well as the conditions under which it is being applied. Interpretations are intended to clarify the intent of specific rules and are not intended to supply consulting information on the application of the Code. Requests for interpretation should be sent to the address above.

If the request is suitable for processing, it will be sent to the Interpretations Committee. After consideration by the committee, which may involve many exchanges of correspondence, the inquirer will be notified of its decision. Decisions are published regularly and may be ordered or accessed online at no cost at **http://standards.ieee.org/nesc**.

The NESC as written is a voluntary standard. However, some editions and some parts of the Code have been adopted, with and without changes, by some state and local jurisdictional authorities. To determine the legal status of the National Electrical Safety Code in any particular state or locality within a state, the authority having jurisdiction should be contacted.

The revision cycle for the 2007 edition of the NESC will be fully electronic. Change proposals and comments will no longer be accepted in hard copy, but will be submitted to the NESC Secretary online via the Internet. For information on how this electronic revision process will take place and for updates and complete information on the NESC, please visit the National Electrical Safety Code Zone on the IEEE Standards Web site at **http://standards.ieee.org/nesc**.

Standards Committee Membership

At the time this Code was approved, Accredited Standards Committee C2 had the following membership:

O. Chuck Amrhyn, *Chair* **Frank A. Denbrock,** *Vice Chair*
Susan L. Vogel, *Secretary*

Organization Represented	*Name*
Alliance for Telephone Industry Solutions	Lawrence M. Slavin
American Insurance Services Group, Inc.	Vacant
American Public Power Association	Michael J. Hyland
	Ron Lunt *(Alt.)*
American Public Transit Association	George S. Pristach
Association of American Railroads	Vacant
Association of Edison Illuminating Companies	John J. Schlee, Jr.
Bonneville Power Administration, US Department of Energy	Jerry L. Reding
Edison Electric Institute	Stephen A. Olinick
	John W. Troglia *(Alt.)*
	David G. Komassa *(Alt.)*
Electronic Industries Association	Percy E. Pool
Institute of Electrical and Electronics Engineers, Inc.	Frank A. Denbrock
	Jerome G. Hanson *(Alt.)*
	Vernon R. Lawson *(Alt.)*
International Association of Government Labor Officials	Bernard O'Neill
International Brotherhood of Electrical Workers	James R. Tomaseski
International Municipal Signal Association	Warren Farrell
National Association of Regulatory Utility Commissioners	B. J. Washburn
National Cable Television Association	F. N. Wilkenloh
	Rex Bullinger *(Alt.)*
National Electrical Contractors Association	O. L. Davis
	Brooke H. Stauffer *(Alt.)*
National Electrical Manufacturers Association	Chris K. Durland
	Lawrence F. Miller *(Alt.)*
National Safety Council	P. Schmidt
National Society of Professional Engineers	William F. Fuller
Rural Electrification Administration, US Department of Agriculture	G. J. Bagnall
Tennessee Valley Authority	Clayton L. Clem
	Nancy A. Knowles *(Alt.)*
Western Area Power Administration, US Department of Energy	Oliver W. Perkins
Individual Member	O. Chuck Amrhyn
Individual Member	Allen L. Clapp
Liaison Representative to Canadian Electrical Code	Susan L. Vogel
Canadian Standards Association Liaison Representative	David Singleton

Subcommittee 1
Purpose, Scope, Application, Definitions, and References
(Sections 1, 2, and 3)

Allen L. Clapp, *Chair* **Charles C. Bleakley,** *Secretary*

Name	Representing
O. Chuck Amrhyn	(Main)
Charles C. Bleakley	(SC7)
David E. Bouchard	*(Alt.)*
Frank A. Denbrock	(SC5)
Allen L. Clapp	*(Alt.)*
D. J. Christofersen	(SC3)
Robert E. Sipler, Jr.	*(Alt.)*
John B. Dagenhart	(SC2)
Donald E. Hooper	(Interpretations)
Ewell T. Robeson	*(Alt.)*
Henry J. Kientz	(SC8)
James R. Tomaseski	*(Alt.)*
David G. Komassa	(SC4)
Eric K. Engdahl	*(Alt.)*

Subcommittee 2
Grounding Methods
(Section 9)

John B. Dagenhart, *Chair* **Ewell T. Robeson,** *Secretary*

Name	Representing
J. Boksiner	ATIS
Harvey L. Bowles	RUS
A. C. Channaiah	SEEX
Terry Page	*(Alt.)*
John B. Dagenhart	IEEE
M. M. Dixon	EEI
L. E. Gaunt	*(Alt.)*
William J. Turner	*(Alt.)*
N. Maxwell	NARUC
Robert Molde	EEI
Michael A. Moore	APPA
Percy E. Pool	EIA
Ewell T. Robeson	EEI
James R. Tomaseski	IBEW
Donald W. Zipse	IEEE

Subcommittee 3
Electric Supply Stations
(Sections 10–19)

D. J. Christofersen, *Chair* **Robert E. Sipler, Jr.,** *Secretary*

Name	Representing
D. J. Christofersen	IEEE
A. L. Comans	AEIC
W. Bruce Dietzman	IEEE
Gary R. Engmann	IEEE
C. C. Hamerla	EEI
S. J. Kollmann	EEI
Brian Winoski	*(Alt.)*
N. Burch Lewis	SEEX
K. C. Labbe	*(Alt.)*
Thomas McNamara	NEMA
Robert E. Sipler, Jr.	NARUC
James R. Tomaseski	IBEW
Philip Young	EEI
George Zaczek	APPA

AAR—Association of American Railroads
AEIC—Association of Edison Illuminating Companies
AISG—American Insurance Services Group, Inc.
AISI—American Iron and Steel Institute
APPA—American Public Power Association
APTA—American Public Transit Association
ATIS—Alliance for Telephone Industry Solutions
AWPA—American Wood Preserves Assocation
BPA—Bonneville Power Admin., US Dept. of Energy
EEI—Edison Electric Institute
EIA—Electronic Industries Association
IAGLO—Int'l. Assoc. of Government Labor Officials
IBEW—International Brotherhood of Electrical Workers
IEEE—Institute of Electrical and Electronics Engineers, Inc.
IMSA—International Municipal Signal Association
NARUC—National Association of Regulatory Utility Commissioners
NCTA—National Cable Television Association
NECA—National Electrical Contractors Association
NEMA—National Electrical Manufacturers Association
NSC—National Safety Council
NSPE—National Society of Professional Engineers
RUS—Rural Utilities Services, US Dept. of Agriculture
SEEX—Southeastern Electric Exchange
TVA—Tennessee Valley Authority
WAPA—Western Area Power Administration, US Dept. of Energy

Subcommittee 4
Overhead Lines—Clearances
(Section 20, 21, 22, and 23)

David G. Komassa, *Chair* | **Eric K. Engdahl,** *Secretary*

O. Chuck Amrhyn	Self	H. N. Johnson, Jr.	SEEX
R. J. Bednarz	Self	D. C. Young	*(Alt.)*
Charles C. Bleakley	AEIC	David G. Komassa	EEI
James. L. Bohlk	RUS	Stephen A. Olinick	EEI
Allen L. Clapp	Self	Robert G. Oswald	IEEE
Charles Crawford	NSPE	Oliver W. Perkins	WAPA
Thomas R. Crowell	EEI	Jerry L. Reding	BPA
Phillip S. Givens	*(Alt.)*	Joseph Renowden	Emeritus
Darrell D. Davidchik	EIA	James Ruehl	NARUC
Eric K. Engdahl	EEI	Mathew C. Schwarz	IEEE
Kevin Drzewiecki	*(Alt.)*	Lawrence M. Slavin	ATIS
Mickey B. Gunter	EEI	R. H. Stevens	Self
Dennis Henry	ATIS	James R. Tomaseski	IBEW
Donald E. Hooper	Self	B. J. Washburn	NARUC
Jan Howard	APPA	F. N. Wilkenloh	NCTA

Subcommittee 5
Overhead Lines—Strength and Loading
(Sections 24, 25, 26, and 27)

Frank A. Denbrock, *Chair* | **Allen L. Clapp,** *Secretary*

Julian Ajello	NARUC	Leon Kempner, Jr.	BPA
Nelson G. Bingel	AWPA	Robert O. Kluge	EEI
Allen L. Clapp	Self	Michael Madore	APPA
Clayton L. Clem	TVA	Robert C. Peters	IEEE
Darrell. D. Davidchik	EIA	T. A. Pinkham	IEEE
Frank A. Denbrock	IEEE	Joseph Rempe	EEI
Nicholas J. DeSantis	EEI	Andrew Schwalm	*(Alt.)*
Thomas Diamantis	*(Alt.)*	E. Wade Shultz	SEEX
Bruce Freimark	EEI	R. Corzine	*(Alt.)*
William. F. Fuller	NSPE	Camille G. Rubeiz	AISI
Douglas Hanson	WAPA	Brian Lacoursiere	*(Alt.)*
Jerome G. Hanson	Self	Richard Aichinger	*(Alt.)*
Edward Harrel	EEI	Lawrence M. Slavin	ATIS
Donald G. Heald	RUS	Richard J. Standford	EEI
Richard W. Hensel	IEEE	Jerry C. Wong	EEI
Walter D. Jones	IEEE		

Subcommittee 7
Underground Lines
(Sections 30–39)

Charles C. Bleakley, *Chair*

David E. Bouchard, *Secretary*
F. William Koch, *Secretary (1987-2000)*

O. Chuck Amrhyn	Self	James D. Mars	Self
Charles C. Bleakley	SEEX	Dennis B. Miller	EEI
Mickey B. Gunter	*(Alt.)*	George S. Pristach	APTA
David E. Bouchard	IEEE	Ewell T. Robeson	IEEE
L. G. Clemons	EEI	Les Shankland	APPA
Donna L. Chaffin	*(Alt.)*	Lawrence M. Slavin	ATIS
R. A. Fernandez	EIA	John C. Spence	EEI
Dennis Henry	ATIS	Monte Szendre	NECA
Trung Hiu	RUS	James R. Tomaseski	IBEW
David H. Kendall	NEMA	F. N. Wilkenloh	NCTA

Subcommittee 8
Work Rules
(Sections 40–44)

Henry J. Kientz, *Chair*

James R. Tomaseski, *Secretary*

George V. Causer	WAPA	J. D. Mitchell	IEEE
J. M. Degen	ATIS	Thomas Verdecchio	*(Alt.)*
J. F. Doering	Self	R. C. Mooney	AIEC
C. W. Grose	Emeritus	Stephen Poholski	NECA
J. A. Dahmer	EEI	Thomas Ryder	EIA
Maury Dunn	*(Alt.)*	Lawrence Schweitzer	IEEE
Robert A. Fass	EEI	Les Shankland	APPA
James Spadafore	*(Alt.)*	R. D. Shores	EIA
Dennis Henry	ATIS	Steven Theis	NUCA
Stephen L. Keys	IEC	James R. Tomaseski	IBEW
James C. Tuggle	*(Alt.)*	G. Tootle	SEEX
Robert W. Baird	*(Alt.)*	D. M. Wallis	OSHA
Henry J. Kientz	Emeritus	J. E. Pipkin	*(Alt.)*

Executive Subcommittee

O. Chuck Amrhyn, *Chair*

Frank A. Denbrock, *Vice Chair*

O. Chuck Amrhyn	Self	Lawrence M. Slavin	ATIS
Frank A. Denbrock	IEEE	B. J. Washburn	NARUC
Michael J. Hyland	APPA	James R. Tomaseski	IBEW
Stephen A. Olinick	EEI		

Noelle D. Humenick
IEEE Standards Project Editor

Contents

SECTION PAGE

Letter Symbols for Units xxxiii

Sec. 1. Introduction to the National Electrical Safety Code 1
- 010. Purpose 1
- 011. Scope 1
- 012. General Rules 1
- 013. Application 1
  - A. New Installations and Extensions 1
  - B. Existing Installations 2
- 014. Waiver 2
  - A. Emergency Installations 2
  - B. Temporary Overhead Installations 2
- 015. Intent 2
- 016. Effective Date 3
- 017. Units of Measure 3

Sec. 2. Definitions of Special Terms 4

Sec. 3. References 14

Sec. 9. Grounding Methods for Electric Supply and Communications Facilities 16
- 90. Purpose 16
- 91. Scope 16
- 92. Point of Connection of Grounding Conductor 16
  - A. Direct Current Systems That Are to Be Grounded 16
    - 1. 750 V and below 16
    - 2. Over 750 V 16
  - B. Alternating Current Systems That Are to Be Grounded 16
    - 1. 750 V and below 16
    - 2. Over 750 V 16
    - 3. Separate Grounding Conductor 17
  - C. Messenger Wires and Guys 17
    - 1. Messenger Wires 17
    - 2. Guys 17
    - 3. Common Grounding of Messengers and Guys on the Same Supporting Structure 17
  - D. Current in Grounding Conductor 17
  - E. Fences 18
- 93. Grounding Conductor and Means of Connection 18
  - A. Composition of Grounding Conductors 18
  - B. Connection of Grounding Conductors 19
  - C. Ampacity and Strength 19
    - 1. System Grounding Conductors for Single-Grounded Systems 19
    - 2. System Grounding Conductors for Multi-Grounded Alternating Current Systems 19
    - 3. Grounding Conductors for Instrument Transformers 19
    - 4. Grounding Conductors for Primary Surge Arresters 19
    - 5. Grounding Conductors for Equipment, Messenger Wires, and Guys 19
    - 6. Fences 19
    - 7. Bonding of Equipment Frames and Enclosures 20
    - 8. Ampacity Limit 20

SECTION PAGE

9. Strength ... 20
D. Guarding and Protection ... 20
E. Underground ... 20
F. Common Grounding Conductor for Circuits, Metal Raceways, and Equipment ... 21
94. Grounding Electrodes ... 21
A. Existing Electrodes ... 21
1. Metallic Water Piping System ... 21
2. Local Systems ... 21
3. Steel Reinforcing Bars in Concrete Foundations and Footings ... 21
B. Made Electrodes ... 21
1. General ... 21
2. Driven Rods ... 22
3. Buried Wire, Strips, or Plates ... 22
4. Pole-Butt Plates and Wire Wraps ... 22
5. Concentric Neutral Cable ... 23
6. Concrete-Encased Electrodes ... 23
95. Method of Connection to Electrode ... 23
A. Ground Connections ... 23
B. Point of Connection to Piping Systems ... 24
C. Contact Surfaces ... 24
96. Ground Resistance Requirements ... 24
A. Supply Stations ... 24
B. Single-Grounded (Unigrounded or Delta) Systems ... 24
C. Multi-Grounded Systems ... 24
97. Separation of Grounding Conductors ... 25
98. Number 98 not used in this edition. ... 26
99. Additional Requirements for Communication Apparatus ... 26
A. Electrode ... 26
B. Electrode Connection ... 26
C. Bonding of Electrodes ... 26

Part 1. Rules for the Installation and Maintenance of Electric Supply Stations and Equipment

Sec. 10. Purpose and Scope of Rules ... 27
100. Purpose ... 27
101. Scope ... 27

Sec. 11. Protective Arrangements in Electric Supply Stations ... 28
110. General Requirements ... 28
A. Enclosure of Equipment ... 28
1. Types of Enclosures ... 28
2. Safety Clearance Zone ... 28
B. Rooms and Spaces ... 28
1. Construction ... 28
2. Use ... 28
3. Ventilation ... 30
4. Moisture and Weather ... 30
C. Electric Equipment ... 30
111. Illumination ... 30
A. Under Normal Conditions ... 30
B. Emergency Lighting ... 30
C. Fixtures ... 30
D. Attachment Plugs and Receptacles for General Use ... 30

SECTION PAGE

E. Receptacles in Damp or Wet Locations 30
112. Floors, Floor Openings, Passageways, and Stairs 32
A. Floors 32
B. Passageways 32
C. Railings 33
D. Stair Guards 33
E. Top Rails 33
113. Exits 33
A. Clear Exits 33
B. Double Exits 33
C. Exit Doors 33
114. Fire-Extinguishing Equipment 33

Sec. 12. Installation and Maintenance of Equipment 34
120. General Requirements 34
121. Inspections 34
A. In-Service Equipment 34
B. Idle Equipment 34
C. Emergency Equipment 34
D. New Equipment 34
122. Guarding Shaft Ends, Pulleys, Belts, and Suddenly Moving Parts 34
A. Mechanical Transmission Machinery 34
B. Suddenly Moving Parts 34
123. Protective Grounding 34
A. Protective Grounding or Physical Isolation of Non-Current-Carrying Metal Parts 34
B. Grounding Method 34
C. Provision for Grounding Equipment During Maintenance 35
D. Grounding Methods for Direct-Current Systems over 750 V 35
124. Guarding Live Parts 35
A. Where Required 35
B. Strength of Guards 35
C. Types of Guards 35
1. Location or Physical Isolation 35
2. Shields or Enclosures 35
3. Supplemental Barriers or Guards Within Electric Supply Stations 36
4. Mats 36
5. Live Parts Below Supporting Surfaces for Persons 36
6. Insulating Covering on Conductors or Parts 36
125. Working Space About Electric Equipment 45
A. Working Space (600 V or Less) 45
1. Clear Spaces 45
2. Access and Entrance to Working Space 45
3. Working Space 45
4. Headroom Working Space 45
5. Front Working Space 45
B. Working Space Over 600 V 45
126. Equipment for Work on Energized Parts 46
127. Classified Locations 46
A. Coal-Handling Areas 46
B. Flammable and Combustible Liquids 46
C. Flammable Liquid Storage Area 47
D. Loading and Unloading Facilities 47

SECTION PAGE

E. Gasoline-Dispensing Stations ... 47
F. Boilers ... 47
G. Gaseous Hydrogen Systems for Supply Equipment ... 47
H. Liquid Hydrogen Systems ... 48
I. Sulfur ... 48
J. Oxygen ... 48
K. Liquefied Petroleum Gas (LPG) ... 48
L. Natural Gas (Methane) ... 48
128. Identification ... 48
129. Mobile Hydrogen Equipment ... 48

Sec. 13. Rotating Equipment ... 49
130. Speed Control and Stopping Devices ... 49
A. Automatic Overspeed Trip Device for Prime Movers ... 49
B. Manual Stopping Devices ... 49
C. Speed Limit for Motors ... 49
D. Number 130D not used in this edition. ... 49
E. Adjustable-Speed Motors ... 49
F. Protection of Control Circuits ... 49
131. Motor Control ... 49
132. Number 132 not used in this edition. ... 49
133. Short-Circuit Protection ... 49

Sec. 14. Storage Batteries ... 50
140. General ... 50
141. Location ... 50
142. Ventilation ... 50
143. Racks ... 50
144. Floors in Battery Areas ... 50
145. Illumination for Battery Areas ... 50
146. Service Facilities ... 50
147. Number 147 not used in this edition. ... 50

Sec. 15. Transformers and Regulators ... 51
150. Current-Transformer Secondary Circuits Protection When Exceeding 600 V ... 51
151. Grounding Secondary Circuits of Instrument Transformers ... 51
152. Location and Arrangement of Power Transformers and Regulators ... 51
A. Outdoor Installations ... 51
B. Indoor Installations ... 51
153. Short-Circuit Protection of Power Transformers ... 51

Sec. 16. Conductors ... 53
160. Application ... 53
161. Electrical Protection ... 53
A. Overcurrent Protection Required ... 53
B. Grounded Conductors ... 53
C. Insulated Power Cables ... 53
162. Mechanical Protection and Support ... 53
163. Isolation ... 53
164. Conductor Terminations ... 53
A. Insulation ... 53
B. Metal-Sheathed or Shielded Cable ... 53

Sec. 17. Circuit Breakers, Reclosers, Switches, and Fuses ... 54
170. Arrangement ... 54

SECTION PAGE

171. Application ... 54
172. Circuit Breakers, Reclosers, and Switches Containing Oil ... 54
173. Switches and Disconnecting Devices ... 54
  A. Capacity ... 54
  B. Provisions for Disconnecting ... 54
  C. Visible Break Switch ... 54
174. Disconnection of Fuses ... 54

Sec. 18. Switchgear and Metal-Enclosed Bus ... 55
180. Switchgear Assemblies ... 55
  A. General Requirements for All Switchgear ... 55
  B. Metal-Enclosed Power Switchgear ... 55
  C. Dead-Front Power Switchboards ... 56
  D. Motor Control Centers ... 56
  E. Control Switchboards ... 56
181. Metal-Enclosed Bus ... 56
  A. General Requirements for All Types of Bus ... 56
  B. Isolated-Phase Bus ... 56

Sec. 19. Surge Arresters ... 58
190. General Requirements ... 58
191. Indoor Locations ... 58
192. Grounding Conductors ... 58
193. Installation ... 58

Part 2. Safety Rules for the Installation and Maintenance of Overhead Electric Supply and Communication Lines

Sec. 20. Purpose, Scope, and Application of Rules ... 59
200. Purpose ... 59
201. Scope ... 59
202. Application of Rules ... 59

Sec. 21. General Requirements ... 60
210. Referenced Sections ... 60
211. Number 211 not used in this edition. ... 60
212. Induced Voltages ... 60
213. Accessibility ... 60
214. Inspection and Tests of Lines and Equipment ... 60
  A. When In Service ... 60
    1. Initial Compliance With Rules ... 60
    2. Inspection ... 60
    3. Tests ... 60
    4. Record of Defects ... 60
    5. Remedying Defects ... 60
  B. When Out of Service ... 60
    1. Lines Infrequently Used ... 60
    2. Lines Temporarily Out of Service ... 60
    3. Lines Permanently Abandoned ... 60
215. Grounding of Circuits, Supporting Structures, and Equipment ... 61
  A. Methods ... 61
  B. Circuits ... 61
    1. Common Neutral ... 61
    2. Other Neutrals ... 61
    3. Other Conductors ... 61
    4. Surge Arresters ... 61

SECTION PAGE

5. Use of Earth as Part of Circuit 61
C. Non-Current-Carrying Parts 61
1. General 61
2. Guys 61
3. Multiple Messengers on the Same Structure 61
216. Arrangement of Switches 62
A. Accessibility 62
B. Indicating Open or Closed Position 62
C. Locking 62
D. Uniform Position 62
217. General 62
A. Supporting Structures 62
1. Protection of Structures 62
2. Readily Climbable Supporting Structures 62
3. Identification 63
4. Obstructions 63
5. Decorative Lighting 63
B. Unusual Conductor Supports 63
218. Tree Trimming 63
A. General 63
B. At Line Crossings, Railroad Crossings, and Limited-Access Highway Crossings 63

Sec. 22. Relations Between Various Classes of Lines and Equipment 64
220. Relative Levels 64
A. Standardization of Levels 64
B. Relative Levels: Supply and Communication Conductors 64
1. Preferred Levels 64
2. Special Construction for Supply Circuits, the Voltage of Which Is 600 V or Less and Carrying Power Not in Excess of 5 kW 64
C. Relative Levels: Supply Lines of Different Voltage Classifications (as classified in Table 235-5) 65
1. At Crossings or Conflicts 65
2. On Structures Used Only by Supply Conductors 65
D. Identification of Overhead Conductors 65
E. Identification of Equipment on Supporting Structures 65
221. Avoidance of Conflict 65
222. Joint Use of Structures 65
223. Communications Protective Requirements 66
A. Where Required 66
B. Means of Protection 66
224. Communication Circuits Located Within the Supply Space and Supply Circuits Located Within the Communication Space 66
A. Communication Circuits Located in the Supply Space 66
B. Supply Circuits Used Exclusively in the Operation of Communication Circuits 67
225. Electric Railway Construction 67
A. Trolley-Contact Conductor Fastenings 67
B. High-Voltage Contact Conductors 67
C. Third Rails 67
D. Prevention of Loss of Contact at Railroad Crossings at Grade 67
E. Guards Under Bridges 68

Sec. 23. Clearances 69
230. General 69

SECTION PAGE

A. Application ... 69
1. Permanent and Temporary Installations ... 69
2. Emergency Installations ... 69
B. Measurement of Clearance and Spacing ... 69
C. Supply Cables ... 69
D. Covered Conductors ... 70
E. Neutral Conductors ... 70
F. Fiber-Optic Cable ... 70
1. Fiber-optic—supply cable ... 70
2. Fiber-optic—communication cable ... 70
G. Alternating- and Direct-Current Circuits ... 70
H. Constant-Current Circuits ... 70
I. Maintenance of Clearances and Spacings ... 71

231. Clearances of Supporting Structures From Other Objects ... 71
A. From Fire Hydrants ... 71
B. From Streets, Roads, and Highways ... 71
C. From Railroad Tracks ... 71

232. Vertical Clearances of Wires, Conductors, Cables, and Equipment Above Ground, Roadway, Rail, or Water Surfaces ... 71
A. Application ... 71
B. Clearance of Wires, Conductors, Cables, Equipment, and Support Arms Mounted on Supporting Structures ... 72
1. Clearance to Wires, Conductors, and Cables ... 72
2. Clearance to Unguarded Rigid Live Parts of Equipment ... 72
3. Clearance to Support Arms and Equipment Cases ... 72
4. Street and Area Lighting ... 72
C. Additional Clearances for Wires, Conductors, Cables, and Unguarded Rigid Live Parts of Equipment ... 72
1. Voltages Exceeding 22 kV ... 72
D. Alternate Clearances for Voltages Exceeding 98 kV AC to Ground or 139 kV DC to Ground ... 73
1. Sag Conditions of Line Conductors ... 73
2. Reference Heights ... 73
3. Electrical Component of Clearance ... 73
4. Limit ... 73

233. Clearances Between Wires, Conductors, and Cables Carried on Different Supporting Structures ... 85
A. General ... 85
1. Conductor Movement Envelope ... 85
2. Clearance Envelope ... 85
B. Horizontal Clearance ... 89
1. Clearance Requirements ... 89
2. Alternate Clearances for Voltages Exceeding 98 kV AC to Ground or 139 kV DC to Ground ... 89
C. Vertical Clearance ... 89
1. Clearance Requirements ... 89
2. Voltages Exceeding 22 kV ... 89
3. Alternate Clearances for Voltage Exceeding 98 kV AC to Ground or 139 kV DC to Ground ... 90

234. Clearance of Wires, Conductors, Cables, and Equipment From Buildings, Bridges, Rail Cars, Swimming Pools, and Other Installations ... 96
A. Application ... 96
1. Vertical and Horizontal Clearances (No Wind Displacement) ... 96
2. Horizontal Clearances (With Wind Displacement) ... 96

SECTION PAGE

3. Transition Between Horizontal and Vertical Clearances 96
B. Clearances of Wires, Conductors, and Cables From Other Supporting Structures 96
C. Clearances of Wires, Conductors, Cables, and Rigid Live Parts From Buildings, Signs, Billboards, Chimneys, Radio and Television Antennas, Tanks, and Other Installations Except Bridges 99
1. Vertical and Horizontal Clearances 99
2. Guarding of Supply Conductors and Rigid Live Parts 99
3. Supply Conductors Attached to Buildings or Other Installations 99
4. Communication Conductors Attached to Buildings or Other Installations 100
5. Ladder Space 100
D. Clearance of Wires, Conductors, Cables, and Unguarded Rigid Live Parts From Bridges 107
1. Vertical and Horizontal Clearances 107
2. Guarding Trolley-Contact Conductors Located Under Bridges 109
E. Clearance of Wires, Conductors, Cables, or Unguarded Rigid Live Parts Installed Over or Near Swimming Areas With No Wind Displacement 109
1. Swimming Pools 109
2. Beaches and Waterways Restricted to Swimming 109
3. Waterways Subject to Water Skiing 109
F. Clearances of Wires, Conductors, Cables, and Rigid Live Parts From Grain Bins 114
1. Grain Bins Loaded by Permanently Installed Augers, Conveyers, or Elevator Systems 114
2. Grain Bins Loaded by Portable Augers, Conveyers or Elevators (With No Wind Displacement) 115
G. Additional Clearances for Voltages Exceeding 22 kV for Wires, Conductors, Cables, and Unguarded Rigid Live Parts of Equipment 117
H. Alternate Clearances for Voltages Exceeding 98 kV AC to Ground or 139 kV DC to Ground 117
1. Sag Conditions of Line Conductors 117
2. Reference Distances 117
3. Electrical Component of Clearance 117
4. Limit 118
I. Clearance of Wires, Conductors, and Cables to Rail Cars 118
1. Rules 232B1 and 232C1 118
2. Rule 232D 118
J. Clearance of Equipment Mounted on Supporting Structures 118
1. Clearance to Unguarded Rigid Live Parts of Equipment 118
2. Clearance to Equipment Cases 118
235. Clearance for Wires, Conductors, or Cables Carried on the Same Supporting Structure 121
A. Application of Rule 121
1. Multiconductor Wires or Cables 121
2. Conductors Supported by Messengers or Span Wires 121
3. Line Conductors of Different Circuits 121
B. Horizontal Clearance Between Line Conductors 121
1. Fixed Supports 121
2. Suspension Insulators 126
3. Alternate Clearances for Different Circuits Where One or Both Circuits Exceed 98 kV AC to Ground or 139 kV DC to Ground 126
C. Vertical Clearance Between Line Conductors 126
1. Basic Clearance for Conductors of Same or Different Circuits 127

SECTION PAGE

2. Additional Clearances ... 127
3. Alternate Clearances for Different Circuits Where One or Both Exceed 98 kV AC, or 139 kV DC to Ground ... 133
4. Communication Worker Safety Zone ... 133

D. Diagonal Clearance Between Line Wires, Conductors, and Cables Located at Different Levels on the Same Supporting Structure ... 134
E. Clearances in Any Direction From Line Conductors to Supports, and to Vertical or Lateral Conductors, Span, or Guy Wires Attached to the Same Support ... 138

1. Fixed Supports ... 138
2. Suspension Insulators ... 138
3. Alternate Clearances for Voltages Exceeding 98 kV AC to Ground or 139 kV DC to Ground ... 138

F. Clearances Between Circuits of Different Voltage Classifications Located in the Supply Space on the Same Support Arm ... 139
G. Conductor Spacing: Vertical Racks ... 140
H. Clearance and Spacing Between Communication Conductors, Cables, and Equipment ... 141
I. Clearances in Any Direction From Supply Line Conductors to Communication Antennas in the Supply Space Attached to the Same Supporting Structure ... 141

236. Climbing Space ... 141

A. Location and Dimensions ... 141
B. Portions of Supporting Structures in Climbing Space ... 141
C. Support Arm Location Relative to Climbing Space ... 142
D. Location of Equipment Relative to Climbing Space ... 142
E. Climbing Space Between Conductors ... 142
F. Climbing Space on Buckarm Construction ... 144
G. Climbing Space Past Longitudinal Runs Not on Support Arms ... 144
H. Climbing Space Past Vertical Conductors ... 144
I. Climbing Space Near Ridge-Pin Conductors ... 144

237. Working Space ... 145

A. Location of Working Spaces ... 145
B. Dimensions of Working Spaces ... 145

1. Along the Support Arm ... 145
2. At Right Angles to the Support Arm ... 145
3. Vertically ... 145

C. Location of Vertical and Lateral Conductors Relative to Working Spaces ... 146
D. Location of Buckarms Relative to Working Spaces ... 146

1. Standard Height of Working Space ... 146
2. Reduced Height of Working Space ... 146

E. Guarding of Energized Equipment ... 146
F. Working Clearances From Energized Equipment ... 146

238. Vertical Clearance Between Certain Communications and Supply Facilities Located on the Same Structure ... 147

A. Equipment ... 147
B. Clearances in General ... 147
C. Clearances for Span Wires or Brackets ... 147
D. Clearance of Drip Loops of Luminaire or Traffic Signal Brackets ... 147
E. Communication Worker Safety Zone ... 147

239. Clearance of Vertical and Lateral Facilities From Other Facilities and Surfaces on the Same Supporting Structure ... 149

A. General ... 149

SECTION PAGE

B. Location of Vertical or Lateral Conductors Relative to Climbing Spaces, Working Spaces, and Pole Steps 149
C. Conductors Not in Conduit 149
D. Guarding and Protection Near Ground 149
E. Requirements for Vertical and Lateral Supply Conductors on Supply Line Structures or Within Supply Space on Jointly Used Structures 149
1. General Clearances 149
2. Special Cases 149
F. Requirements for Vertical and Lateral Communication Conductors on Communication Line Structures or Within the Communication Space on Jointly Used Structures 150
1. Clearances From Communication Conductors 150
2. Clearances From Supporting Structure Surfaces 150
G. Requirements for Vertical Supply Conductors and Cables Passing Through Communication Space on Jointly Used Line Structures 150
1. Guarding—General 150
2. Cables and Conductors in Conduit or Covering 151
3. Protection Near Trolley, Ungrounded Traffic Signal, or Ungrounded Luminaire Attachments 151
4. Aerial Services 151
5. Clearance From Through Bolts and Other Metal Objects 151
H. Requirements for Vertical Communication Conductors Passing Through Supply Space on Jointly Used Structures 151
1. Metal-Sheathed Communication Cables 151
2. Communication Conductors 151
3. Communication Grounding Conductors 151
4. Clearance From Through Bolts and Other Metal Objects 152
I. Operating Rods 152
J. Additional Rules for Standoff Brackets 152

Sec. 24 Grades of Construction 155
240. General 155
241. Application of Grades of Construction to Different Situations 155
A. Supply Cables 155
B. Order of Grades 155
C. At Crossings 155
1. Grade of Upper Line 155
2. Grade of Lower Line 155
3. Multiple Crossings 155
D. Conflicts 155
242. Grades of Construction for Conductors 155
A. Constant-Current Circuit Conductors 156
B. Railway Feeder and Trolley-Contact Circuit Conductors 156
C. Communication Circuit Conductors Located in the Supply Space 156
D. Fire-Alarm Circuit Conductors 156
E. Neutral Conductors of Supply Circuits 156
F. Surge-Protection Wires 156
243. Grades of Construction for Line Supports 160
A. Structures 160
B. Crossarms and Support Arms 160
C. Pins, Armless Construction Brackets, Insulators, and Conductor Fastenings 160

Sec. 25. Loading for Grades B and C 161

SECTION PAGE

250. General Loading Requirements and Maps ... 161
A. General ... 161
B. Combined Ice and Wind Loading ... 161
C. Extreme Wind Loading ... 161
1. Velocity Pressure Exposure Coefficient k_z ... 163
2. Gust Response Factor, G_{RF} ... 163
251. Conductor Loading ... 170
A. General ... 170
B. Load Components ... 170
1. Vertical Load Component ... 170
2. Horizontal Load Component ... 170
3. Total Load ... 170
252. Loads on Line Supports ... 171
A. Assumed Vertical Loads ... 171
B. Assumed Transverse Loads ... 171
1. Transverse Loads From Conductors and Messengers ... 171
2. Wind Loads on Structures ... 171
3. At Angles ... 171
4. Span Lengths ... 171
C. Assumed Longitudinal Loading ... 171
1. Change in Grade of Construction ... 171
2. Jointly Used Poles at Crossings Over Railroads, Communication Lines, or Limited-Access Highways ... 172
3. Deadends ... 172
4. Unequal Spans and Unequal Vertical Loads ... 172
5. Stringing Loads ... 172
6. Longitudinal Capability ... 172
7. Communication Conductors on Unguyed Supports at Railroad and Limited-Access Highway Crossings ... 172
D. Simultaneous Application of Loads ... 172
253. Overload Factors for Structures, Crossarms, Support Hardware, Guys, Foundations, and Anchors ... 172

Sec. 26. Strength Requirements ... 175
260. General (see also Section 20) ... 175
A. Preliminary Assumptions ... 175
B. Application of Strength Factors ... 175
261. Grades B and C Construction ... 175
A. Supporting Structures ... 175
1. Metal, Prestressed-, and Reinforced-Concrete Structures ... 175
2. Wood Structures ... 176
3. Transverse-Strength Requirements for Structures Where Side Guying Is Required, But Can Only Be Installed at a Distance ... 177
4. Longitudinal-Strength Requirements for Sections of Higher Grade in Lines of a Lower-Grade Construction ... 177
B. Strength of Foundations, Settings, and Guy Anchors ... 177
C. Strength of Guys and Guy Insulators ... 177
1. Metal and Prestressed-Concrete Structures ... 177
2. Wood and Reinforced-Concrete Poles and Structures ... 177
D. Crossarms and Braces ... 178
1. Concrete and Metal Crossarms and Braces ... 178
2. Wood Crossarms and Braces ... 178
3. Crossarms and Braces of Other Materials ... 178
4. Additional Requirements ... 178

SECTION PAGE

E. Insulators 179
F. Strength of Pin-Type or Similar Construction and Conductor Fastenings 179
1. Longitudinal Strength 179
2. Double Pins and Conductor Fastenings 179
3. Single Supports Used in Lieu of Double Wood Pins 179
G. Armless Construction 179
1. General 179
2. Insulating Material 179
3. Other Components 179
H. Open Supply Conductors and Overhead Shield Wires 179
1. Sags and Tensions 179
2. Splices, Taps, and Dead-End Fittings 180
3. Trolley-Contact Conductors 180
I. Supply Cable Messengers 180
J. Open-Wire Communication Conductors 180
K. Communication Cables 180
L. Paired Communication Conductors 180
1. Paired Conductors Supported on Messenger 180
2. Paired Conductors Not Supported on Messenger 181
M. Support Hardware 181
262. Number 262 not used in this edition 183
263. Grade N Construction 183
A. Poles 183
B. Guys 183
C. Crossarm Strength 183
D. Supply-Line Conductors 183
E. Service Drops 183
1. Size of Open-Wire Service Drops 183
2. Tension of Open-Wire Service Drops 184
3. Cabled Service Drops 184
F. Trolley-Contact Conductors 184
G. Communication Conductors 184
H. Street and Area Lighting Equipment 184
I. Insulators 184
264. Guying and Bracing 185
A. Where Used 185
B. Strength 185
C. Point of Attachment 185
D. Guy Fastenings 185
E. Guy Markers and Protection 186
F. Electrolysis 186
G. Anchor Rods 186

Sec. 27. Line Insulation 187
270. Application of Rule 187
271. Material and Marking 187
272. Ratio of Flashover to Puncture Voltage 187
273. Insulation Level 187
274. Factory Tests 187
275. Special Insulator Applications 187
A. Insulators for Constant-Current Circuits 187
B. Insulators for Single-Phase Circuits Directly Connected to Three-Phase Circuits 188

SECTION PAGE

276. Number 276 Not Used in this Edition 188
277. Mechanical Strength of Insulators 188
278. Aerial Cable Systems 188
    A. Electrical Requirements 188
    B. Mechanical Requirements 189
279. Guy and Span Insulators 189
    A. Insulators 189
        1. Properties of Guy Insulators 189
        2. Use of Guy Insulators 189
        3. Corrosion Protection 189
    B. Span-Wire Insulators 189
        1. Properties of Span-Wire Insulators 189
        2. Use of Span-Wire Insulators 190

Sec. 28 Section number 28 not used in this edition. 190

Sec. 29 Section number 29 not used in this edition 190

Part 3. Safety Rules for the Installation and Maintenance of Underground Electric Supply and Communication Lines

Sec. 30. Purpose, Scope, and Application of Rules 191
300. Purpose 191
301. Scope 191
302. Application of Rules 191

Sec. 31. General Requirements Applying to Underground Lines 192
310. Referenced Sections 192
311. Installation and Maintenance 192
312. Accessibility 192
313. Inspection and Tests of Lines and Equipment 192
    A. When In Service 192
        1. Initial Compliance With Safety Rules 192
        2. Inspection 192
        3. Tests 192
        4. Record of Defects 192
        5. Remedying Defects 192
    B. When Out of Service 192
        1. Lines Infrequently Used 192
        2. Lines Temporarily Out of Service 192
        3. Lines Permanently Abandoned 192
314. Grounding of Circuits and Equipment 192
    A. Methods 192
    B. Conductive Parts to Be Grounded 192
    C. Circuits 193
        1. Neutrals 193
        2. Other Conductors 193
        3. Surge Arresters 193
        4. Use of Earth as Part of Circuit 193
315. Communications Protective Requirements 193
    A. Where Required 193
    B. Means of Protection 193
316. Induced Voltage 193

Sec. 32. Underground Conduit Systems 194
320. Location 194

SECTION PAGE

A. Routing ... 194
1. General ... 194
2. Natural Hazards ... 194
3. Highways and Streets ... 194
4. Bridges and Tunnels ... 194
5. Crossing Railroad Tracks ... 194
6. Submarine Crossing ... 194
B. Separation From Other Underground Installations ... 194
1. General ... 194
2. Separations Between Supply and Communication Conduit Systems ... 195
3. Sewers, Sanitary and Storm ... 195
4. Water Lines ... 195
5. Gas and Other Fuel Lines ... 195
6. Steam Lines ... 195
321. Excavation and Backfill ... 195
A. Trench ... 195
B. Quality of Backfill ... 195
322. Ducts and Joints ... 195
A. General ... 195
B. Installation ... 196
1. Restraint ... 196
2. Joints ... 196
3. Externally Coated Pipe ... 196
4. Building Walls ... 196
5. Bridges ... 196
6. In Vicinity of Manholes ... 196
323. Manholes, Handholes, and Vaults ... 196
A. Strength ... 196
B. Dimensions ... 198
C. Manhole Access ... 198
D. Covers ... 198
E. Vault and Utility Tunnel Access ... 198
F. Ladder Requirements ... 199
G. Drainage ... 199
H. Ventilation ... 199
I. Mechanical Protection ... 199
J. Identification ... 199

Sec. 33. Supply Cable ... 200
330. General ... 200
331. Sheaths and Jackets ... 200
332. Shielding ... 200
A. General ... 200
B. Material ... 200
333. Cable Accessories and Joints ... 200

Sec. 34. Cable in Underground Structures ... 201
340. General ... 201
341. Installation ... 201
A. General ... 201
B. Cable in Manholes and Vaults ... 201
1. Supports ... 201
2. Clearance ... 201
3. Identification ... 202
342. Grounding and Bonding ... 202

SECTION		PAGE
	343. Fireproofing	202
	344. Communication Cables Containing Special Supply Circuits	202
Sec. 35.	Direct-Buried Cable	204
	350. General	204
	351. Location and Routing	204
	A. General	204
	B. Natural Hazards	205
	C. Other Conditions	205
	1. Swimming Pools	205
	2. Buildings and Other Structures	205
	3. Railroad Tracks	205
	4. Highways and Streets	206
	5. Submarine Crossings	206
	352. Installation	206
	A. Trenching	206
	B. Plowing	206
	C. Boring	206
	D. Depth of Burial	206
	353. Deliberate Separations—Equal to or Greater Than 300 mm (12 in) From Underground Structures or Other	207
	A. General	207
	B. Crossings	207
	C. Parallel Facilities	207
	D. Thermal Protection	207
	354. Random Separation—Separation Less Than 300 mm (12 in) From Underground Structures or Other Cables	207
	A. General	207
	B. Supply Cables or Conductors	207
	C. Communication Cables or Conductors	207
	D. Supply and Communication Cables or Conductors	208
	1. General	208
	2. Grounded Bare or Semiconducting Jacketed Neutral Supply Cables	208
	3. Insulating Jacketed Grounded Neutral Supply Cables	209
	4. Insulating Jacketed Grounded Neutral Supply Cables in Nonmetallic Duct	209
	E. Supply and Communication Cables or Conductors and Non-Metallic Water and Sewer Lines	209
Sec. 36.	Risers	210
	360. General	210
	361. Installation	210
	362. Pole Risers—Additional Requirements	210
	363. Pad-Mounted Installations	210
Sec. 37.	Supply Cable Terminations	211
	370. General	211
	371. Support at Terminations	211
	372. Identification	211
	373. Clearances in Enclosures or Vaults	211
	374. Grounding	211
Sec. 38.	Equipment	212
	380. General	212
	381. Design	212
	382. Location in Underground Structures	213

SECTION PAGE

383. Installation 213
384. Grounding 213
385. Identification 213

Sec. 39. Installation in Tunnels 214
390. General 214
391. Environment 214

Part 4. Rules for the Operation of Electric Supply and Communications Lines and Equipment

Sec. 40. Purpose and Scope 215
400. Purpose 215
401. Scope 215
402. Referenced Sections 215

Sec. 41. Supply and Communications Systems—Rules for Employers 216
410. General Requirements 216
A. General 216
B. Emergency Procedures and First Aid Rules 216
C. Responsibility 216
411. Protective Methods and Devices 216
A. Methods 216
B. Devices and Equipment 217
C. Inspection and Testing of Protective Devices 217
D. Signs and Tags for Employee Safety 217
E. Identification and Location 217
F. Fall Protection 217

Sec. 42. General Rules for Employees 218
420. Personal General Precautions 218
A. Rules and Emergency Methods 218
B. Qualifications of Employees 218
C. Safeguarding Oneself and Others 218
D. Energized or Unknown Conditions 218
E. Ungrounded Metal Parts 218
F. Arcing Conditions 219
G. Liquid-Cell Batteries 219
H. Tools and Protective Equipment 219
I. Clothing 219
J. Ladders and Supports 219
K. Fall Protection 219
L. Fire Extinguishers 220
M. Machines or Moving Parts 220
N. Fuses 220
O. Cable Reels 220
P. Street and Area Lighting 221
421. General Operating Routines 221
A. Duties of a First-Level Supervisor or Person in Charge 221
B. Area Protection 221
1. Areas Accessible to Vehicular and Pedestrian Traffic 221
2. Areas Accessible to Employees Only 221
3. Locations With Crossed or Fallen Wires 221
C. Escort 221
422. Overhead Line Operating Procedures 222
A. Setting, Moving, or Removing Poles In or Near Energized Electric Supply Lines 222
B. Checking Structures Before Climbing 222

SECTION PAGE

C. Installing and Removing Wires or Cables 222
423. Underground Line Operating Procedures 222
A. Guarding Manhole and Street Openings 222
B. Testing for Gas in Manholes and Unventilated Vaults 222
C. Flames 223
D. Excavation 223
E. Identification 223
F. Operation of Power-Driven Equipment 223

Sec. 43. Additional Rules for Communications Employees 224
430. General 224
431. Approach to Energized Conductors or Parts 224
432. Joint-Use Structures 224
433. Attendant on Surface at Joint-Use Manhole 224
434. Sheath Continuity 224

Sec. 44. Additional Rules for Supply Employees 226
440. General 226
441. Energized Conductors or Parts 226
A. Minimum Approach Distance to Live Parts 226
1. General 226
2. Precautions for Approach—Voltages from 51 V to 300 V 226
3. Precautions for Approach—Voltages from 301 V to 72.5 kV 226
4. Transient Overvoltage Control Above 72.5 kV 227
5. Altitude Correction 227
6. Calculation of Approach Distances 227
B. Additional Approach Requirements 235
C. Live-Line Tool Clear Insulation Length 235
1. Clear Live-Line Tool Length 235
2. Live-Line Conductor Support Tool Length 235
442. Switching Control Procedures 236
A. Designated Person 236
B. Specific Work 236
C. Operations at Stations 236
D. Re-energizing After Work 236
E. Tagging Electric Supply Circuits Associated With Work Activities 236
F. Restoration of Service After Automatic Trip 237
G. Repeating Oral Messages 237
443. Work on Energized Lines and Equipment 237
A. General Requirements 237
B. Requirement for Assisting Employee 238
C. Opening and Closing Switches 238
D. Working Position 238
E. Protecting Employees by Switches and Disconnectors 238
F. Making Connections 238
G. Switchgear 238
H. Current Transformer Secondaries 238
I. Capacitors 239
J. Gas-Insulated Equipment 239
K. Attendant on Surface 239
L. Unintentional Grounds on Delta Circuits 239
444. De-energizing Equipment or Lines to Protect Employees 239
A. Application of Rule 239
B. Employee's Request 239
C. Operating Switches, Disconnectors, and Tagging 239
D. Employee's Protective Grounds 240
E. Proceeding With Work 240
F. Reporting Clear—Transferring Responsibility 240

G. Removal of Tags 240
H. Sequence of Re-energizing 240
445. Protective Grounds 241
A. Installing Grounds 241
1. Current-Carrying Capacity of Grounds 241
2. Initial Connections 241
3. Test for Voltage 241
4. Completing Grounds 241
B. Removing Grounds 241
446. Live Work 241
447 Protection Against Arcing and Other Damage While Installing and Maintaining Insulators and Conductors 242

FIGURES		RULE	PAGE
Figure D-1.	Sag and Apparent Sag	Sec. 2	10
Figure 110-1	Safety Clearance to Electric Supply Station Fences	110A2	29
Figure 124-1.	Clearance From Live Parts	124A1	36
Figure 124-2.	Railing Used as Guards	124C3	44
Figure 233-1.	Use of Clearance Envelope and Conductor Movement Envelopes to Determine Applicable Clearance	233A	86
Figure 233-2.	Conductor Movement Envelope	233A1a	87
Figure 233-3.	Clearance Envelope	233B	89
Figure 234-1(a)	Clearance Diagram for Building	234A3	97
Figure 234-1(b)	Clearance Diagram for Other Structures	234A3	97
Figure 234-1(c)	Transitional Clearance When H is Greater Than V	234	97
Figure 234-2.	Clearances of Service Drop Terminating on Support Mast	234C3d(1)(b)	108
Figure 234-3.	Swimming Pool Clearances	234E3	109
Figure 234-4(a)	Clearance Envelope for Grain Bins Filled by Permanently Installed Augers, Conveyors, or Elevators	234F2c	115
Figure 234-4(b)	Clearance Envelope for Grain Bins Filled by Portable Augers, Conveyors, or Elevators	234F2a	116
Figure 234-5	Rail Car Clearances	234I	119
Figure 235-1.	Clearance Diagram for Energized Conductor	235D	134
Figure 236-1.	Rule 236G, Exception 3	236G	145
Figure 237-1.	Obstruction of Working Space by Buckarm	237D1	146
Figure 250-1.	General Loading Map of United States With Respect to Loading of Overhead Lines	250B	162
Figure 250-2(a)	Basic Wind Speeds (miles per hour)	250C	166
Figure 250-2(b)	Basic Wind Speeds	250C	167
Figure 250-2(c)	Western Gulf of Mexico Hurricane Coastline	250C	168
Figure 250-2(d)	Eastern Gulf of Mexico and Southeastern US Hurricane Coastline	250C	169
Figure 250-2(e)	Mid and Northern Atlantic Hurricane Coastline	250C	169
Figure 323-1.	Roadway Vehicle Load	323A1	197
Figure 323-2.	Wheel Load Area	323A1	197
Figure 350-1.	Symbols for Identification of Buried Cables	350G	205

TABLES			
Table 110-1	Values for Use with Fig 110-1	11B2	29
Table 111-1.	Illumination Levels	111A	31
Table 124-1.	Clearance From Live Parts	124A1	37
Table 125-1.	Working Space	125A3	45
Table 232-1.	Vertical Clearance of Wires, Conductors, and Cables Above Ground, Roadway, Rail, or Water Surfaces	232B1	74
Table 232-2.	Vertical Clearance of Equipment Cases, Support Arms, Platforms, Braces and Unguarded Rigid Live Parts Above Ground, Roadway, or Water Surfaces	232B2	81
Table 232-3.	Reference Heights	232D2	83
Table 232-4.	Electrical Component of Clearance in Rule 232D3a	232D3a	84
Table 233-1.	Vertical Clearance Between Wires, Conductors, and Cables Carried on Different Supporting Structures	233C1	91
Table 233-2.	Clearance Between Supply Wires, Conductors, and Cables in Rule 233C3b(1)	233C3b(1)	94
Table 233-3.	Reference Heights	233C3a	95
Table 234-1.	Clearance of Wires, Conductors, Cables, and Unguarded Rigid Live Parts Adjacent but Not Attached to Buildings and Other Installations Except Bridges	234C1a	104

TABLES		RULE	PAGE
Table 234-2.	Clearance of Wires, Conductors, Cables, and Unguarded Rigid Live Parts From Bridges	234D1a	110
Table 234-3.	Clearance of Wires, Conductors, Cables or Unguarded Rigid Live Parts Over or Near Swimming Pools	234E1	113
Table 234-4.	Electrical Component of Clearance of Buildings, Bridges, and Other Installations	234H3a	
Table 234-5.	Reference Distances	234H2	121
Table 235-1.	Horizontal Clearance Between Wires, Conductors, or Cables at Supports	235B1a	123
Table 235-2.	Horizontal Clearances Between Line Conductors Smaller Than AWG No. 2 at Supports, Based on Sags	235B1b(1)	124
Table 235-3.	Horizontal Clearances Between Line Conductors AWG No. 2 or Larger at Supports, Based on Sags	235B1b(2)	125
Table 235-4.	Electrical Clearances in Rule 235B3a(1)	235B3a(1)	128
Table 235-5.	Vertical Clearance Between Conductors at Supports	235C1	131
Table 235-6.	Clearance in Any Direction From Line Conductors to Supports and to Vertical or Lateral Conductors, Span, or Guy Wires Attached to the Same Support	235E1	135
Table 235-7.	Clearance in Any Direction From Line Conductors to Supports	235E3b	140
Table 236-1.	Horizontal Clearance Between Conductors Bounding the Climbing Space	236E	143
Table 238-1.	Vertical Clearance Between Supply Conductors and Communications Equipment, Between Communication Conductors and Supply Equipment, and Between Supply and Communications Equipment	238B	148
Table 238-2.	Vertical Clearance of Span Wires and Brackets From Communication Lines	238C	148
Table 239-1.	Clearance of Open Vertical and Lateral Conductors	239E1	153
Table 239-2.	Clearance Between Open Vertical Conductors and Pole Surface	239E2a(1)	154
Table 242-1.	Grades of Construction for Supply Conductors Alone, at Crossing, or on the Same Structures With Other Conductors	242	157
Table 242-2.	Grades of Construction for Communication Conductors Alone, or in Upper Position of Crossing or on Joint Poles	242	159
Table 250-1.	Ice, Wind, and Temperature	250C	162
Table 250-2.	Velocity Pressure Exposure Coefficient k_z, Structure and Wire	250C	163
Table 250-3	Structure and Wire Gust Response Factors, G_{RF}	250C	164
Table 251-1.	Temperatures and Constants	251B3	170
Table 253-1.	Overload Factors for Structures, Crossarms, Support Hardware, Guys, Foundation, and Anchors to Be Used with the Strength Factors of Table 261-1A	253	173
Table 253-2.	Alternate Overload Factors for Wood and Reinforced (Not Prestressed) Concrete Structures to Be Used with the Strength Factors of Table 261-1B	253	174
Table 261-1A.	Strength Factors for Structures, Crossarms, Support Hardware, Guys, Foundations, and Anchors for Use with Overload Factors of Table 253-1	261M	182
Table 261-1B.	Strength Factors for Structures and Crossarms for Use with Overload Factor of Table 253-2	261D2	182
Table 261-2	Dimensions of Crossarm Cross Section of Select Southern Pine and Douglas Fir	261D2	183

Table 263-1.	Sizes for Grade N Supply Line Conductors (AWG for Copper and Aluminum, Stl WG for Steel)	263D1	184
Table 263-2.	Sizes of Service Drops of 750 V or Less	263E1a(1)	185
Table 273-1.	Insulation Level Requirements	273	188
Table 341-1.	Clearance Between Supply and Communications Facilities in Joint-Use Manholes and Vaults	341B2b(5)	202
Table 352-1.	Supply Cable or Conductor Burial Depth	352D2	206
Table 431-1.	Overhead Supply Lines and Equipment Minimum Approach Distances to Exposed Energized Parts	431	225
Table 441-1.	AC Live Work Minimum Approach Distance	441A4	228
Table 441-2.	AC Live Work Minimum Approach Distance With Transient Overvoltage Factor	441A4	229
Table 441-3.	AC Live Work Minimum Approach Distance With Transient Overvoltage Factor	441A4	231
Table 441-4.	DC Live Work Minimum Approach Distance With Transient Overvoltage Factor	441A1	233
Table 441-5.	Altitude Correction Factor	441A4	234
Table 441-6	Maximum Use Voltage for Rubber Insulating Equipment	441A4	234

APPENDIX PAGE

Appendix A **Uniform System of Clearances Adopted in the 1990 Edition**

Introduction 243
The 1990 Changes 243
Clearance Rules and Tables Prior to 1990 249
Clearance Values Prior to 1990 253
Summary—Prior Editions 253
Clearances Subcommittee Activities 253
Summary 253

Appendix B **Bibliography** 254

APPENDIX FIGURES

Figure A1. Clearance at Maximum Sag 248
Figure A2. Basic Clearance 250
Figure A3. Additional Clearance Required for High-Temperature Operation 251
Figure A4. Additional Clearance Required for Long-Span Construction 252

APPENDIX TABLES

Table A-1. 245
Table A-2a. Reference Components of Rule 232 246
Table A-2b. Reference Components of Rule 234 247

Index 257

Revision Procedures 279
Form for Change Proposal 281
Time Schedule 282
Working Group Assignments 283
Tentative Interim Amendments 286

Letter Symbols for Units

This code uses standard symbols for units. They have the following meanings:

A	ampere
C	degree Celsius
ft	foot
ft^2	square foot
ft^3	cubic foot
F	degree Farenheit
g	gram
Hz	hertz
h	hour
in	inch
in^2	square inch
k	kilo (10^3)
kg	kilogram
kPa	kilo pascal
km^2	square kilometer
kV	kilovolt (1000 volts)
kVA	kilovoltampere
kW	kilowatt
m	meter
m^2	square meter
m^3	cubic meter
m	milli (10^{-3})
mA	milliampere
mi	mile (statute)
mm	millimeter
min	minute (time)
N	newton
Pa	pascal
lb	pound
s	second (time)
V	volt
W	watt

Section 1.
Introduction to the
National Electrical Safety Code®

010. Purpose

The purpose of these rules is the practical safeguarding of persons during the installation, operation, or maintenance of electric supply and communication lines and associated equipment.

These rules contain the basic provisions that are considered necessary for the safety of employees and the public under the specified conditions. This code is not intended as a design specification or as an instruction manual.

011. Scope

A. These rules cover supply and communication lines, equipment, and associated work practices employed by a public or private electric supply, communications, railway, or similar utility in the exercise of its function as a utility. They cover similar systems under the control of qualified persons, such as those associated with an industrial complex or utility interactive system.

B. The NESC covers utility facilities and functions up to the service point.

NOTE: The National Electrical Code® (NEC®), NFPA 70-1999[1] covers utilization wiring requirements beyond the service point.

C. NESC rules cover street and area lights (supplied by underground or overhead conductors) under the exclusive control of utilities (including their authorized contractors) or other qualified persons (such as those associated with an industrial complex).

NOTE: Luminaires not under such exclusive control are governed by the requirements of the NEC.

D. NESC rules do not cover installations in mines, ships, railway rolling equipment, aircraft, or automotive equipment, or utilization wiring except as covered in Parts 1 and 3.

012. General Rules

A. All electric supply and communication lines and equipment shall be designed, constructed, operated, and maintained to meet the requirements of these rules.

B. The utilities, authorized contractors, or other entities, as applicable, performing design, construction, operation, or maintenance tasks for electric supply or communication lines or equipment covered by this code shall be responsible for meeting applicable requirements.

C. For all particulars not specified in these rules, construction and maintenance should be done in accordance with accepted good practice for the given local conditions known at the time by those responsible for the construction or maintenance of the communication or supply lines and equipment.

013. Application

A. New Installations and Extensions

1. These rules shall apply to all new installations and extensions, except that they may be waived or modified by the administrative authority. When so waived or modified, safety shall be provided in other ways.

   EXAMPLE: Alternative working methods, such as the use of barricades, guards, or other electrical protective equipment, may be implemented along with appropriate alternative working clearances as a means of providing safety when working near energized conductors.

[1]Information on references can be found in Section 3.

2. Types of construction and methods of installation other than those specified in the rules may be used experimentally to obtain information, if done where
   a. Qualified supervision is provided,
   b. Equivalent safety is provided, and
   c. On joint use facilities, all affected parties agree.

B. Existing Installations
1. Where an existing installation meets, or is altered to meet, these rules, such installation is considered to be in compliance with this edition and is not required to comply with any previous edition.
2. Existing installations, including maintenance replacements, that currently comply with prior editions of the Code, need not be modified to comply with these rules except as may be required for safety reasons by the administrative authority.
3. Where conductors or equipment are added, altered, or replaced on an existing structure, the structure or the facilities on the structure need not be modified or replaced if the resulting installation will be in compliance with either (a) the rules that were in effect at the time of the original installation, or (b) the rules in effect in a subsequent edition to which the installation has been previously brought into compliance, or (c) the rules of this edition in accordance with Rule 013B1.

014. Waiver

The person responsible for an installation may modify or waive rules in the case of emergency or temporary installations.

A. Emergency Installations
1. The clearances required in Section 23 may be decreased for emergency installations. See Rule 230A.
2. The strength of material and construction for emergency installations shall be not less than that required for Grade N construction. See Rule 263.
3. Emergency installations shall be removed, replaced, or relocated, as desired, as soon as practical.

B. Temporary Overhead Installations
When an installation is temporary, or where facilities are temporarily relocated to facilitate other work, the installation shall meet the requirements for non-temporary installation except that the strength of material and construction shall be not less than that required for Grade N construction. See Rule 263.

015. Intent

A. The word "shall" indicates provisions that are mandatory.

B. The word "should" indicates provisions that are normally and generally practical for the specified conditions. However, where the word "should" is used, it is recognized that, in certain instances, additional local conditions not specified herein may make these provisions impractical. When this occurs, the difference in conditions shall be appropriately recognized and Rule 012 shall be met.

C. Footnotes to a table have the force and effect required or allowed by the rule that specifies the use of the table.

D. Exceptions to a rule have the same force and effect required or allowed by the rule to which the exception applies.

E. The word *"RECOMMENDATION"* indicates provisions considered desirable, but that are not intended to be mandatory.

F. The word *"NOTE"* or the word *"EXAMPLE"* used in a rule indicates material provided for information or illustrative purposes only. *"NOTES"* and *"EXAMPLES"* are not mandatory and are not considered to be a part of Code requirements.

G. A *"RECOMMENDATION," "EXCEPTION,"* or *"NOTE"* applies to all text in that rule above its location that is indented to the same level.

016. Effective Date

This edition may be used at any time on or after the publication date. Additionally, this edition shall become effective no later than 180 days following its publication date for application to new installations and extensions where both design and approval were started after the expiration of that period, unless otherwise stipulated by the administrative authority.

NOTE: A period of 180 days is allowed for utilities and regulatory authorities to acquire copies of the new edition and to change regulations, internal standards, and procedures as may be required. There is neither an intention to require or imply that this edition be implemented before 180 days from the publication date, nor an intention to prohibit earlier implementation.

017. Units of Measure

A. Numerical values in the requirements of this code are stated in the metric system[2] and in the customary inch-foot-pound system. In text, the metric value is shown first with the customary inch-foot-pound (inside parentheses) following. Extensive detailed tables are duplicated. The first, marked **m**, contains metric (SI) values; the second, marked **in**, **ft**, or **lb**, contains the inch-foot-pound values. Tensions and wind loads are stated in newtons, the SI unit of force.

The SI values and the customary inch-foot-pound values are not, nor are they intended to be, identical measures. The values shown in each system of measurement have been rounded to convenient numbers in order to simplify measurement and to minimize errors. The values shown in each system are functional equivalents for safety purposes.

The values required in this code have been carefully developed and evaluated to ensure that the intended levels of safety are provided in both systems; neither is distinguishable from the other for safety purposes. The values specified in either system of measurement may be used, or the values of the two systems may be intermixed, as desired.[3]

B. The dimensions of physical items referenced in this code, such as wires and ground rods, are "nominal values" assigned for the purpose of convenient designation. Due to manufacturing limitations or other restraints, other standards may set tolerances, variations, or ranges for the dimensions of such items.

[2]Le Systeme Internationale d'Unites (The International System of Units [or SI] in the modern version of the metric system). For basic information and conversion factors, see IEEE/ASTM SI 10-1997 [B21], listed in Appendix B.

[3]It is recognized that many equivalent utility system components may be purchased in both SI and customary units.

Section 2.
Definitions of Special Terms

The following definitions are for use with the National Electrical Safety Code. For other use, and for definitions not contained herein, see *The Authoritative Dictionary of IEEE Standards Terms,* Seventh Edition.

administrative authority. The governmental authority exercising jurisdiction over application of this code.

ampacity. The current-carrying capacity, expressed in amperes, of an electric conductor under stated thermal conditions.

anchorage. A secure point of attachment to which the fall protection system is connected.

automatic. Self-acting, operating by its own mechanism when actuated by some impersonal influence—as, for example, a change in current strength; not manual; without personal intervention. Remote control that requires personal intervention is not automatic, but manual.

backfill (noun). Materials such as sand, crushed stone, or soil, that are placed to fill an excavation.

ballast section (railroads). The section of material, generally trap rock, that provides support under railroad tracks.

bonding. The electrical interconnecting of conductive parts, designed to maintain a common electrical potential.

cable. A conductor with insulation, or a stranded conductor with or without insulation and other coverings (single-conductor cable), or a combination of conductors insulated from one another (multiple-conductor cable).

cable jacket. A protective covering over the insulation, core, or sheath of a cable.

cable sheath. A conductive protective covering applied to cables.
NOTE: A cable sheath may consist of multiple layers, of which one or more is conductive.

cable terminal. A device that provides insulated egress for the conductors. *Syn:* **termination**.

circuit. A conductor or system of conductors through which an electric current is intended to flow.

circuit breaker. A switching device capable of making, carrying, and breaking currents under normal circuit conditions and also making, carrying for a specified time, and breaking currents under specified abnormal conditions such as those of short circuit.

clearance. The clear distance between two objects measured surface to surface.

climbing. The vertical movement (ascending and descending) and horizontal movement to access or depart the worksite.

common use. Simultaneous use by two or more utilities of the same kind.

communication lines. *See:* **lines.**

conductor.

1. A material, usually in the form of a wire, cable, or bus bar, suitable for carrying an electric current.

2. **bundled conductor.** An assembly of two or more conductors used as a single conductor and employing spacers to maintain a predetermined configuration. The individual conductors of this assembly are called subconductors.

3. **covered conductor.** A conductor covered with a dielectric having no rated insulating strength or having a rated insulating strength less than the voltage of the circuit in which the conductor is used.

4. **fiber-optic conductor.** *See:* **fiber-optic cable—communication** or **fiber-optic cable—supply.**

5. **grounded conductor.** A conductor that is intentionally grounded, either solidly or through a noninterrupting current-limiting device.

6. **grounding conductor.** A conductor that is used to connect the equipment or the wiring system with a grounding electrode or electrodes.

7. **insulated conductor.** A conductor covered with a dielectric (other than air) having a rated insulating strength equal to or greater than the voltage of the circuit in which it is used.

8. **lateral conductor.** A wire or cable extending in a general horizontal direction at an angle to the general direction of the line conductors, and entirely supported on one structure.

9. **line conductor.** (Overhead supply or communication lines.) A wire or cable intended to carry electric currents, extending along the route of the line, supported by poles, towers, or other structures, but not including vertical or lateral conductors.

10. **open conductor.** A type of electric supply or communication line construction in which the conductors are bare, covered, or insulated and without grounded shielding, individually supported at the structure either directly or with insulators. *Syn:* open wire.

conductor shielding. An envelope that encloses the conductor of a cable and provides an equipotential surface in contact with the cable insulation.

conduit. A structure containing one or more ducts.

NOTE: Conduit may be designated as iron-pipe conduit, tile conduit, etc. If it contains only one duct it is called *single-duct conduit;* if it contains more than one duct it is called *multiple-duct conduit,* usually with the number of ducts as a prefix, for example, *two-duct multiple conduit.*

conduit system. Any combination of duct, conduit, conduits, manholes, handholes, and vaults joined to form an integrated whole.

current-carrying part. A conducting part intended to be connected in an electric circuit to a source of voltage. Non-current-carrying parts are those not intended to be so connected.

de-energized. Disconnected from all sources of electrical supply by open switches, disconnectors, jumpers, taps, or other means.

NOTE: De-energized conductors or equipment could be electrically charged or energized through various means, such as induction from energized circuits, portable generators, lightning, etc.

designated person. A qualified person designated to perform specific duties under the conditions existing. *Syn:* designated employee.

disconnecting or isolating switch. A mechanical switching device used for changing the connections in a circuit or for isolating a circuit or equipment from a source of power.

NOTE: It is required to carry normal load current continuously as well as abnormal or short-circuit current for short intervals, as specified. It is also required to open or close circuits either when negligible current is broken or made, or when no significant change in the voltage across the terminals of each of the switch poles occurs. *Syn:* disconnector, isolator.

duct. A single enclosed raceway for conductors or cable.

effectively grounded. Intentionally connected to earth through a ground connection or connections of sufficiently low impedance and having sufficient current-carrying capacity to limit the buildup of voltages to levels below that which may result in undue hazard to persons or to connected equipment.

electric supply equipment. Equipment that produces, modifies, regulates, controls, or safeguards a supply of electric energy. *Syn:* supply equipment.

electric supply lines. *See:* **lines.**

electric supply station. Any building, room, or separate space within which electric supply equipment is located and the interior of which is accessible, as a rule, only to qualified persons. This includes generating stations and substations, including their associated generator, storage battery, transformer, and switchgear rooms or enclosures, but does not include facilities such as pad-mounted equipment and installations in manholes and vaults.

1. **generating station.** A plant wherein electric energy is produced by conversion from some other form of energy (for example, chemical, nuclear, solar, mechanical, or hydraulic) by means of suitable apparatus. This includes all generating station auxiliaries and other associated equipment required for the operation of the plant. Not included are stations producing power exclusively for use with communications systems.
2. **substation.** An enclosed assemblage of equipment, e.g., switches, circuit breakers, buses, and transformers, under the control of qualified persons, through which electric energy is passed for the purpose of switching or modifying its characteristics.

enclosed. Surrounded by case, cage, or fence designed to protect the contained equipment and limit the likelihood, under normal conditions, of dangerous approach or accidental contact by persons or objects.

energized. Electrically connected to a source of potential difference, or electrically charged so as to have a potential significantly different from that of earth in the vicinity. *Syn:* live.

equipment. A general term including fittings, devices, appliances, fixtures, apparatus, and similar terms used as part of or in connection with an electric supply or communications system.

exposed. Not isolated or guarded.

fall arrest system. The assemblage of equipment, such as a line-worker's body belt, aerial belt, or full body harness in conjunction with a connecting means, with or without an energy absorbing device, and an anchorage to limit the forces a worker can experience during a fall.

fall prevention system. A system, which may include a positioning device system, intended to prevent a worker from falling from an elevation.

fall protection program. A program intended to protect workers from injury due to falls from elevations.

fall protection system (hardware). Consists of either a fall prevention system or a fall arrest system.

fiber-optic cable—communication. A fiber optic cable meeting the requirements for a communication line and located in the communication space of overhead or underground facilities.

fiber-optic cable—supply. A fiber-optic cable located in the supply space of overhead or underground facilities.

fireproofing (of cables). The application of a fire-resistant covering.

generating station. *See:* **electric supply station.**

grounded. Connected to or in contact with earth or connected to some extended conductive body that serves instead of the earth.

grounded effectively. *See:* **effectively grounded**.

grounded system. A system of conductors in which at least one conductor or point is intentionally grounded, either solidly or through a noninterrupting current-limiting device.

guarded. Covered, fenced, enclosed, or otherwise protected, by means of suitable covers or casings, barrier rails or screens, mats or platforms, designed to limit the likelihood, under normal conditions, of dangerous approach or accidental contact by persons or objects.
NOTE: Wires that are insulated but not otherwise protected are not normally considered to be guarded. See exceptions under applicable rules.

handhole. An access opening, provided in equipment or in a below-the-surface enclosure in connection with underground lines, into which personnel reach but do not enter, for the purpose of installing, operating, or maintaining equipment or cable or both.

harness. A component with a design of straps that is fastened about the worker in a manner so as to contain the torso and distribute the fall arrest forces over at least the upper thighs, pelvis, chest, and shoulders with means for attaching it to other components and subsystems.
NOTE: Wherever the word "harness" is used in this code, it refers to full body harness.

in service. Lines and equipment are considered in service when connected to the system and intended to be capable of delivering energy or communication signals, regardless of whether electric loads or signaling apparatus are presently being served from such facilities.

insulated. Separated from other conducting surfaces by a dielectric (including air space) offering a high resistance to the passage of current.
NOTE: When any object is said to be insulated, it is understood to be insulated for the conditions to which it is normally subjected. Otherwise, it is, within the purpose of these rules, uninsulated.

insulation (as applied to cable). That which is relied upon to insulate the conductor from other conductors or conducting parts or from ground.

insulation shielding. An envelope that encloses the insulation of a cable and provides an equipotential surface in contact with the cable insulation.

insulator. Insulating material in a form designed to support a conductor physically and electrically separate it from another conductor or object.

isolated. Not readily accessible to persons unless special means for access are used.

isolated by elevation. Elevated sufficiently so that persons may safely walk underneath.

isolator. *See:* **disconnecting or isolating switch.**

jacket. A protective covering over the insulation, core, or sheath of a cable.

joint use. Simultaneous use by two or more kinds of utilities.

lanyard. A flexible line or webbing, rope, wire rope, or strap that generally has a connector at each end for connecting the line-worker's body belt, aerial belt, or full body harness to an energy absorbing device, lifeline, or anchorage.

limited access highways. As used herein, limited access highways are fully controlled highways where access is controlled by a governmental authority for purposes of improving traffic flow and safety. Fully controlled access highways have no grade crossings and have carefully designed access connections.

lines.

1. **communication lines.** The conductors and their supporting or containing structures that are used for public or private signal or communications service, and which operate at potentials not exceeding 400 V to ground or 750 V between any two points of the circuit, and the transmitted power of which does not exceed 150 W. When operating at not more than 90 V ac or 150 V dc, no limit is placed on the transmitted power of the system. Under specified conditions, communication cables may include communication circuits exceeding the preceding limitation where such circuits are also used to supply power solely to communications equipment.

NOTE: Telephone, telegraph, railroad-signal, data, clock, fire, police-alarm, cable-television, and other systems conforming with the above are included. Lines used for signaling purposes, but not included under the above definition, are considered as supply lines of the same voltage and are to be so installed.

2. **electric supply lines.** Those conductors used to transmit electric energy and their necessary supporting or containing structures. Signal lines of more than 400 V are always supply lines within the meaning of the rules, and those of less than 400 V may be considered as supply lines, if so run and operated throughout. *Syn:* supply lines.

line-worker's body belt. A belt that consists of a belt strap and D-rings and which may include a cushion section or a tool saddle.

live. *See:* **energized.**

manhole. A subsurface enclosure that personnel may enter used for the purpose of installing, operating, and maintaining submersible equipment and cable.

manhole cover. A removable lid that closes the opening to a manhole or similar subsurface enclosure.

manhole grating. A grid that provides ventilation and a protective cover for a manhole opening.

manual. Capable of being operated by personal intervention.

minimum approach distance. The closest distance a qualified employee is permitted to approach either an energized or a grounded object, as applicable for the work method being used.

multigrounded/multiple grounded system. A system of conductors in which a neutral conductor is intentionally grounded solidly at specified intervals. A multigrounded or multiple grounded system may or may not be effectively grounded. *See:* **effectively grounded.**

neutral conductor. A system conductor other than a phase conductor that provides a return path for current to the source. Not all systems have a neutral conductor. An example is an ungrounded delta system containing only three energized phase conductors.

out of service. Lines and equipment are considered out of service when disconnected from the system and when not intended to be capable of delivering energy or communications signals.

pad-mounted equipment. A general term describing enclosed equipment, the exterior of which enclosure is at ground potential, positioned on a surface-mounted pad.

positioning device system. A system of equipment or hardware that, when used with its line-worker's body belt or full body harness, allows a worker to be supported on an elevated vertical surface, such as a pole or tower, and work with both hands free.

positioning strap. A strap with snaphook(s) to connect to the D-rings of a line-worker's body belt or full body harness.

prestressed-concrete structures. Concrete structures that include metal tendons that are tensioned and anchored either before or after curing of the concrete.

pulling iron. An anchor secured in the wall, ceiling, or floor of a manhole or vault to attach rigging used to pull cable.

pulling tension. The longitudinal force exerted on a cable during installation.

qualified. Having been trained in and having demonstrated adequate knowledge of the installation, construction, or operation of lines and equipment and the hazards involved, including identification of and exposure to electric supply and communication lines and equipment in or near the workplace. An employee who is undergoing on-the-job training and who, in the course of such training, has demonstrated an ability to perform duties safely at his or her level of training, and who is under the direct supervision of a qualified person, is considered to be a qualified person for the performance of those duties.

qualified climber. A worker who, by reason of training and experience, understands the methods and has routinely demonstrated proficiency in climbing techniques and familiarity with the hazards associated with climbing.

raceway. Any channel designed expressly and used solely for holding conductors.

random separation. Installed with no deliberate separation.

readily climbable supporting structure. A supporting structure having sufficient handholds and footholds that the structure can be climbed easily by an average person without using a ladder, special tools or devices, or extraordinary physical effort. A supporting structure with a space of 2.45 m (8 ft) or more between handholds and footholds starting not more than 1.80 m (6 ft) above grade is not readily climbable.

NOTE 1: Round poles, diagonal braces on towers, and guy wires are not considered to be readily climbable.

NOTE 2: Diagonal braces on towers are not considered to be handholds or footholds except that, at their point of attachment to a vertical structural member, they may create a foothold.

remotely operable (as applied to equipment). Capable of being operated from a position external to the structure in which it is installed or from a protected position within the structure.

roadway. The portion of highway, including shoulders, for vehicular use. *See also:* **shoulder**; **traveled way.**

NOTE: A divided highway has two or more roadways.

rural districts. All places not urban. This may include thinly settled areas within city limits.

sag.

1. The distance measured vertically from a conductor to the straight line joining its two points of support. Unless otherwise stated in the rule, the sag referred to is the sag at the midpoint of the span. See Fig D-1.
2. **initial unloaded sag.** The sag of a conductor prior to the application of any external load.

3. **final sag.** The sag of a conductor under specified conditions of loading and temperature applied, after it has been subjected for an appreciable period to the loading prescribed for the loading district in which it is situated, or equivalent loading, and the loading removed. Final sag shall include the effect of inelastic deformation (creep).

4. **final unloaded sag.** The sag of a conductor after it has been subjected for an appreciable period to the loading prescribed for the loading district in which it is situated, or equivalent loading, and the loading removed. Final unloaded sag shall include the effect of inelastic deformation (creep).

5. **total sag.** The distance measured vertically from the conductor to the straight line joining its two points of support, under conditions of ice loading equivalent to the total resultant loading for the district in which it is located.

6. **maximum total sag.** The total sag at the midpoint of the straight line joining the two points of support of the conductor.

7. **apparent sag of a span.** The maximum distance between the wire in a given span and the straight line between the two points of support of the wire, measured perpendicularly from the straight line. See Fig D-1.

8. **sag of a conductor at any point in a span.** The distance measured vertically from the particular point in the conductor to a straight line between its two points of support.

9. **apparent sag at any point in the span.** The distance, at the particular point in the span, between the wire and the straight line between the two points of support of the wire, measured perpendicularly from the straight line.

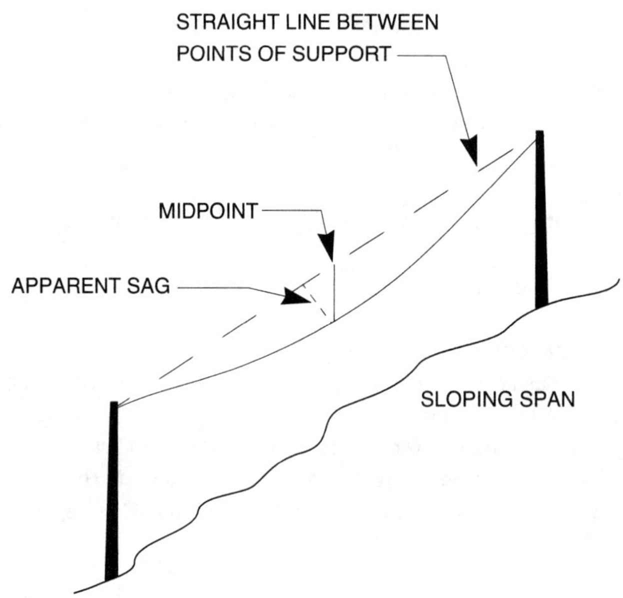

Fig D-1
Sag and Apparent Sag

separation. The distance between two objects, measured surface to surface, and usually filled with a solid or liquid material.

service drop. The overhead conductors between the electric supply or communication line and the building or structure being served.

service point. The point of connection between the facilities of the serving utility and the premises wiring.

shield wire/conductor (also referred to as static wire or ground wire). A grounded conductor installed to shield transmission or distribution phase conductors from direct lightning strikes.

shoulder. The portion of the roadway contiguous with the traveled way for accommodation of stopped vehicles for emergency use and for lateral support of base and surface course.

side-wall pressure. The crushing force exerted on a cable during installation.

single grounded system/unigrounded system. A system of conductors in which one conductor is intentionally grounded solidly at a specific location, typically at the source.

spacer cable. A type of electric supply-line construction consisting of an assembly of one or more covered conductors, separated from each other and supported from a messenger by insulating spacers.

span length. The horizontal distance between two adjacent supporting points of a conductor.

span wire. An auxiliary suspension wire that serves to support one or more trolley contact conductors or a light fixture and the conductors that connect it to a supply system.

structure conflict. A line so situated with respect to a second line that the overturning of the first line will result in contact between its supporting structures or conductors and the conductors of the second line, assuming that no conductors are broken in either line.

substation. *See:* electric supply station.

supply equipment. *See:* **electric supply equipment.**

supply station. *See:* **electric supply station.**

supporting structure. The main supporting unit (usually a pole or tower).

susceptiveness. The characteristics of a communication circuit, including its connected apparatus, that determine the extent to which it is adversely affected by inductive fields.

switch. A device for opening and closing or for changing the connection of a circuit. In these rules, a switch is understood to be manually operable, unless otherwise stated.

switchboard. A type of switchgear assembly that consists of one or more panels with electric devices mounted thereon, and associated framework.

tag. Accident prevention tag (DANGER, PEOPLE AT WORK, etc.) of a distinctive appearance used for the purpose of personnel protection to indicate that the operation of the device to which it is attached is restricted.

termination. *See:* **cable terminal.**

transferring (as applied to fall protection). The act of moving from one distinct object to another (e.g., between an aerial device and a structure).

transformer vault. An isolated enclosure either above or below ground with fire-resistant walls, ceiling, and floor, in which transformers and related equipment are installed, and which is not continuously attended during operation. *See also:* **vault.**

transitioning (as applied to fall protection). The act of moving from one location to another on equipment or a structure.

traveled way. The portion of the roadway for the movement of vehicles, exclusive of shoulders and full-time parking lanes.

ungrounded system. A system of conductors in which no conductor or point is intentionally grounded, either solidly or through a noninterrupting current-limiting device.

unigrounded system. *See:* **single grounded system/unigrounded system.**

unloaded tension.

1. **initial.** The longitudinal tension in a conductor prior to the application of any external load.

2. **final.** The longitudinal tension in a conductor after it has been subjected for an appreciable period to the loading prescribed for the loading district in which it is situated, or equivalent loading, and the loading removed. Final unloaded tension shall include the effect of inelastic deformation (creep).

urban districts. Thickly settled areas (whether in cities or suburbs) or where congested traffic often occurs. A highway, even though in thinly settled areas, on which the traffic is often very heavy, is considered as urban.

utility. An organization responsible for the installation, operation, or maintenance of electric supply or communications systems.

utility interactive system. An electric power production system that is operating in parallel with and capable of delivering energy to a utility electric supply system.

utilization equipment. Equipment, devices, and connected wiring that utilize electric energy for mechanical, chemical, heating, lighting, testing, or similar purposes and are not a part of supply equipment, supply lines, or communication lines.

vault. A structurally solid enclosure, including all sides, top, and bottom, above or below ground where entry is limited to personnel qualified to install, maintain, operate, or inspect the equipment or cable enclosed. The enclosure may have openings for ventilation, personnel access, cable entrance, and other openings required for operation of equipment in the vault.

voltage.

1. The effective (rms) potential difference between any two conductors or between a conductor and ground. Voltages are expressed in nominal values unless otherwise indicated. The nominal voltage of a system or circuit is the value assigned to a system or circuit of a given voltage class for the purpose of convenient designation. The operating voltage of the system may vary above or below this value.

2. **voltage of circuit not effectively grounded.** The highest nominal voltage available between any two conductors of the circuit.

NOTE: If one circuit is directly connected to and supplied from another circuit of higher voltage (as in the case of an autotransformer), both are considered to be of the higher voltage, unless the circuit of the lower voltage is effectively grounded, in which case its voltage is not determined by the circuit of higher voltage. Direct connection implies electric connection as distinguished from connection merely through electromagnetic or electrostatic induction.

3. **voltage of a constant-current circuit.** The highest normal full-load voltage of the circuit.

4. **voltage of an effectively grounded circuit.** The highest nominal voltage available between any conductor of the circuit and ground unless otherwise indicated.

5. **voltage to ground of:**
   a. **a grounded circuit.** The highest nominal voltage available between any conductor of the circuit and that point or conductor of the circuit that is grounded.
   b. **an ungrounded circuit.** The highest nominal voltage available between any two conductors of the circuit concerned.
6. **voltage to ground of a conductor of:**
   a. **a grounded circuit.** The nominal voltage between such conductor and that point or conductor of the circuit that is grounded.
   b. **an ungrounded circuit.** The highest nominal voltage between such conductor and any other conductor of the circuit concerned.

wire gages. Throughout these rules the American Wire Gage (AWG), formerly known as Brown & Sharpe (B&S), is the standard gage for copper, aluminum, and other conductors, excepting only steel conductors, for which the Steel Wire Gage (Stl WG) is used.
NOTE: The Birmingham Wire Gage is obsolete.

worksite (as applied to fall protection). The location on the structure or equipment where, after the worker has completed the climbing (horizontally and vertically), the worker is in position to perform the assigned work or task.

Section 3.
References

The following standards form a part of the National Electrical Safety Code to the extent indicated in the rules herein.[4]

ANSI C29.1-1988, American National Standard Test Methods for Electrical Power Insulators. [Rules 272, 273, and 277 NOTE 1a][5]

ANSI C29.2-1992, American National Standard for Wet-Process Porcelain and Toughened Glass Insulators (Suspension Type). [Rules 272 and 441B4c]

ANSI C29.3-1986 (R1995), American National Standard for Wet-Process Porcelain Insulators (Spool Type). [Rule 272]

ANSI C29.4-1989 (R1995), American National Standard for Wet-Process Porcelain Insulators (Strain Type). [Rule 272]

ANSI C29.5-1984 (R1995), American National Standard for Low- and Medium-Voltage Pin Type Wet-Process Porcelain Insulators. [Rule 272]

ANSI C29.6-1996, American National Standard for High-Voltage Pin Type Wet-Process Porcelain Insulators. [Rule 272]

ANSI C29.7-1996, American National Standard for High-Voltage Line-Post Type Wet-Process Porcelain Insulators. [Rules 272 and 277 NOTE 2]

ANSI C84.1-1995 (R2001), American National Standard for Voltage Ratings for Electric Power Systems and Equipment (60 Hz). [Rule 441]

ANSI O5.1-1992, American National Standard Specifications and Dimensions for Wood Poles. [Rule 261]

ANSI Z535.1-1998, American National Standard for Safety Color Code. [Rule 110A1 NOTE, Rule 217A1c NOTE, Rule 323C4 NOTE, Rule 381G2 NOTE, and Rule 411D]

ANSI Z535.2-1998, American National Standard for Environmental and Facility Safety Signs. [Rule 110A1 NOTE, Rule 217A1c NOTE, Rule 323C4 NOTE, Rule 381G2 NOTE, and Rule 411D]

ANSI Z535.3-1998, American National Standard for Criteria for Safety Symbols. [Rule 110A1 NOTE, Rule 217A1c NOTE, Rule 323C4 NOTE, Rule 381G2 NOTE, and Rule 411D]

ANSI Z535.4-1998, American National Standard for Product Safety Signs and Labels. [Rule 110A1 NOTE, Rule 217A1c NOTE, Rule 323C4 NOTE, Rule 381G2 NOTE, and Rule 411D]

ANSI Z535.5-1998, American National Standard for Accident Prevention Tags (for Temporary Hazards). [Rule 110A1 NOTE, Rule 217A1c NOTE, Rule 323C4 NOTE, Rule 381G2 NOTE, and Rule 411D]

[4]The standards listed here were the editions used in this revision of the code. In some cases, newer editions may be in effect. Contact the publisher for information about availability.

[5]ANSI publications are available from the Sales Department, American National Standards Institute, 11 West 42nd Street, 13th Floor, New York, NY 10036, USA (http://www.ansi.org/).

ANSI/SIA A92.2-1990, American National Standard for Vehicle Mounted Elevating and Rotating Aerial Devices. [Rule 446]

ASCE 7-1998, ASCE Standard for Minimum Design Loads for Buildings and Other Structures [250C][6]

ASME B15.1-2000, ASME Standard for Mechanical Power Transmission Apparatus. [Rule 122]

ASTM D 178-1993 (R1998), ASTM Standard Specification for Rubber Insulating Matting.[7] [Rule 124]

IEEE Std 4-1995, IEEE Standard Techniques for High-Voltage Testing.[8] [Rule 441]

IEEE 100, The Authoritative Dictionary of IEEE Standards Terms, Seventh Edition. [Section 2, General]

IEEE Std 516-1995, IEEE Guide for Maintenance Methods on Energized Power-Lines. [Rules 441 and 446]

IEEE Std 1313-1993, IEEE Standard for Power Systems—Insulation Coordination. [Rule 124][9]

NFPA 30-2000, Flammable and Combustible Liquids Code.[10] [Rule 127]

NFPA 58-2001, Storage and Handling of Liquefied Petroleum Gases. [Rule 127]

NFPA 59-2001, Storage and Handling of Liquefied Petroleum Gases at Utility Gas Plants. [Rule 127]

NFPA 59A-2001, Production, Storage, and Handling of Liquefied Natural Gas (LNG). [Rule 127]

NFPA 70-1999, National Electrical Code (NEC). [Rules 011, 124, and 127]

NFPA 496-1998, Standard for Purged and Pressurized Enclosures for Electrical Equipment. [Rule 127]

NFPA 8503-1997, Standard for the Installation and Operation of Pulverized Fuel Systems. [Rule 127]

[6]ASCE publications are available from ASCE Publications, 1801 Alexander Bell Dr., Reston, VA 20191, (800) 548-ASCE (2723) (http://www.asce.org).

[7]ASTM publications are available from the American Society for Testing and Materials, 100 Barr Harbor Drive, West Conshohocken, PA 19428-2959, USA (http://www.astm.org/).

[8]IEEE publications are available from the Institute of Electrical and Electronics Engineers, 445 Hoes Lane, P.O. Box 1331, Piscataway, NJ 08855-1331, USA (http://standards.ieee.org/).

[9]IEEE Std 1313-1993 has been withdrawn; however, copies can be obtained from Global Engineering, 15 Inverness Way East, Englewood, CO 80112-5704, USA, tel. (303) 792-2181 (http://global.ihs.com/).

[10]NFPA publications are available from Publications Sales, National Fire Protection Association, 1 Batterymarch Park, P.O. Box 9101, Quincy, MA 02269-9101, USA.

Section 9.
Grounding Methods for Electric Supply and Communications Facilities

090. Purpose

The purpose of Section 9 of this code is to provide practical methods of grounding, as one of the means of safeguarding employees and the public from injury that may be caused by electrical potential.

091. Scope

Section 9 of this code covers methods of protective grounding of supply and communication conductors and equipment. The rules requiring grounding are in other parts of this code.

These rules do not cover the grounded return of electric railways nor those lightning protection wires that are normally independent of supply or communication wires or equipment.

092. Point of Connection of Grounding Conductor

A. Direct Current Systems That Are to Be Grounded

1. 750 V and below

Connection shall be made only at supply stations. In three-wire dc systems, the connection shall be made to the neutral.

2. Over 750 V

Connection shall be made at both the supply and load stations. The connection shall be made to the neutral of the system. The ground or grounding electrode may be external to or remotely located from each of the stations.

One of the two stations may have its grounding connection made through surge arresters provided the other station neutral is effectively grounded as described above.

EXCEPTION: Where the stations are not geographically separated as in back-to-back converter stations, the neutral of the system should be connected to ground at one point only.

B. Alternating Current Systems That Are to Be Grounded

1. 750 V and below

The point of the grounding connection on a wye-connected three-phase four-wire system, or on a single-phase three-wire system, shall be the neutral conductor. On other one-, two-, or three-phase systems with an associated lighting circuit or circuits, the point of grounding connection shall be on the common circuit conductor associated with the lighting circuits.

The point of grounding connection on a three-phase three-wire system, whether derived from a delta-connected or an ungrounded wye-connected transformer installation not used for lighting, may be any of the circuit conductors, or it may be a separately derived neutral.

The grounding connections shall be made at the source, and at the line side of all service equipment.

2. Over 750 V

a. Nonshielded (Bare or Covered Conductors or Insulated Nonshielded Cables)

Grounding connection shall be made at the neutral of the source. Additional connections may be made, if desired, along the length of the neutral, where this is one of the system conductors.

b. Shielded

(1) Surge-Arrester Cable-Shielding Interconnection

Cable-shielding grounds shall be bonded to surge-arrester grounds, where provided, at points where underground cables are connected to overhead lines.

(2) Cable Without Insulating Jacket

Connection shall be made to the neutral of the source transformer and at cable termination points.

(3) Cable With Insulating Jacket

Additional bonding and connections between the cable insulation shielding or sheaths and the system ground are recommended. In multi-grounded shielded cable systems, the shielding (including sheath) shall be grounded at each cable joint exposed to personnel contact. Where multi-grounded shielding cannot be used for electrolysis or sheath-current reasons, the shielding sheaths and splice-enclosure devices shall be insulated for the voltage that may appear on them during normal operation.

Bonding transformers or reactors may be substituted for direct ground connection at one end of the cable.

3. Separate Grounding Conductor

If a separate grounding conductor is used as an adjunct to a cable run underground, it shall be connected either directly or through the system neutral to the source transformers, source transformer accessories, and cable accessories where these are to be grounded. This grounding conductor shall be located in the same direct burial or duct bank run (or the same duct if this is of magnetic material) as the circuit conductors.

EXCEPTION: The grounding conductor for a circuit that is installed in a magnetic duct need not be in the same duct if the duct containing the circuit is bonded to the separate grounding conductor at both ends.

C. Messenger Wires and Guys

1. Messenger Wires

Messenger wires required to be grounded shall be connected to grounding conductors at poles or structures at maximum intervals as listed below:

a. Where messenger wires are adequate for system grounding conductors (Rules 093C1, 093C2, and 093C5), four connections in each 1.6 km (1 mi).

b. Where messenger wires are not adequate for system grounding conductors, eight connections in each 1.6 km (1 mi), exclusive of service grounds.

2. Guys

Guys that are required to be grounded shall be connected to one or more of the following:

a. A grounded metallic supporting structure.

b. An effective ground on a nonmetallic supporting structure.

c. A line conductor that has at least four ground connections in each mile of line in addition to the ground connections at individual services.

3. Common Grounding of Messengers and Guys on the Same Supporting Structure

a. Where messengers and guys on the same supporting structure are required to be grounded, they shall be bonded together and grounded by connection to:

(1) One grounding conductor that is grounded at that structure, or to

(2) Separate grounding conductors or grounded messengers that are bonded together and grounded at that structure, or to

(3) One or more grounded line conductors or grounded messengers that are (a) bonded together at this structure or elsewhere and (b) multi-grounded elsewhere at intervals as specified in Rules 092C1 and 092C2.

b. At common crossing structures, messengers and guys that are required to be grounded shall be bonded together at that structure and grounded in accordance with Rule 092C3a.

EXCEPTION: This rule does not apply to guys that are connected to an effectively grounded overhead static wire.

D. Current in Grounding Conductor

Ground connection points shall be so arranged that under normal circumstances there will be no objectionable flow of current over the grounding conductor. If an objectionable flow of current occurs over a grounding conductor due to the use of multi-grounds, one or more of the following should be used:

1. Abandon one or more grounds.
2. Change location of grounds.
3. Interrupt the continuity of the conductor between ground connections.

4. Subject to the approval of the administrative authority, take other effective means to limit the current.

   The system ground of the source transformer shall not be removed.

   The temporary currents set up under abnormal conditions while the grounding conductors are performing their intended protective functions are not considered objectionable. The conductor shall have the capability of conducting anticipated fault current without thermal overloading or excessive voltage buildup. Refer to Rule 093C.

E. Fences

Fences that are required to be grounded by other parts of this code shall be designed to limit touch, step, and transferred voltages in accordance with industry practices.

NOTE: IEEE Std 80-2000 [B22][12] is one source that may be utilized to provide guidance in meeting these requirements.

The grounding connections shall be made either to the grounding system of the enclosed equipment or to a separate ground.

1. Fences shall be grounded at each side of a gate or other opening.
2. Gates shall be bonded to the grounding conductor, jumper, or fence.
3. A buried bonding jumper shall be used to bond across a gate or other opening in the fence, unless a nonconducting fence section is used.
4. If barbed wire strands are used above the fence fabric, the barbed wire strands shall be bonded to the grounding conductor, jumper, or fence.
5. When fence posts are of conducting material, the grounding conductor shall be connected to the fence post or posts, as required, with suitable connecting means.
6. When fence posts are of nonconducting material, suitable bonding connection shall be made to the fence mesh strands and the barbed wire strands at each grounding conductor point.

093. Grounding Conductor and Means of Connection

A. Composition of Grounding Conductors

In all cases, the grounding conductor shall be made of copper or other metals or combinations of metals that will not corrode excessively during the expected service life under the existing conditions and, if practical, shall be without joint or splice. If joints are unavoidable, they shall be so made and maintained as to not materially increase the resistance of the grounding conductor and shall have appropriate mechanical and corrosion-resistant characteristics. For surge arresters and ground detectors, the grounding conductor or conductors shall be as short, straight, and free from sharp bends as practical. Metallic electrical equipment cases or the structural metal frame of a building or structure may serve as part of a grounding conductor to an acceptable grounding electrode.

In no case shall a circuit-opening device be inserted in the grounding conductor or connection except where its operation will result in the automatic disconnection from all sources of energy of the circuit leads connected to the equipment so grounded.

EXCEPTION 1: For dc systems over 750 V, grounding conductor circuit-opening devices shall be permitted for changing between a remote electrode and a local ground through surge arresters.

EXCEPTION 2: Temporary disconnection of grounding conductors for testing purposes, under competent supervision, shall be permitted.

EXCEPTION 3: Disconnection of a grounding conductor from a surge arrester is allowed when accomplished by means of a surge-arrester disconnector.

NOTE: The base of the surge arrester may remain at line potential following operation of the disconnector.

[12]The numbers in brackets correspond to those bibliographical items listed in Appendix B.

B. Connection of Grounding Conductors

Connection of the grounding conductor shall be made by a means matching the characteristics of both the grounded and grounding conductors, and shall be suitable for the environmental exposure. These means include brazing, welding, mechanical and compression connections, ground clamps, and ground straps. Soldering is acceptable only in conjunction with lead sheaths.

C. Ampacity and Strength

For bare grounding conductors, the short time ampacity is the current that the conductor can carry for the time during which the current flows without melting or affecting the design characteristics of the conductor. For insulated grounding conductors, the short time ampacity is the current that the conductor can carry for the applicable time without affecting the design characteristics of the insulation. Where grounding conductors at one location are paralleled, the increased total current capacity may be considered.

1. System Grounding Conductors for Single-Grounded Systems

   The system grounding conductor or conductors for a system with single-system grounding electrode or set of electrodes, exclusive of grounds at individual services, shall have a short time ampacity adequate for the fault current that can flow in the grounding conductors for the operating time of the system-protective device. If this value cannot be readily determined, continuous ampacity of the grounding conductor or conductors shall be not less than the full-load continuous current of the system supply transformer or other source of supply.

2. System Grounding Conductors for Multi-Grounded Alternating Current Systems

   The system grounding conductors for an ac system with grounds at more than one location exclusive of grounds at individual services shall have continuous total ampacities at each location of not less than one-fifth that of the conductors to which they are attached. (See also Rule 093C8.)

3. Grounding Conductors for Instrument Transformers

   The grounding conductor for instrument cases and secondary circuits for instrument transformers shall not be smaller than AWG No. 12 copper or shall have equivalent short time ampacity.

4. Grounding Conductors for Primary Surge Arresters

   The grounding conductor or conductors shall have adequate short time ampacity under conditions of excess current caused by or following a surge. Individual arrester grounding conductors shall be no smaller than AWG No. 6 copper or AWG No. 4 aluminum.

   EXCEPTION: Arrester grounding conductors may be copper-clad or aluminum-clad steel wire having not less than 30% of the conductivity of solid copper or aluminum wire of the same diameter, respectively.

   Where flexibility of the grounding conductor, such as adjacent to the base of the arrester, is vital to its proper operation, a suitably flexible conductor shall be employed.

5. Grounding Conductors for Equipment, Messenger Wires, and Guys

   a. Conductors

      The grounding conductors for equipment, raceways, cable, messenger wires, guys, sheaths, and other metal enclosures for wires shall have short time ampacities adequate for the available fault current and operating time of the system fault-protective device. If no overcurrent or fault protection is provided, the ampacity of the grounding conductor shall be determined by the design and operating conditions of the circuit, but shall be not less than that of AWG No. 8 copper. Where the adequacy and continuity of the conductor enclosures and their attachment to the equipment enclosures is assured, this path can constitute the equipment grounding conductor.

   b. Connections

      Connections of the grounding conductor shall be to a suitable lug, terminal, or device not disturbed in normal inspection, maintenance, or operation.

6. Fences

   The grounding conductor for fences required to be grounded by other parts of this code shall meet the requirements of Rule 093C5 or shall be steel wire not smaller than Stl WG No. 5.

7. Bonding of Equipment Frames and Enclosures

   Where required, a low-impedance metallic path shall be provided to conduct fault current back to the grounded terminal of the local supply. Where the supply is remote, the metallic path shall interconnect the equipment frames and enclosures with all other nonenergized conducting components within reach and shall additionally be connected to ground as outlined in Rule 093C5. Short time ampacities of bonding conductors shall be adequate for the duty involved.

8. Ampacity Limit

   No grounding conductor need have greater ampacity than either:

   a. The phase conductors that would supply the ground fault current, or
   b. The maximum current that can flow through it to the ground electrode or electrodes to which it is attached. For a single grounding conductor and connected electrode or electrodes, this would be the supply voltage divided by the electrode resistance (approximately).

9. Strength

   All grounding conductors shall have mechanical strength suitable for the conditions to which they may reasonably be subjected.

   Furthermore, unguarded grounding conductors shall have a tensile strength not less than that of AWG No. 8 soft-drawn copper, except as noted in Rule 093C3.

D. Guarding and Protection

1. The grounding conductors for single-grounded systems and those exposed to mechanical damage shall be guarded. However, grounding conductors need not be guarded where not readily accessible to the public nor where grounding multi-grounded circuits or equipment.
2. Where guarding is required, grounding conductors shall be protected by guards suitable for the exposure to which they may reasonably be subjected. The guards should extend for not less than 2.45 m (8 ft) above the ground or platform from which the grounding conductors are accessible to the public.
3. Where guarding is not required, grounds shall be protected by being substantially attached closely to the surface of the pole or other structure in areas of exposure to mechanical damage and, where practical, on the portion of the structure having least exposure.
4. Guards used for grounding conductors of lightning-protection equipment shall be of nonmetallic materials if the guard completely encloses the grounding conductor or is not bonded at both ends to the grounding conductor.

E. Underground

1. Grounding conductors laid directly underground shall be laid slack or shall be of sufficient strength to allow for earth movement or settling that is normal at the particular location.
2. Direct-buried uninsulated joints or splices in grounding conductors shall be made with methods suitable for the application and shall have appropriate corrosion resistance, required permanence, appropriate mechanical characteristics, and required ampacity. The number of joints or splices should be the minimum practical.
3. Grounding cable insulation shielding systems shall be interconnected with all other accessible grounded power supply equipment in manholes, handholes, and vaults.

   EXCEPTION: Where cathodic protection or shield cross-bonding is involved, interconnection may be omitted.

4. Looped magnetic elements such as structural steel, piping, reinforcing bars, etc., should not separate grounding conductors from the phase conductors of circuits they serve.
5. Metals used for grounding, in direct contact with earth, concrete, or masonry, shall have been proven suitable for such exposure.

   NOTE 1: Under present technology, aluminum has not generally been proven suitable for such use.

   NOTE 2: Metals of different galvanic potentials that are electrically interconnected may require protection against galvanic corrosion.

6. Sheath Transposition Connections (Cross-Bonding)
   a. Where cable insulating shields or sheaths, which are normally connected to ground, are insulated from ground to minimize shield circulating currents, they shall be insulated from personnel contact at accessible locations. Transposition connections and bonding jumpers shall be insulated for nominal 600 V service, unless the normal shielding voltage exceeds this level, in which case the insulation shall be ample for the working voltage to ground.
   b. Bonding jumpers and connecting means shall be sized and selected to carry the available fault current without damaging jumper insulation or sheath connections.

F. Common Grounding Conductor for Circuits, Metal Raceways, and Equipment

Where the ampacity of a supply system grounding conductor is also adequate for equipment grounding requirements, this conductor may be used for the combined purpose. Equipment referred to includes the frames and enclosures of supply system control and auxiliary components, conductor raceways, cable shields, and other enclosures.

094. Grounding Electrodes

The grounding electrode shall be permanent and adequate for the electrical system involved. A common electrode or electrode system shall be employed for grounding the electrical system and the conductor enclosures and equipment served by that system. This may be accomplished by interconnecting these elements at the point of connection of grounding conductor, Rule 092.

Grounding electrodes shall be one of the following:

A. Existing Electrodes

Existing electrodes consist of conducting items installed for purposes other than grounding:

1. Metallic Water Piping System

   Extensive metallic underground cold water piping systems may be used as grounding electrodes.

   NOTE: Such systems normally have very low resistance to earth and have been extensively used in the past. They are the preferred electrode type where they are readily accessible.

   EXCEPTION: Water systems with nonmetallic, non-current-carrying pipe or insulating joints are not suitable for use as grounding electrodes.

2. Local Systems

   Isolated buried metallic cold water piping connecting to wells having sufficiently low measured resistance to earth may be used as grounding electrodes.

   NOTE: Care should be exercised to ensure that all parts that might become disconnected are effectively bonded together.

3. Steel Reinforcing Bars in Concrete Foundations and Footings

   The reinforcing bar system of a concrete foundation or footing that is not insulated from direct contact with earth, and that extends at least 900 mm (3 ft) below grade, constitutes an effective and acceptable type of grounding electrode. Where steel supported on this foundation is to be used as a grounding conductor (tower, structure, etc.), it shall be interconnected by bonding between anchor bolts and reinforcing bars or by cable from the reinforcing bars to the structure above the concrete.

   The normally applied steel ties are considered to provide adequate bonding between bars of the reinforcing cage.

   NOTE: Where reinforcing bars in concrete are not suitably connected to a metal structure above the concrete, and the latter structure is subjected to grounding discharge currents (even connected to another electrode), there is likelihood of damage to the intervening concrete from ground-seeking current passing through the semiconducting concrete.

B. Made Electrodes

1. General

   Where made electrodes are used, they shall, as far as practical, penetrate permanent moisture level and below the frostline. Made electrodes shall be of metal or combinations of metals that do not corrode excessively under the existing conditions for the expected service life.

All outer surfaces of made electrodes shall be conductive, that is, not having paint, enamel, or other covering of an insulating type.

2. Driven Rods
   a. Driven rods may be sectional; the total length shall be not less than 2.45 m (8 ft). Iron or steel rods shall have a cross-sectional dimension of not less than 15 mm (5/8 in). Copper-clad, stainless steel, or stainless steel-clad rods shall have a cross-sectional dimension of not less than 12 mm (1/2 in).
   b. Longer rods or multiple rods may be used to reduce the ground resistance. Spacing between multiple rods should be not less than 1.8 m (6 ft).

      EXCEPTION: Other dimensions or configurations may be used if their suitability is supported by a qualified engineering study.
   c. Driven depth shall be not less than 2.45 m (8 ft). The upper end shall be flush with or below the ground level unless suitably protected.

      EXCEPTION 1: Where rock bottom is encountered, driven depth may be less than 2.45 m (8 ft), or other types of electrode may be employed.

      EXCEPTION 2: When contained within pad-mounted equipment, vaults, manholes, or similar enclosures, the driven depth may be reduced to 2.3 m (7.5 ft).
3. Buried Wire, Strips, or Plates

   In areas of high soil resistivity or shallow bedrock, or where lower resistance is required than attainable with driven rods, one or more of the following electrodes may be more useful:
   a. Wire

      Bare wires 4 mm (0.162 in) in diameter or larger, conforming to Rule 093E5, buried in earth at a depth not less than 450 mm (18 in) and not less than 30 m (100 ft) total in length, laid approximately straight, constitute an acceptably made electrode. (This is frequently designated a counterpoise.) The wire may be in a single length or may be several lengths connected at ends or at some point away from the ends. The wire may take the form of a network with many parallel wires spaced in two-dimensional array, referred to as a grid.

      EXCEPTION 1: Where rock bottom is encountered, burial depth may be less than 450 mm (18 in).

      EXCEPTION 2: Other lengths or configurations may be used if their suitability is supported by a qualified engineering study.
   b. Strips

      Strips of metal not less than 3.0 m (10 ft) in total length and with total (two sides) surface not less than 0.47 m^2 (5 ft^2) buried in soil at a depth not less than 450 mm (18 in) constitute an acceptably made electrode. Ferrous metal electrodes shall be not less than 6 mm (0.25 in) in thickness and nonferrous metal electrodes not less than 1.5 mm (0.06 in).

      NOTE: Strip electrodes are frequently useful in rocky areas where only irregularly shaped pits are practical to excavate.
   c. Plates or Sheets

      Metal plates or sheets having not less than 0.185 m^2 (2 ft^2) of surface exposed to the soil, and at a depth of not less than 1.5 m (5 ft), constitute an acceptable made electrode. Ferrous metal electrodes shall be not less than 6 mm (0.25 in) in thickness and nonferrous metal electrodes not less than 1.5 mm (0.06 in).
4. Pole-Butt Plates and Wire Wraps
   a. General

      In areas of very low soil resistivity there are two constructions, described in specifications b and c below, that may provide effective grounding electrode functions although they are inadequate in most other locations. Where these have been proven to have adequately low earth resistance by the application of Rule 96, two such electrodes may be counted as one made electrode and ground for application of Rules 92C1a, 92C2b, 96C, and 97C; however, these types shall not be the sole grounding electrode at transformer locations.
   b. Pole-Butt Plates

      Subject to the limitations of Rule 94B4a, a pole-butt plate on the base of a wooden pole, possibly folded up around the base of the pole butt, may be considered an acceptable electrode in locations where the limitations of Rule 96 are met. The plates shall be not less than 6 mm (1/4 in) thick if of ferrous metal and not less than 1.5 mm (0.06 in) thick if of nonferrous metal. Further, the plate area exposed to the soil shall be not less than 0.046 m^2 (0.5 ft^2).

c. Wire Wrap

Subject to the limitations of Rule 94B4a, made electrodes may be wire attached to the pole previous to the setting of the pole. The wire shall be of copper or other metals that will not corrode excessively under the existing conditions and shall have a continuous bare or exposed length below ground level of not less than 3.7 m (12 ft), shall extend to the bottom of the pole, and shall not be smaller than AWG No. 6.

5. Concentric Neutral Cable

Systems employing extensive [30 m (100 ft) minimum length] buried bare concentric neutral cable in contact with the earth may employ the concentric neutral as a grounding electrode. The concentric neutral may be covered with a semi-conducting jacket that has a radial resistivity not exceeding 100 m·Ω and that will remain essentially stable in service. The radial resistivity of the jacket material is that value calculated from measurements on a unit length of cable, of the resistance between the concentric neutral and a surrounding conducting medium. Radial resistivity equals resistance of unit length times the surface area of jacket divided by the average thickness of the jacket over the neutral conductors. All dimensions are to be expressed in meters.

6. Concrete-Encased Electrodes

A metallic wire, rod, or structural shape, meeting Rule 093E5 and encased in concrete, that is not insulated from direct contact with earth, shall constitute an acceptable ground electrode. The concrete depth below grade shall be not less than 300 mm (1 ft), and a depth of 750 mm (2.5 ft) is recommended. Wire shall be no smaller than AWG No. 4 if copper, or 9 mm (3/8 in) diameter or AWG No. 1/0 if steel. It shall be not less than 6.1 m (20 ft) long, and shall remain entirely within the concrete except for the external connection. The conductor should be run as straight as practical.

The metal elements may be composed of a number of shorter lengths arrayed within the concrete and connected together (for example, the reinforcing system in a structural footing).

EXCEPTION: Other wire length or configurations may be used if their suitability is supported by a qualified engineering study.

NOTE 1: The lowest resistance per unit wire length will result from a straight wire installation.

NOTE 2: The outline of the concrete need not be regular, but may conform to an irregular or rocky excavation.

NOTE 3: Concrete-encased electrodes are frequently more practical or effective than driven rods or strips or plates buried directly in earth.

095. Method of Connection to Electrode

A. Ground Connections

The grounding connection shall be as accessible as practical and shall be made to the electrode by methods that provide the required permanence, appropriate mechanical characteristics, corrosion resistance, and required ampacity such as:

1. An effective clamp, fitting, braze, or weld.
2. A bronze plug that has been tightly screwed into the electrode.
3. For steel-framed structures, employing a concrete-encased reinforcing bar electrode, a steel rod similar to the reinforcing bar shall be used to join, by welding, a main vertical reinforcing bar to an anchor bolt. The bolt shall be substantially connected to the baseplate of the steel column supported on that footing. The electrical system may then be connected (for grounding) to the building frame by welding or by a bronze bolt tapped into a structural member of that frame.
4. For nonsteel frame structures employing a concrete-encased rod or wire electrode, an insulated copper conductor of size meeting the requirements of Rule 093C (except not smaller than AWG No. 4) shall be connected to the steel rod or wire using a cable clamp suitable for steel cable. This clamp and all the bared portion of the copper conductor, including ends of exposed strands within the concrete, shall be completely covered with mastic or sealing compound before concrete is poured. The copper conductor end shall be brought to or out of the concrete surface at the required location for connection to the electrical system. If the copper wire is carried beyond the surface of the concrete, it shall be no smaller than AWG No. 2.

Alternately, the copper wire may be brought out of the concrete at the bottom of the hole and carried external to the concrete for surface connection.

B. Point of Connection to Piping Systems

1. The point of connection of a grounding conductor to a metallic water piping system shall be as near as is practical to the water-service entrance to the building or near the equipment to be grounded and shall be accessible. If a water meter is between the point of connection and the underground water pipe, the metallic water piping system shall be made electrically continuous by bonding together all parts between the connection and the pipe entrance that may become disconnected, such as meters and service unions.
2. Made grounds or grounded structures should be separated by 3.0 m (10 ft) or more from pipelines used for the transmission of flammable liquids or gases operating at high pressure [1030 kPa (150 lb/in^2) or greater] unless they are electrically interconnected and cathodically protected as a single unit. Grounds within 3.0 m (10 ft) of such pipelines should be avoided or shall be coordinated so that hazardous ac conditions will not exist and cathodic protection of the pipeline will not be nullified.

   RECOMMENDATION: It is recommended that calculations or tests be used to determine the required separation of ground electrodes for high-voltage direct-current (HVDC) systems from flammable liquid or high-pressure gas pipelines.

   NOTE: Ground electrodes for HVDC systems over 750 V may require greater separation.

C. Contact Surfaces

If any coating of nonconducting material, such as enamel, rust, or scale, is present on electrode contact surfaces at the point of connection, such a coating shall be thoroughly removed where required to obtain the requisite good connection. Special fittings so designed as to make such removal of nonconducting coatings unnecessary may also be used.

096. Ground Resistance Requirements

Grounding systems shall be designed to minimize hazard to personnel and shall have resistances to ground low enough to permit prompt operation of circuit protective devices. Grounding systems may consist of buried conductors and grounding electrodes.

A. Supply Stations

Supply stations may require extensive grounding systems consisting of multiple buried conductors, grounding electrodes, or interconnected combinations of both. Grounding systems shall be designed to limit touch, step, mesh, and transferred potentials in accordance with industry practices.

NOTE: IEEE Std 80-2000 [B21] is one source that may be utilized to provide guidance in meeting these requirements.

B. Single-Grounded (Unigrounded or Delta) Systems

Individual made electrodes shall, where practical, have a resistance to ground not exceeding 25 Ω. If a single electrode resistance exceeds 25 Ω, two electrodes connected in parallel shall be used.

C. Multi-Grounded Systems

The neutral, which shall be of sufficient size and ampacity for the duty involved, shall be connected to a made or existing electrode at each transformer location and at a sufficient number of additional points with made or existing electrodes to total not less than four grounds in each 1.6 km (1 mi) of the entire line, not including grounds at individual services.

EXCEPTION: Where underwater crossings are encountered, the requirement of made electrodes to total not less than four grounds in each 1.6 km (1 mi) of the entire line does not apply for the underwater portion if the neutral is of sufficient size and capacity for the duty involved and the requirements of Rule 092B2 are met.

NOTE: Multi-grounded systems extending over a substantial distance are more dependent on the multiplicity of grounding electrodes than on the resistance to ground of any individual electrode. Therefore, no specific values are imposed for the resistance of individual electrodes.

RECOMMENDATION: This rule may be applied to shield wire(s) grounded at the source and which meet the multigrounding requirements of this rule.

097. Separation of Grounding Conductors

A. Except as permitted in Rule 097B, grounding conductors from equipment and circuits of each of the following classes shall be run separately to the grounding electrode for each of the following classes:

1. Surge arresters of circuits over 750 V and frames of any equipment operating at over 750 V.
2. Lighting and power circuits under 750 V.
3. Shield wires of power circuits.
4. Lightning rods, unless attached to a grounded metal supporting structure.

Alternatively, the grounding conductors shall be run separately to a sufficiently heavy ground bus or system ground cable that is well connected to ground at more than one place.

B. The grounding conductors of the equipment classes detailed in Rules 097A1, 097A2, and 097A3 may be interconnected utilizing a single grounding conductor, provided:

1. There is a direct-earth grounding connection at each surge-arrester location, and
2. The secondary neutral or the grounded secondary phase conductor is common with or connected to a primary neutral or a shield wire meeting the grounding requirements of Rule 097C.

C. Primary and secondary circuits utilizing a single conductor as a common neutral shall have at least four ground connections on such conductor in each 1.6 km (1 mi) of line, exclusive of ground connections at customers' service equipment.

D. Ungrounded or Single-Grounded Systems and Multi-Grounded Systems

1. Ungrounded or Single-Grounded Systems

   Where the secondary neutral is not interconnected with the primary surge-arrester grounding conductor as in Rule 097B, interconnection may be made through a spark gap or device that performs an equivalent function. The gap or device shall have a 60 Hz breakdown voltage of at least twice the primary circuit voltage but not necessarily more than 10 kV. At least one other grounding connection on the secondary neutral shall be provided with its grounding electrode located at a distance of not less than 6.1 m (20 ft) from the surge-arrester grounding electrode in addition to customer's grounds at each service entrance. The primary grounding conductor, or the secondary grounding conductor, shall be insulated for 600 V.

2. Multi-Grounded Systems

   On multi-grounded systems, the primary and secondary neutrals should be interconnected according to Rule 097B. However, where it is necessary to separate the neutrals, interconnection of the neutrals shall be made through a spark gap or a device that performs an equivalent function. The gap or device shall have a 60 Hz breakdown voltage not exceeding 3 kV. At least one other grounding connection on the secondary neutral shall be provided with its grounding electrode located at a distance not less than 1.80 m (6 ft) from the primary neutral and surge-arrester grounding electrode in addition to the customer's grounds at each service entrance.Where the primary and secondary neutrals are not directly interconnected, the primary grounding conductor, or the secondary grounding conductor, or both, shall be insulated for 600 V.

   NOTE 1: A difference of voltage can exist where primary and secondary neutrals are not directly interconnected. For example, where metallic equipment is bonded to the secondary grounding conductor and is installed on the same pole, the primary grounding conductor would be insulated.

   NOTE 2: Cooperation of all communications and supply utilities, customers of these utilities, and others may be necessary to obtain effective isolation between primary and secondary neutrals.

E. Where separate electrodes are used for system isolation, separate grounding conductors shall be used. Where multiple electrodes are used to reduce grounding resistance, they may be bonded together and connected to a single grounding conductor.

F. Made electrodes used for grounding surge arresters of ungrounded supply systems operated at potentials exceeding 15 kV phase to phase should be located at least 6.1 m (20 ft) from buried communication cables. Where lines with lesser separations are to be constructed, reasonable advance notice should be given to the owners or operators of the affected systems.

G. Communication Systems

Where both electric supply systems and communication systems are grounded on a joint use structure, either a single grounding conductor should be used for both systems or the electric supply and communication grounding conductors should be bonded together, except where separation is required by Rule 097A. Where the electric supply utility is maintaining isolation between primary and secondary neutrals, the communication system ground shall be connected only to the primary grounding conductor.

098. Number 098 not used in this edition.

099. Additional Requirements for Communication Apparatus

Where required to be grounded by other parts of this code, communication apparatus shall be grounded in the following manner.

See *NOTE* 2 in Rule 097D2.

A. Electrode

The grounding conductor shall be connected to an acceptable grounding electrode as follows:

1. Where available and where the supply service is grounded to an acceptable electrode, as described in Rule 094, to the grounded metallic supply service conduit, service equipment enclosure, grounding electrode conductors, or grounding electrode conductors' metal enclosure.
2. Where the grounding means of Rule 099A1 is not available, to a grounding electrode as described in Rule 094A.
3. Where the grounding means of Rule 099A1 or 099A2 are not available, to a grounding electrode as described in Rule 094B.

   EXCEPTION: A variance to Rule 094B2 is allowed for this application. Iron or steel rods may have a cross-sectional dimension of not less than 13 mm (.50 in) and a length of not less than 1.50 m (5 ft). The driven depth shall be 1.50 m (5 ft), subject to *EXCEPTION 1* of Rule 094B2.

B. Electrode Connection

The grounding conductor shall preferably be made of copper (or other material that will not corrode excessively under the prevailing conditions of use) and shall be not less than AWG No. 14 in size. The grounding conductor shall be attached to the electrode by means of a bolted clamp or other suitable methods.

C. Bonding of Electrodes

A bond not smaller than AWG No. 6 copper or equivalent shall be placed between the communication grounding electrode and the supply system neutral grounding electrode where separate electrodes are used in or on the same building or structure being served.

RECOMMENDATION: If water piping is used as a bonding means, care must be taken to assure that the metallic path is continuous between electrodes.

NOTE: See NEC Article 800-40(d) for corresponding NEC requirements.

Part 1.
Rules for the Installation and Maintenance of Electric Supply Stations and Equipment

Section 10.
Purpose and Scope of Rules

100. Purpose

The purpose of Part 1 of this code is the practical safeguarding of persons during the installation, operation, or maintenance of electric supply stations and their associated equipment.

101. Scope

Part 1 of this code covers the electric supply conductors and equipment, along with the associated structural arrangements in electric supply stations, that are accessible only to qualified personnel. It also covers the conductors and equipment employed primarily for the utilization of electric power when such conductors and equipment are used by the utility in the exercise of its function as a utility.

Section 11.
Protective Arrangements in Electric Supply Stations

110. General Requirements

A. Enclosure of Equipment

1. Types of Enclosures

Rooms and spaces in which electric supply conductors or equipment are installed shall be so arranged with fences, screens, partitions, or walls to form an enclosure as to limit the likelihood of entrance of unauthorized persons or interference by them with equipment inside. Entrances not under observation of an authorized attendant shall be kept locked.

A safety sign shall be displayed at each entrances. For fenced electric supply stations, a safety sign shall be displayed on each side.

NOTE: ANSI Z535.1-1998, ANSI Z535.2-1998, ANSI Z535.3-1998, ANSI Z535.4-1998, and ANSI Z535.5-1998 contain information regarding safety signs.

Metal fences, when used to enclose electric supply stations having energized electric conductors or equipment, shall have a height not less than 2.13 m (7 ft) overall and shall be grounded in accordance with Section 9.

The requirements for fence height may be satisfied with any one of the following:

a. Fence fabric, 2.13 m (7 ft) or more in height.

b. A combination of 1.80 m (6 ft) or more of fence fabric and an extension utilizing three or more strands of barbed wire to achieve an overall height of the fence of not less than 2.13 m (7 ft).

c. Other types of construction, such as nonmetallic material, that present equivalent barriers to climbing or other unauthorized entry.

2. Safety Clearance Zone

Fences or walls, when installed as barriers for unauthorized personnel, shall be located such that exposed live parts are outside the safety clearance zone as illustrated in Figure 110-1 and shown in Table 110-1.

EXCEPTION 1: Where a fence, partition, or wall with no openings through which sticks or other objects can be inserted is utilized, live parts complying with the requirements of this code may be installed within the safety clearance zone if they are below the horizontal line projected from the top of the fence or wall.

EXCEPTION 2: The safety clearance zone requirement is not applicable to internal fences within an electric supply station perimeter fence.

B. Rooms and Spaces

All rooms and spaces in which electric supply equipment is installed shall comply with the following requirements:

1. Construction

They shall be as much as practical noncombustible.

NOTE: This rule is not intended to prevent wood poles from being used to support conductors or equipment in electric supply stations.

2. Use

They should be as much as practical free from combustible materials, dust, and fumes and shall not be used for manufacturing or for storage.

EXCEPTION 1: Equipment or material essential for maintenance of the installed equipment may be stored if guarded or separated from live parts as required by Rule 124.

EXCEPTION 2: Materials related to station, transmission, or distribution construction or maintenance work may be stored in the station if located in an area separated from the station electric supply equipment by a fence meeting the requirements of Rule 110A.

EXCEPTION 3: Stored material related to station, transmission, and distribution construction and maintenance work in progress may be temporarily located in a storage space meeting all of the following requirements:

(1) Guarded or separated from live parts as required by Rule 124.

(2) Station exits continue to meet the requirements of Rule 113.

(3) Station working space continues to meet the requirements of Rule 125.

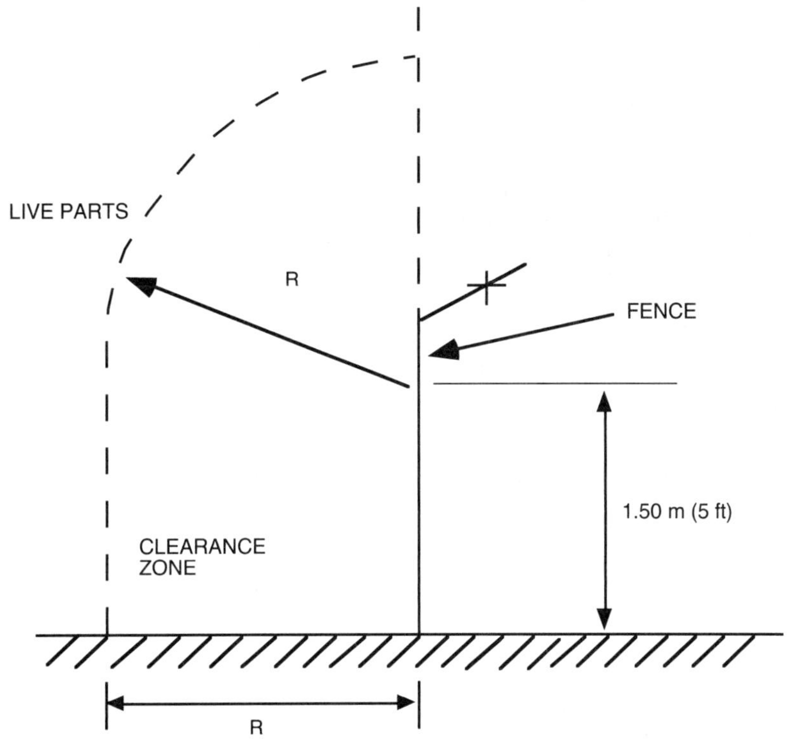

Fig 110-1
Safety Clearance to Electric Supply Station Fences

Table 110-1
Values for Use with Fig 110-1

Nominal Voltage Between Phases	Typical BIL	Dimension "R"	
		m	ft
151–7200	95	3.0	10.0
13 800	110	3.1	10.1
23 000	150	3.1	10.3
34 500	200	3.2	10.6
46 000	250	3.3	10.9
69 000	350	3.5	11.6
115 000	550	4.0	13.0
138 000	650	4.2	13.7
161 000	750	4.4	14.3
230 000	825	4.5	14.9
230 000	900	4.7	15.4
345 000	1050	5.0	16.4
345 000	1175	5.3	17.3
345 000	1300	5.5	18.3
500 000	1550	6.0	19.8
500 000	1800	6.6	21.5
765 000	2050	7.1	23.4

(4) Access is limited to qualified personnel.

(5) The storage location and content is such that the risk of fire does not unreasonably jeopardize station operation.

(For battery areas, see Section 14; for guarding, see Rule 124; for auxiliary equipment in classified locations, see Rule 127.)

3. Ventilation

There should be sufficient ventilation to maintain operating temperatures within ratings, arranged to minimize accumulation of airborne contaminants under any operating conditions.

4. Moisture and Weather

They should be dry. In outdoor stations or stations in wet tunnels, subways or other moist or high-humidity locations, the equipment shall be suitably designed to withstand the prevailing atmospheric conditions.

C. Electric Equipment

All stationary equipment shall be supported and secured in a manner consistent with reasonably expected conditions of service. Consideration shall be given to the fact that certain heavy equipment, such as transformers, can be secured in place by their weight. However, equipment that generates dynamic forces during operation may require appropriate additional measures.

111. Illumination

A. Under Normal Conditions

Rooms and spaces shall have means for artificial illumination. Illumination levels not less than those listed in Table 111-1 are recommended for safety to be maintained on the task.

B. Emergency Lighting

1. A separate emergency source of illumination with automatic initiation, from an independent generator, storage battery, or other suitable source, shall be provided in every attended station.
2. Emergency lighting of 11 lux (1 footcandle) shall be provided in exit paths from all areas of attended stations. Consideration must be given to the type of service to be rendered, whether of short or long duration. The minimum duration shall be 1-1/2 h. It is recommended that emergency circuit wiring shall be kept independent of all other wiring and equipment.

C. Fixtures

Arrangements for permanent fixtures and plug receptacles shall be such that portable cords need not be brought into dangerous proximity to live or moving parts. All lighting shall be controlled and serviced from safely accessible locations.

D. Attachment Plugs and Receptacles for General Use

Portable conductors shall be attached to fixed wiring only through separable attachment plugs that will disconnect all poles by one operation. Receptacles installed on two- or three-wire single-phase, ac branch circuits shall be of the grounding type. Receptacles connected to circuits having different voltages, frequencies, or types of current (ac or dc) on the same premises shall be of such design that attachment plugs used on such circuits are not interchangeable.

E. Receptacles in Damp or Wet Locations

All 120 V ac permanent receptacles shall either be provided with ground-fault interrupter (GFI) protection or be on a grounded circuit that is tested at such intervals as experience has shown to be necessary.

Table 111–1
Illumination Levels

Location	lux	footcandles
Generating Station		
Air-conditioning equipment, air preheater and fan floor, ash sluicing	55	5
Auxiliaries, battery areas, boiler feed pumps, tanks, compressors, gage area	110	10
Boiler platforms	55	5
Burner platforms	110	10
Cable room, circulator, or pump bay	55	5
Chemical laboratory	270	25
Coal conveyor, crusher, feeder, scale area, pulverizer, fan area, transfer tower	55	5
Condensers, deaerator floor, evaporator floor, heater floors	55	5
Control rooms Vertical face of switchboards Simplex or section of duplex operator:		
Type A — Large centralized control room 1.68 m (66 in) above floor	270	25
Type B — Ordinary control room 1.68 m (66 in) above floor	160	15
Section of duplex facing away from operator	160	15
Bench boards (horizontal level)	270	25
Area inside duplex switchboards	55	5
Rear of all switchboard panels (vertical)	55	5
Dispatch boards		
Horizontal plane (desk level)	270	25
Vertical face of board [1.22 m (48 in) above floor, facing operator]:		
System load dispatch room	270	25
Secondary dispatch room	160	15
Hydrogen and carbon dioxide manifold area	110	10
Precipitators	55	5
Screen house	110	10
Soot or slag blower platform	55	5
Steam headers and throttles	55	5
Switchgear, power	110	10
Telephone equipment room	110	10
Tunnels or galleries, piping	55	5
Turbine bay sub-basement	110	10
Turbine room	160	15
Visitors' gallery	110	10
Water treating area	110	10

Table 111–1 ***(Continued)***
Illumination Levels

Location	lux	footcandles
Generating Station (Exterior)		
Catwalks	22	2
Cinder dumps	2.2	0.2
Coal-storage area	2.2	0.2
Coal unloading		
Dock (loading or unloading zone)	55	5
Barge storage area	5.5	0.5
Car dumper	5.5	0.5
Tipple	55	5
Conveyers	22	2
Entrances Generating or service building		
Main	110	10
Secondary	22	2
Gate house		
Pedestrian entrance	110	10
Conveyor entrance	55	5
Fence	2.2	0.2
Fuel-oil delivery headers	55	5
Oil storage tanks	11	1
Open yard	2.2	0.2
Platforms—boiler, turbine deck	55	5
Roadway		
Between or along buildings	11	1
Not bordered by buildings	5.5	0.5
Substation		
General horizontal	22	2
Specific vertical (on disconnects)	22	2

112. Floors, Floor Openings, Passageways, and Stairs

A. Floors

Floors shall have even surfaces and afford secure footing. Slippery floors or stairs should be provided with antislip covering.

B. Passageways

Passageways, including stairways, shall be unobstructed and shall, where practical, provide at least 2.13 m (7 ft) head room. Where the preceding requirements are not practical, the obstructions should be painted, marked, or indicated by safety signs, and the area properly lighted.

NOTE: ANSI Z535.1-1998, ANSI Z535.2-1998, ANSI Z535.3-1998, ANSI Z535.4-1998, and ANSI Z535.5-1998 contain information regarding safety signs.

C. Railings

All floor openings without gratings or other adequate cover and raised platforms and walkways in excess of 300 mm (1 ft) in height shall be provided with railings. Openings in railings for units such as fixed ladders, cranes, and the like shall be provided with adequate guards such as grates, chains, or sliding pipe sections.

D. Stair Guards

All stairways consisting of four or more risers shall be provided with handrails.

NOTE: For additional information, see ANSI A1264.1-1995 [B5].

E. Top Rails

All top rails shall be kept unobstructed for a distance of 75 mm (3 in) in all directions except from below at supports.

113. Exits

A. Clear Exits

Each room or space and each working space about equipment shall have a means of exit, which shall be kept clear of all obstructions.

B. Double Exits

If the plan of the room or space and the character and arrangement of equipment are such that an accident would be likely to close or make inaccessible a single exit, a second exit shall be provided.

C. Exit Doors

Exit doors shall swing out and be equipped with panic bars, pressure plates, or other devices that are normally latched but open under simple pressure.

EXCEPTION: This rule does not apply to exit doors in buildings and rooms containing low-voltage, nonexplosive equipment, and to gates in fences for outdoor equipment installations.

114. Fire-Extinguishing Equipment

Fire-extinguishing equipment approved for the intended use shall be conveniently located and conspicuously marked.

Section 12.
Installation and Maintenance of Equipment

120. General Requirements

A. All electric equipment shall be constructed, installed, and maintained so as to safeguard personnel as far as practical.

B. The rules of this section are applicable to both ac and dc supply stations.

121. Inspections

A. In-Service Equipment

Electric equipment shall be inspected and maintained at such intervals as experience has shown to be necessary. Equipment or wiring found to be defective shall be put in good order or permanently disconnected.

B. Idle Equipment

Infrequently used equipment or wiring shall be inspected and tested before use to determine its fitness for service. Idle equipment energized but not connected to a load shall be inspected and maintained at such intervals as experience has shown to be necessary.

C. Emergency Equipment

Equipment and wiring maintained for emergency service shall be inspected and tested at such intervals as experience has shown to be necessary to determine its fitness for service.

D. New Equipment

New equipment shall be inspected and tested before being placed in service. New equipment shall be tested in accordance with standard industry practices.

122. Guarding Shaft Ends, Pulleys, Belts, and Suddenly Moving Parts

A. Mechanical Transmission Machinery

The methods for safeguarding pulleys, belts, and other equipment used in the mechanical transmission of power shall be in accordance with ANSI/ASME B15.1-2000.

B. Suddenly Moving Parts

Parts of equipment that move suddenly in such a way that persons in the vicinity are likely to be injured by such movement shall be guarded or isolated.

123. Protective Grounding

A. Protective Grounding or Physical Isolation of Non-Current-Carrying Metal Parts

All electric equipment shall have the exposed non-current-carrying metal parts, such as frames of generators and switchboards, cases of transformers, switches, and operating levers, effectively grounded or physically isolated. All metallic guards including rails, screen fences, etc., about electric equipment shall be effectively grounded.

B. Grounding Method

All grounding that is intended to be a permanent and effective protective measure, such as surge-arrester grounding, grounding of circuits, equipment, or wire raceways, shall be made in accordance with the methods specified in Section 9 of this code.

NOTE: For additional information, see IEEE Std 80-2000 [B22].

C. Provision for Grounding Equipment During Maintenance
Electric equipment or conductors normally operating at more than 600 V between conductors, on or about which work is occasionally done while isolated from a source of electric energy by disconnecting or isolating switches only, shall be provided with some means for grounding, such as switches, connectors, or a readily accessible means for connecting a portable grounding conductor. When necessary, grounding may be omitted on conductors normally operating at 25 kV or less and not influenced by higher voltage conductors, where visible openings in the source of supply are available, and are properly tagged in the open position. (See Part 4 of this code.)

D. Grounding Methods for Direct-Current Systems over 750 V
On dc systems greater than 750 V, the dc system shall be grounded in accordance with the methods specified in Section 9 of this code.

124. Guarding Live Parts

A. Where Required

1. Guards shall be provided around all live parts operating above 150 V to ground without an adequate insulating covering, unless their location gives sufficient horizontal or vertical clearance or a combination of these clearances to limit the likelihood of accidental human contact. Clearances from live parts to any permanent supporting surface for workers shall equal or exceed either of those shown in Table 124-1 and illustrated in Fig 124-1.
2. Parts over or near passageways through which material may be carried, or in or near spaces such as corridors, storerooms, and boiler rooms used for nonelectrical work, shall be guarded or given clearances in excess of those specified such as may be necessary to secure reasonable safety. The guards shall be substantial and completely shield or enclose the live parts without openings. In spaces used for nonelectrical work, guards should be removable only by means of tools or keys.
3. Each portion of parts of indeterminate potential, such as telephone wires exposed to induction from high-voltage lines, ungrounded neutral connections, ungrounded frames, ungrounded parts of insulators or surge arresters, or ungrounded instrument cases connected directly to a high-voltage circuit, shall be guarded in accordance with Rule 124A1 on the basis of the maximum voltage that may be present on the surface of that portion. The vertical clearance above grade of the bottom of such part shall be not less than 2.60 m (8.5 ft) unless it is enclosed or guarded in accordance with Rule 124C.

B. Strength of Guards
Guards shall be sufficiently strong and shall be supported rigidly and securely enough to limit the likelihood of them being displaced or dangerously deflected by a person slipping or falling against them.

C. Types of Guards

1. Location or Physical Isolation
Parts having clearances equal to or greater than specified in Table 124-1 are guarded by location. Parts are guarded by isolation when all entrances to enclosed spaces, runways, fixed ladders, and the like are kept locked, barricaded, or roped off, and safety signs are posted at all entrances.
NOTE: ANSI Z535.1-1998, ANSI Z535.2-1998, ANSI Z535.3-1998, ANSI Z535.4-1998, and ANSI Z535.5-1998 contain information regarding safety signs.
2. Shields or Enclosures
Guards less than 100 mm (4 in) outside of the guard zone shall completely enclose the parts from contact up to the heights listed in column 2 of Table 124-1. They shall be not closer to the live parts than listed in column 4 of Table 124-1, except when suitable insulating material is used with circuits of less than 2500 V to ground. If more than 100 mm (4 in) outside the guard zone, the guards shall extend at least 2.60 m (8.5 ft) above the floor. Covers or guards, which must at any time be removed while the parts they guard are live, shall be so arranged that they cannot readily be brought into contact with live parts.

3. Supplemental Barriers or Guards Within Electric Supply Stations
   If the vertical distance in Table 124-1 cannot be obtained, railings or fences may be used. Railings or fences, if used, shall be not less than 1.07 m (3.5 ft) high and shall be located at a horizontal distance of at least 900 mm (3 ft) [and preferably not more than 1.20 m (4 ft)] from the nearest point of the guard zone that is less than 2.60 m (8.5 ft) above the floor or grade (see Fig 124-2).
4. Mats
   Mats of rubber or other suitable insulating material complying with ASTM D 178-88 may be used at switchboards, switches, or rotating machinery as supplementary protection.
5. Live Parts Below Supporting Surfaces for Persons
   The supporting surfaces for persons above live parts shall be without openings. Toe boards at least 150 mm (6 in) high and handrails shall be provided at all edges.
6. Insulating Covering on Conductors or Parts
   Conductors and parts may be considered as guarded by insulation if they have either of the following:
   a. Insulation covering of a type and thickness suitable for the voltage and conditions under which they are expected to be operated, and if operating above 2500 V to ground, having metallic insulation shielding or semiconducting shield in combination with suitable metallic drainage that is grounded to an effective ground.
      EXCEPTION: Nonshielded insulated conductors listed by a qualified testing laboratory shall be permitted for use up to 8000 V (phase to phase) when the conductors meet the requirements of the NEC, Article 310-6.
   b. Barriers or enclosures that are electrically and mechanically suitable for the conditions under which they are expected to be operated.

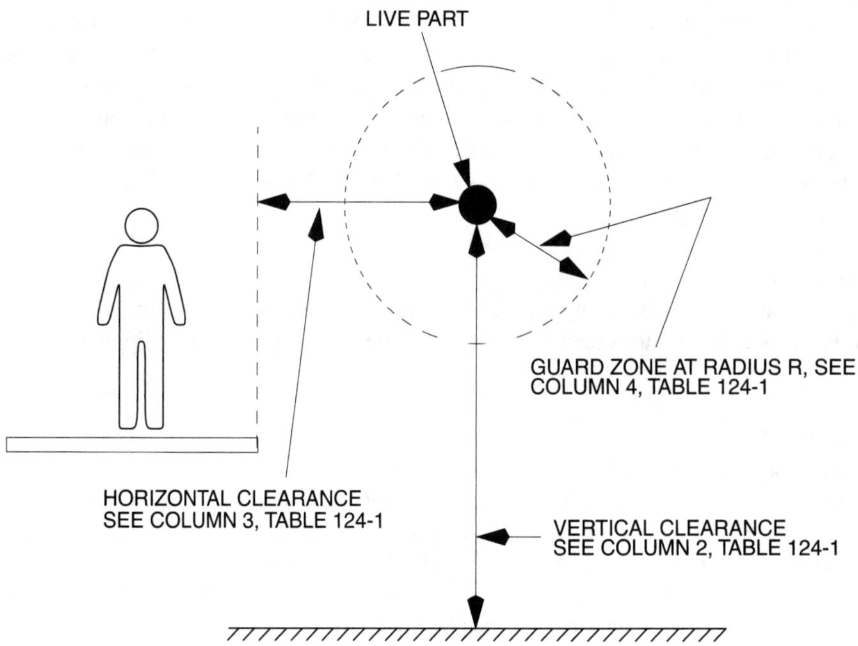

Fig 124-1
Clearance From Live Parts

m

Table 124-1
Clearances From Live Parts
Part A—Low, Medium, and High Voltages (based on BIL factors)

Maximum Design Voltage between Phases (1)	Basic Impulse Insulation Level [5] (BIL)	Vertical clearance of unguarded parts (2)[1]	Horizontal clearance of unguarded parts (3)[1]	Clearance guard to live parts (4)[1]
kV	kV	m	m	mm
0.151–0.6	—	2.64	1.02	50
2.4	—	2.67	1.02	76
7.2	95	2.69	1.02	101
15	95	2.69	1.02	101
15	110	2.74	1.07	152
25	125	2.77	1.09	177
25	150	2.82	1.14	228
35	200	2.90	1.22	304
48	250	3.00	1.32	406
72.5	250	3.00	1.32	406
72.5	350	3.18	1.50	584
121	350	3.18	1.50	584
121	550	3.53	1.85	939
145	350	3.18	1.50	584
145	550	3.53	1.85	939
145	650	3.71	2.03	1117
169	550	3.53	1.85	939
169	650	3.71	2.03	1117
169	750	3.91	2.24	1320
242	550	3.53	1.85	939
242	650	3.71	2.03	1117
242	750	3.91	2.24	1320
242	900	4.19	2.51	1600
242	1050	4.52	2.84	1930

m

Table 124-1 ***(Continued)***
PART B — Extra-High Voltages (based on switching-surge factors)[2]

Maximum design voltage between phases (1)	Switching-surge factor per unit (A) [4]	Switching surge line to ground (B) [4]	Vertical clearance of unguarded parts (2) [1]	Horizontal clearance of unguarded parts (3) [1]	Clearance guard to live parts (4) [1]
		kV	m	m	m
362 000	2.2 or below	650	4.7	3.0	2.13
	2.3	680	4.9	3.2	2.30
	2.4	709	5.0	3.4	2.45
	2.5	739	5.2	3.6	2.60
	2.6	768	5.4	3.7	2.80
	2.7	798	5.6	3.9	3.0
	2.8	828	5.8	4.1	3.2
	2.9	857	6.0	4.3	3.4
	3.0	887	6.1	4.5	3.6
550 000	1.8 or below	808	5.7	4.1	3.2
	1.9	853	5.9	4.3	3.4
	2.0	898	6.2	4.6	3.6
	2.1	943	6.6	4.9	4.0
	2.2	988	6.9	5.2	4.3
	2.3	1033	7.2	5.5	4.6
	2.4	1078	7.5	5.8	4.9
	2.5	1123	7.9	6.2	5.3
	2.6	1167	8.2	6.6	5.6
	2.7	1212	8.6	7.0	6.0
800 000	1.5	980	6.8	5.1	4.2
	1.6	1045	7.3	5.6	4.7
	1.7	1110	7.8	6.1	5.2
	1.8	1176	8.3	6.6	5.7
	1.9	1241	8.8	7.2	6.2
	2.0	1306	9.4	7.7	6.8
	2.1	1372	10.0	8.3	7.4
	2.2	1437	10.6	8.9	8.0
	2.3	1502	11.2	9.5	8.6
	2.4	1567	11.8	10.0	9.2

m

Table 124-1 *(Continued)*
PART C — Extra-High Voltages (based on BIL factors)[2]

Maximum design voltage between phases (1)	Basic impulse insulation level[5] (BIL) (C)[4]	Vertical clearance of unguarded parts (2)[1]	Horizontal clearance of unguarded parts (3)[1]	Clearance guard to live parts (4)[1]
	kV	m	m	m
362 000	1050	4.7	3.0	2.13
362 000	1300	5.2	3.6	2.60
550 000	1550	5.7	4.1	3.2
550 000	1800	6.2	4.6	3.6
800 000	2050	6.8	5.2	4.2

m

Table 124-1 *(Continued)*

PART D — High Voltage Direct Current (Based on Transient Overvoltage)

Maximum design voltage conductor to ground	Transient overvoltage per unit (A)[4]	Transient overvoltage line to grnd (B)[4]	Vertical clearance of unguarded parts	Horizontal clearance of unguarded parts	Clearance guard to live parts[1]
kV		kV	m	m	m
250	1.5 or below	375	3.81	2.13	1.22
	1.6	400	3.89	2.22	1.30
	1.7	425	3.97	2.30	1.38
	1.8	450	4.05	2.38	1.46
400	1.5 or below	600	4.54	2.87	1.95
	1.6	640	4.67	3.00	2.08
	1.7	680	4.88	3.21	2.29
	1.8	720	5.12	3.45	2.53
500	1.5 or below	750	5.29	3.62	2.70
	1.6	800	5.60	3.92	3.01
	1.7	850	5.96	4.29	3.37
	1.8	900	6.35	4.67	3.76
600	1.5 or below	900	6.35	4.67	3.76
	1.6	960	6.73	5.06	4.14
	1.7	1020	7.13	5.45	4.54
	1.8	1080	7.57	5.89	4.97

m

Table 124-1 ***(Continued)***

PART D — High Voltage Direct Current (Based on Transient Overvoltage)

Maximum design voltage conductor to ground	Transient overvoltage per unit (A) [4]	Transient overvoltage line to grnd (B) [4]	Vertical clearance of unguarded parts	Horizontal clearance of unguarded parts	Clearance guard to live parts[1]
kV		kV	m	m	m
750	1.5 or below	1125	7.90	6.23	5.31
	1.6	1200	8.44	6.76	5.85
	1.7	1275	9.06	7.38	6.47
	1.8	1350	9.72	8.04	7.13

[1] Interpolate for intermediate values. The clearances in column 4 of this table are solely for guidance in installing guards without definite engineering design and are not to be considered as a requirement for such engineering design. For example, the clearances in the tables above are not intended to refer to the clearances between live parts and the walls of the cells, compartments, or similar enclosing structures. They do not apply to the clearances between bus bars and supporting structures nor to clearances between the blade of a disconnecting switch and its base. However, where surge-protective devices are applied to protect the live parts, the vertical clearances, column 2 of Table 124-1 Part A may be reduced provided the clearance is not less than 2.6 m plus the electrical clearance between energized parts and ground as limited by the surge-protective devices.

[2] Clearances shall satisfy either switching-surge or BIL duty requirements, whichever are greater.

[3] Switching-Surge Factor—an expression of the maximum switching-surge crest voltage in terms of the maximum operating line-to-neutral crest voltage of the power system.

[4] The values of columns A, B, and C are power system design factors that shall correlate with selected clearances. Adequate data to support these design factors should be available.

[5] The selection of station BIL shall be coordinated with surge-protective devices when BIL is used to determine clearance. BIL—Basic Impulse Insulation Level—For definition and application, see IEEE Std 1313-1993.

ft

Table 124-1
Clearances From Live Parts
Part A—Low, Medium, and High Voltages (based on BIL factors)

Maximum Design Voltage between Phases (1)	Basic Impulse Insulation Level [5] (BIL)	Vertical clearance of unguarded parts (2)[1]		Horizontal clearance of unguarded parts (3)[1]		Clearance guard to live parts (4)[1]	
kV	kV	ft	in	ft	in	ft	in
0.151–0.6	—	8	8	3	4	0	2
2.4	—	8	9	3	4	0	3
7.2	95	8	10	3	4	0	4
15	95	8	10	3	4	0	4
15	110	9	0	3	6	0	6
25	125	9	1	3	7	0	7
25	150	9	3	3	9	0	9
35	200	9	6	4	0	1	0
48	250	9	10	4	4	1	4
72.5	250	9	10	4	4	1	4
72.5	350	10	5	4	11	1	11
121	350	10	5	4	11	1	11
121	550	11	7	6	1	3	1
145	350	10	5	4	11	1	11
145	550	11	7	6	1	3	1
145	650	12	2	6	8	3	8
169	550	11	7	6	1	3	1
169	650	12	2	6	8	3	8
169	750	12	10	7	4	4	4
242	550	11	7	6	1	3	1
242	650	12	2	6	8	3	8
242	750	12	10	7	4	4	4
242	900	13	9	8	3	5	3
242	1050	14	10	9	4	6	4

ft

Table 124-1 *(Continued)*

PART B — Extra-High Voltages (based on switching-surge factors)[2]

Maximum design voltage between phases (1)	Switching surge factor [3] per unit (A) [4]	Switching surge line to ground (B) [4]	Vertical clearance of unguarded parts (2) [1]		Horizontal clearance of unguarded parts (3) [1]		Clearance guard to live parts (4) [1]	
		kv	ft	in	ft	in	ft	in
362 000	2.2 or below	650	15	6	10	0	7	0
	2.3	680	16	0	10	6	7	6
	2.4	709	16	6	11	0	8	0
	2.5	739	17	2	11	8	8	8
	2.6	768	17	9	12	3	9	3
	2.7	798	18	4	12	10	9	10
	2.8	828	18	11	13	5	10	5
	2.9	857	19	7	14	1	11	1
	3.0	887	20	2	14	8	11	8
550 000	1.8 or below	808	18	10	13	4	10	4
	1.9	853	19	6	14	0	11	0
	2.0	898	20	6	15	0	12	0
	2.1	943	21	6	16	0	13	0
	2.2	988	22	6	17	0	14	0
	2.3	1033	23	7	18	1	15	1
	2.4	1078	24	8	19	2	16	2
	2.5	1123	25	10	20	4	17	4
	2.6	1167	27	0	21	6	18	6
	2.7	1212	28	4	22	10	19	10
800 000	1.5	980	22	4	16	10	13	10
	1.6	1045	23	11	18	5	15	5
	1.7	1110	25	6	20	0	17	1
	1.8	1176	27	3	21	9	18	9
	1.9	1241	29	0	23	6	2	6
	2.0	1306	30	10	25	4	22	4
	2.1	1372	32	9	27	3	24	3
	2.2	1437	34	8	29	3	26	2
	2.3	1502	36	9	31	3	28	3
	2.4	1567	38	9	33	3	30	3

ft

Table 124-1 *(Continued)*

PART C — Extra-High Voltages (based on BIL factors)[2]

Maximum design voltages between phases (1)	Basic impulse insulation level [5] (BIL) (C) [4]	Vertical clearance of unguarded parts (2) [1]		Horizontal clearance of unguarded parts (3) [1]		Clearance guard to live parts (4) [1]	
	kV	ft	in	ft	in	ft	in
362 000	1050	15	6	10	0	7	0
362 000	1300	17	2	11	8	8	8
550 000	1550	18	10	13	4	10	4
550 000	1800	20	6	15	0	12	0
800 000	2050	22	5	16	11	13	11

ft

Table 124-1 *(Continued)*

PART D — High Voltage Direct Current (Based on Transient Overvoltage)

Maximum design voltage conductor to ground	Transient overvoltage per unit (A) [4]	Transient overvoltage line to grnd (B) [4]	Vertical clearance of unguarded parts		Horizontal clearance of unguarded parts		Clearance guard to live parts[1]	
kV		kV	ft	in	ft	in	ft	in
250	1.5 or below	375	12	6	7	0	4	0
	1.6	400	12	9	7	3	4	3
	1.7	425	13	0	7	7	4	6
	1.8	450	13	3	7	10	4	9
400	1.5 or below	600	14	11	9	5	6	5
	1.6	640	15	4	9	10	6	10
	1.7	680	16	0	10	6	7	6
	1.8	720	16	10	11	4	8	4
500	1.5 or below	750	17	4	11	11	8	10
	1.6	800	18	4	12	10	9	11
	1.7	850	19	7	14	1	11	1
	1.8	900	20	10	15	4	12	4
600	1.5 or below	900	20	10	15	4	12	4
	1.6	960	22	1	16	7	13	7
	1.7	1020	23	5	17	11	14	11
	1.8	1080	24	10	19	4	16	4
750	1.5 or below	1125	25	11	20	5	17	5
	1.6	1200	27	8	22	2	19	2

ft

Table 124-1 ***(Continued)***

PART D — High Voltage Direct Current (Based on Transient Overvoltage)

Maximum design voltage conductor to ground	Transient overvoltage per unit (A) [4]	Transient overvoltage line to grnd (B) [4]	Vertical clearance of unguarded parts		Horizontal clearance of unguarded parts		Clearance guard to live parts[1]	
kV		kV	ft	in	ft	in	ft	in
	1.7	1275	29	9	24	3	21	3
	1.8	1350	31	11	26	5	23	5

[1] Interpolate for intermediate values. The clearances in column 4 of this table are solely for guidance in installing guards without definite engineering design and are not to be considered as a requirement for such engineering design. For example, the clearances in the tables above are not intended to refer to the clearances between live parts and the walls of the cells, compartments, or similar enclosing structures. They do not apply to the clearances between bus bars and supporting structures nor to clearances between the blade of a disconnecting switch and its base. However, where surge-protective devices are applied to protect the live parts, the vertical clearances, column 2 of Table 124-1 Part A may be reduced provided the clearance is not less than 8.5 ft plus the electrical clearance between energized parts and ground as limited by the surge-protective devices.

[2] Clearances shall satisfy either switching-surge or BIL duty requirements, whichever are greater.

[3] Switching-Surge Factor—an expression of the maximum switching-surge crest voltage in terms of the maximum operating line-to-neutral crest voltage of the power system.

[4] The values of columns A, B, and C are power system design factors that shall correlate with selected clearances. Adequate data to support these design factors should be available.

[5] The selection of station BIL shall be coordinated with surge-protective devices when BIL is used to determine clearance. BIL—Basic Impulse Insulation Level—For definition and application, see IEEE Std 1313-1993.

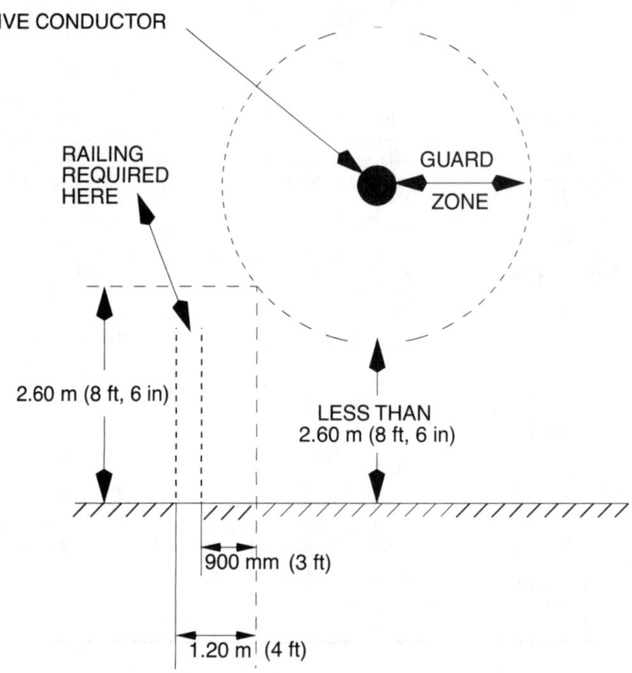

Fig 124-2
Railing Used as Guards

125. Working Space About Electric Equipment

A. Working Space (600 V or Less)

Access and working space shall be provided and maintained about electric equipment to permit ready and safe operation and maintenance of such equipment.

1. Clear Spaces

   Working space required by this section shall not be used for storage. When normally enclosed energized parts are exposed for inspection or servicing, the working space, if in a passageway or general open space, shall be guarded.

2. Access and Entrance to Working Space

   At least one entrance shall be provided to give access to the working space about electric equipment.

3. Working Space

   The working space in the direction of access to energized parts operating at 600 V or less that require examination, adjustment, servicing, or maintenance while energized shall be not less than indicated in Table 125-1. In addition to the dimensions shown in Table 125-1, the working space shall be not less than 750 mm (30 in) wide in front of the electric equipment. Distances shall be measured from the energized parts if such are exposed or from the enclosure front or opening if such are enclosed. Concrete, brick, or tile walls shall be considered grounded.

4. Headroom Working Space

   The headroom of working spaces about switchboards or control centers shall be not less than 2.13 m (7 ft).

5. Front Working Space

   In all cases where there are energized parts normally exposed on the front of switchboards or motor control centers, the working space in front of such equipment shall not be less than 900 mm (3 ft).

Table 125-1
Working Space

Voltage to ground	Condition	Clear Distance					
		(mm) 1	(ft) 1	(mm) 2	(ft) 2	(mm) 3	(ft) 3
0–150		900	3	900	3	900	3
151–600		900	3	1070	3-1/2	1200	4

Where the *Conditions* are as follows:

1. Exposed energized parts on one side and no energized or grounded parts on the other side of the working space, or exposed energized parts on both sides effectively guarded by suitable wood or other insulating materials. Insulated wire or insulated bus bars operating at not over 300 V shall not be considered energized parts.
2. Exposed energized parts on one side and grounded parts on the other side.
3. Exposed energized parts on both sides of the work space (not guarded as provided in Condition 1) with the operator between.

EXCEPTION: Working space shall not be required in back of assemblies, such as dead-front switchboards or motor control centers where there are no renewable or adjustable parts such as fuses or switches on the back and where all connections are accessible from locations other than the back.

B. Working Space Over 600 V

Working space shall be in accordance with Table 124-1 clearances for guarding.

126. Equipment for Work on Energized Parts

When it is necessary for personnel to move themselves, material, or tools within the guard zone of unguarded energized parts, protective equipment shall be provided.

This protective equipment shall be periodically inspected, tested, and kept in a safe condition. Protective equipment shall be rated for not less than the voltage involved. Refer to Section 3 for a list of specifications for equipment.

127. Classified Locations

Electrical installations in classified areas shall meet the requirements of the National Electrical Code (NEC) (NFPA 70-1999), Articles 500 through 517. Areas classified in accordance with NEC Article 500 shall comply with the requirements of that Article and A. through L. below. Areas classified with the optional Zone method in accordance with NEC Article 505 shall comply with the requirements of that article.

A. Coal-Handling Areas

1. Unventilated spaces inside or above coal-storage silos or bunkers, or other enclosed coal-storage and coal-handling spaces where methane may accumulate in explosive or ignitable mixtures as defined in Article 500-5 of the NEC, are Class I, Division 1, Group D locations. Electric equipment in other locations in which flammable gases or vapors may exist shall be in accordance with the NEC, Article 500-5, or the locations shall be adequately ventilated
2. Adequate ventilation exists when the method of ventilation will limit the likelihood of accumulation of significant quantities of vapor-air concentrations from exceeding 25% of the lower flammable limit.
3. Tunnels beneath stockpiles or surge piles; spaces inside, above, or below coal-storage silos or bunkers; or other enclosed coal-storage or coal-handling spaces or areas shall be Class II, Group F, Division 1 or Division 2 locations as determined by the NEC.
4. Enclosed sections where only wet coal is handled, or enclosed sections so cut off as to be free from dangerous amounts of coal dust, are not classified. Coal shall be considered to be wet if enough water sprays are installed and maintained to limit the atmospheric concentration of total entrapped volatiles to 8% or less. (See ASTM D 3175 for coal and coke dusts).
5. Locations having completely dust-tight pulverized fuel systems designed and installed in compliance with NFPA 8503-1997 shall not be considered classified.
6. Portable lamps for use in fuel bunkers or bins shall be suitable for Class II, Division 1 locations.
7. Sparking electric tools shall not be used where combustible dust or dust clouds are present.
8. An equipment grounding conductor shall be carried with the power conductors and serve to ground the frames of all equipment supplied from that circuit. The origin of the grounding conductor shall be:
   a. Ungrounded delta or wye-transformer frame ground.
   b. Grounded delta or wye-transformer grounded secondary connection.
   c. Resistance grounded wye—the grounded side of the grounding resistor.
9. Ungrounded systems should be equipped with a ground-fault indicating device to give both a visual and audible alarm upon the occurrence of a ground fault in the system.

B. Flammable and Combustible Liquids

1. Flammable liquid shall mean a liquid having a flash point below 38 °C (100 °F) and having a vapor pressure not exceeding 275 kPa (40 lb/in^2) (absolute) at 38 °C (100 °F) and shall be known as a Class I liquid.
2. Combustible liquid shall mean a liquid having a flash point greater than or equal to 38 °C (100 °F) and having a vapor pressure not exceeding 275 kPa (40 lb/in^2) (absolute) at 38 °C (100 °F).
3. Class I liquids are subdivided as follows:
   a. Class IA includes those having flash points below 23 °C (73 °F) and having a boiling point below 38 °C (100 °F).
   b. Class IB includes those having flash points below 23 °C (73 °F).
   c. Class IC includes those having flash points at or above 23 °C (73 °F) and below 38 °C (100 °F).

4. Combustible liquids are subdivided as follows:
   a. Class II includes those having flash points equal to or greater than 38 °C (100 °F) but less than 60 °C (140 °F).
   b. Class IIIA includes those having flash points equal to or greater than 60 °C (140 °F) but less than 93 °C (200 °F).
   c. Class IIIB includes those having flash points greater than or equal to 93 °C (200 °F).

C. Flammable Liquid Storage Area

Electric wiring and equipment located in flammable liquid storage areas shall be installed in accordance with applicable sections of NFPA 30-2000 and the NEC.

D. Loading and Unloading Facilities for Flammable and Combustible Liquids

Electric equipment located in the area shall be installed in accordance with applicable sections of NFPA 30-2000 and the NEC.

E. Gasoline-Dispensing Stations

Electric equipment installed in areas used for dispensing flammable liquids shall be installed in accordance with applicable sections of NFPA 30A-2000 and the NEC.

F. Boilers

When storing, handling, or burning fuel oils that have flash points below 38 °C (100 °F) the installation shall conform to NFPA 30-2000 and the NEC.

NOTE: Attention must be given to electrical installations in areas where flammable vapors or gases may be present in the atmosphere. Typical locations are burner areas, fuel-handling equipment areas, fuel-storage areas, pits, sumps, and low spots where fuel leakage or vapors may accumulate. The NEC, Article 500 provides for classifying such areas and defines requirements for electrical installations in the areas so classified. The burner front piping and equipment shall be designed and constructed to eliminate hazardous concentrations of flammable gases that exist continuously, intermittently, or periodically under normal operating conditions. Providing the burners are thoroughly purged before removal for cleaning, burner front maintenance operations will not cause hazardous concentrations of flammable vapors to exist frequently. With such provisions, the burner front is not normally classified more restrictively than Class I, Division 2.

G. Gaseous Hydrogen Systems for Supply Equipment

1. Outdoor storage areas shall not be located beneath electric power lines.
2. Safety considerations at specific storage areas.

   Electric equipment shall be suitable for Class I, Division 2 locations:
   a. Within 4.6 m (15 ft) of outdoor storage spaces
   b. Within adequately ventilated separate buildings or special rooms for storing hydrogen
   c. Within 7.6 m (25 ft) of a hydrogen storage space in an adequately ventilated building used for other purposes
3. Space around elements of the generator hydrogen seal oil system shall not be considered classified for electrical installation except where external venting is not provided in the bearing drain system.

4. Spaces around the hydrogen piping system beyond the point where the hydrogen storage system connects to distribution piping shall not be considered classified for electrical installations, outside the boundaries established in 127G2a and 127G2c.

H. Liquid Hydrogen Systems
1. Electric wiring and equipment located within 900 mm (3 ft) of a point where connections are regularly made and disconnected shall be in accordance with the NEC, Article 501, Class I, Group B, Division 1 locations.
2. Except as provided in Paragraph 1, electric wiring and equipment located within 7.6 m (25 ft) of a point where connections are regularly made and disconnected or within 7.6 m (25 ft) of a liquid hydrogen storage container, shall be in accordance with the NEC, Article 501, Class I, Group B, Division 2 locations. When equipment approved for Class I, Group B atmospheres is not commercially available, the equipment may be (1) purged or ventilated in accordance with NFPA 496-1998, (2) intrinsically safe, or (3) approved for Class I, Group C atmospheres. This requirement does not apply to electric equipment that is installed on mobile supply trucks or tank cars from which the storage container is filled.

I. Sulfur
1. Electric wiring and equipment located in areas where sulfur dust is in suspension in explosive or ignitable mixtures during normal operations shall be suitable for Class II, Division 1, Group G.

J. Oxygen
1. Bulk oxygen installations are not defined as classified locations.

K. Liquefied Petroleum Gas (LPG)
Electric equipment and wiring installed in areas used for handling, storage, or utilization of LPG shall be installed in accordance with applicable sections of NFPA 58-1998, NFPA 59-1998, and the NEC.

L. Natural Gas (Methane)
Electric equipment and wiring installed in areas used for handling, storage, or utilization of natural gas shall be installed in accordance with applicable sections of NFPA 59A-1990 and the NEC.

NOTE: NFPA 497M-1997 [B41] and API RP500, 7 January 1998 [B9], provide additional guidelines for classifying these areas.

128. Identification

Electric equipment and devices shall be identified for safe use and operation. The identification shall be as nearly uniform as practical throughout any one station. Identification marks shall not be placed on removable covers or doors that could be interchanged.

129. Mobile Hydrogen Equipment

Mobile hydrogen supply units being used to replenish a hydrogen system shall be bonded both to the grounding system and to the grounded parts of the hydrogen system.

Section 13.
Rotating Equipment

Rotating equipment includes generators, motors, motor generators, and rotary converters.

130. Speed Control and Stopping Devices

A. Automatic Overspeed Trip Device for Prime Movers

When harmful overspeed can occur, prime movers driving generating equipment shall be provided with automatic overspeed trip devices in addition to their governors.

B. Manual Stopping Devices

An operator-initiated stopping device shall be provided for any machine that drives an electric power generator or rotary uninterruptible power supply (motor-generator). The operator-initiated stopping device shall be accessible to the operator during normal operation. Manual controls to be used in emergency for machinery and electric equipment shall be located so as to provide protection to the operator in the event of such emergency.

C. Speed Limit for Motors

Machines of the following types shall be provided with speed-limiting devices unless their inherent characteristics or the load and the mechanical connection thereto are such as to safely limit the speed.

1. Separately excited dc motors
2. Series motors

D. Number 130D not used in this edition.

E. Adjustable-Speed Motors

Adjustable-speed motors, controlled by means of field regulation, shall, in addition to the provisions of Rule 130C, be so equipped and connected that the field cannot be weakened sufficiently to permit dangerous speed.

F. Protection of Control Circuits

Where speed-limiting or stopping devices and systems are electrically operated, the control circuits by which such devices are actuated shall be protected from mechanical damage. Such devices and systems should be of the automatic tripping type.

131. Motor Control

All motors arranged such that an unexpected starting of the motor might create an exposure of personnel to injury shall have the motor control circuit designed to block unintended re-energization of the motor after a power supply interruption of a duration sufficient for moving equipment to become stationary. The motor control shall be such that an operator must take some action to restart the motor, or automatic restarting shall be preceded by warning signals and a time delay sufficient for personnel action to limit the likelihood of injury. This requirement does not apply to those motors with an emergency use and where the opening of the circuit may cause less safe conditions.

132. Number 132 not used in this edition.

133. Short-Circuit Protection

Means shall be provided to automatically disconnect an electric motor from the supply source in the event of high-magnitude short-circuit currents within the motor.

Section 14.
Storage Batteries

140. General

The provisions of this section are intended to apply to all stationary installations of storage batteries. For operating precautions, see Part 4 of this code.

Space shall be provided around batteries for safe inspection, maintenance, testing, and cell replacement and space left above the cells to allow for operation of lifting equipment when required, addition of water, and taking measurements.

141. Location

Storage batteries shall be located within a protective enclosure or area accessible only to qualified persons. A protective enclosure can be a battery room, control building, or a case, cage, or fence that will protect the contained equipment and limit the likelihood of inadvertent contact with energized parts.

142. Ventilation

The battery area shall be ventilated, either by a natural or powered ventilation system to limit hydrogen accumulation to less than an explosive mixture. Failure of a continuously operated or automatically controlled powered ventilation system required by design to limit hydrogen accumulation to less than an explosive mixture shall be annunciated.

143. Racks

Racks refer to frames designed to support cells or trays. Racks shall be firmly anchored, preferably to the floor. Anchoring to both walls and floors is not recommended. Racks made of metal shall be grounded.

144. Floors in Battery Areas

Floors of battery areas should be of an acid-resistive material, painted with acid-resistive paint, or otherwise protected. Provision should be made to contain spilled electrolyte.

145. Illumination for Battery Areas

Lighting fixtures shall be protected from physical damage by guards or isolation. Receptacles and lighting switches should be located outside of battery areas.

146. Service Facilities

A. Proper eye protection and clothing shall be provided in the battery area during battery maintenance and installation and shall consist of the following:

1. Goggles or face shield
2. Acid-resistant gloves
3. Protective aprons and overshoes
4. Portable or stationary water facilities or neutralizing agent for rinsing eyes and skin

B. Safety signs inside and outside of a battery room or in the vicinity of a battery area, prohibiting smoking, sparks, or flame shall be provided.

NOTE: ANSI Z535.1-1998, ANSI Z535.2-1998, ANSI Z535.3-1998, ANSI Z535.4-1998, and ANSI Z535.5-1998 contain information regarding safety signs.

147. Number 147 not used in this edition.

Section 15.
Transformers and Regulators

150. Current-Transformer Secondary Circuits Protection When Exceeding 600 V

Secondary circuits, when in the vicinity of primary circuits exceeding 600 V should, except for short lead lengths at the terminals of the transformer, have the secondary wiring adequately protected by means of conduit, covering, or some other protection. Any metallic covering used shall be effectively grounded, giving appropriate consideration to circulating currents. Current transformers shall have provision for shorting the secondary winding.

151. Grounding Secondary Circuits of Instrument Transformers

The secondary circuits of instrument transformers shall be effectively grounded where functional requirements permit.

152. Location and Arrangement of Power Transformers and Regulators

A. Outdoor Installations

1. Power transformers and regulators shall be so installed that all energized parts are enclosed or guarded so as to limit the likelihood of inadvertent contact, or the energized parts shall be physically isolated in accordance with Rule 124. The case shall be effectively grounded or guarded.
2. The installation of liquid-filled transformers shall utilize one or more of the following methods to minimize fire hazards. The method to be applied shall be according to the degree of the fire hazard. Recognized methods are the use of less flammable liquids, space separation, fire-resistant barriers, automatic extinguishing systems, absorption beds, and enclosures.

   The amount and characteristics of liquid contained should be considered in the selection of space separation, fire-resistant barriers, automatic extinguishing systems, absorption beds, and enclosures that confine thc liquid of a ruptured transformer tank, all of which are recognized as safeguards.

B. Indoor Installations

1. Transformers and regulators 75 kVA and above containing an appreciable amount of flammable liquid and located indoors shall be installed in ventilated rooms or vaults separated from the balance of the building by fire walls. Doorways to the interior of the building shall be equipped with fire doors and shall have means of containing the liquid.
2. Transformers or regulators of the dry type or containing a nonflammable liquid or gas may be installed in a building without a fireproof enclosure. When installed in a building used for other than station purposes, the case or the enclosure shall be so designed that all energized parts are enclosed in the case that is grounded in accordance with Rule 123. As an alternate, the entire unit may be enclosed so as to limit the likelihood of inadvertent contact by persons with any part of the case or wiring. When installed, the pressure-relief vent of a unit containing a nonbiodegradable liquid shall be furnished with a means for absorbing toxic gases.
3. Transformers containing less flammable liquid may be installed in a supply station building in such a way as to minimize fire hazards. The amount of liquid contained, the type of electrical protection, and tank venting shall be considered in the selection of space separation from combustible materials or structures, liquid confinement, fire-resistant barriers or enclosures, or extinguishing systems.

153. Short-Circuit Protection of Power Transformers

Power transformers shall be provided with means to disconnect automatically the source of supply of current for a high magnitude short circuit (fault) within the transformer.

The devices for automatically disconnecting the source of supply may be a circuit breaker, circuit switcher, fuse, thyristor blocking, or other reasonable methods either locally or remotely connected to the transformer. This includes disconnecting the generator electric field source together with the source of mechanical energy upon detection of a fault in either the generator step-up or station auxiliary transformer. Removing a single phase rather than all three phases to extinguish short-circuit current is acceptable.

EXCEPTION: Transformers other than power transformers are exempt from this rule. This includes instrument transformers, neutral grounding transformers, regulating transformers, and other transformers specifically for control, protection, or metering.

Section 16.
Conductors

160. Application

Conductors shall be suitable for the location, use, and voltage. Conductors shall have ampacity that is adequate for the application.

161. Electrical Protection

A. Overcurrent Protection Required

Conductors and insulation shall be protected against excessive heating by the design of the system and by overcurrent, alarm, indication, or trip devices.

B. Grounded Conductors

Conductors normally grounded for the protection of persons shall be arranged without overcurrent protection or other means that could interrupt their continuity to ground.

C. Insulated Power Cables

Insulated power cable circuits shall be provided with short-circuit protection that will isolate the short circuit from the supply.

162. Mechanical Protection and Support

A. All conductors shall be adequately supported to withstand forces caused by the maximum short-circuit current to which they may be subjected. Where support conductors extend outside the electric supply station, such conductors and their supports shall comply with the grades of construction, strength, and loading requirements of Part 2 of this code.

B. Where conductors, conductor insulation, or conductor supports may be subjected to mechanical damage, casing, armor, or other means shall be employed to limit the likelihood of damage or disturbance.

163. Isolation

All nonshielded insulated conductors of more than 2500 V to ground and bare conductors of more than 150 V to ground shall be isolated by elevation or guarded in accordance with Rule 124.

Nonshielded, insulated, and jacketed conductors may be installed in accordance with Rule 124C6.

164. Conductor Terminations

A. Insulation

Ends and joints of insulated conductors, unless otherwise adequately guarded, shall have insulating covering equivalent to that of other portions of the conductor.

B. Metal-Sheathed or Shielded Cable

Insulation of the conductors, where leaving the metal sheath or shield, shall be protected from mechanical damage, moisture, and excessive electrical stress.

Section 17.
Circuit Breakers, Reclosers, Switches, and Fuses

170. Arrangement

Circuit breakers, reclosers, switches, and fuses shall be so installed as to be accessible only to persons qualified for operation and maintenance. Walls, barriers, latched doors, location, isolation, or other means shall be provided to protect persons from energized parts or arcing. Conspicuous markings (such as numbers/letters/symbols) shall be provided on each device and at any remote operating points so as to facilitate identification by employees authorized to operate the device. No device identification shall be duplicated within the same supply station. When the contact parts of a switching device are not normally visible, the device shall be equipped with an indicator to show all normal operating positions.

171. Application

Circuit breakers, circuit switchers, reclosers, switches, and fuses should be utilized with due regard to their assigned ratings of voltage and continuous and momentary currents. Devices that are intended to interrupt fault current shall be capable of safely interrupting the maximum short-circuit current they are intended to interrupt, and for the circumstances under which they are designed to operate. The interrupting capacity should be reviewed prior to each significant system change.

172. Circuit Breakers, Reclosers, and Switches Containing Oil

Circuit-interrupting devices containing flammable liquids shall be adequately segregated from other equipment and buildings to limit damage in the event of an explosion or fire. Segregation may be provided by spacing, by fire-resistant barrier walls, or by metal cubicles. Gas-relief vents should be equipped with oil-separating devices or piped to a safe location. Means shall be provided to control oil that could be discharged from vents or by tank rupture. This may be accomplished by absorption beds, pits, drains, or by any combination thereof. Buildings or rooms housing this equipment shall be of fire-resistant construction.

173. Switches and Disconnecting Devices

A. Capacity

Switches shall be of suitable voltage and ampere rating for the circuit in which they are installed. Switches used to break load current shall be marked with the current that they are rated to interrupt.

B. Provisions for Disconnecting

Switches and disconnectors shall be so arranged that they can be locked in the open and closed positions, or plainly tagged where it is not practical to install locks. (See Part 4 of this code). For devices that are operated remotely and automatically, the control circuit shall be provided with a positive disconnecting means near the apparatus to limit the likelihood of accidental operation of the mechanism.

C. Visible Break Switch

A visible break switch or disconnector shall be inserted in each ungrounded conductor between electric supply equipment or lines and sources of energy of more than 600 V, if the equipment or lines may have to be worked on without protective grounding while the sources may be energized.

Where metal-clad switchgear equipment is used, the withdrawn position of the circuit breaker, where clearly indicated, constitutes a visible break for this purpose.

174. Disconnection of Fuses

Fuses in circuits of more than 150 V to ground or more than 60 A shall be classified as disconnecting fuses or be so arranged that before handling:

A. The fuses can be disconnected from all sources of electric energy, or

B. The fuses can be conveniently removed by means of insulating handles.

Fuses can be used to disconnect from the source when they are so rated.

Section 18.
Switchgear and Metal-Enclosed Bus

180. Switchgear Assemblies

A. General Requirements for All Switchgear

1. To minimize movement, all switchgear shall be secured in a manner consistent with conditions of service and applicable manufacturer's instructions.
2. Cable routed to switchgear shall be supported to minimize forces applied to conductor terminals.
3. Piping containing liquids, or corrosive or hazardous gases, shall not be routed in the vicinity of switchgear unless suitable barriers are installed to protect the switchgear from damage in the event of a pipe failure.
4. Switchgear shall not be located where foreign flammable or corrosive gases or liquids routinely and normally are discharged. Companion equipment such as transformers and switchgear are not considered foreign.
5. Switchgear should not be installed in a location that is still specifically under active construction, especially where welding and burning are required directly overhead. Special precautions should be observed to minimize impingement of slag, metal filings, moisture, dust, or hot particles.
   EXCEPTION: Switchgear may be installed in a general construction area if suitable temporary protection is provided to minimize the risks associated with general construction activities.
6. Precautions shall be taken to protect energized switchgear from damage when maintenance is performed in the area.
7. Switchgear enclosure surfaces shall not be used as physical support for any item unless specifically designed for that purpose.
8. Enclosure interiors shall not be used as storage areas unless specifically designed for that purpose.
9. Metal instrument cases shall be grounded and enclosed in covers that are metal and grounded, or are of insulating material.

B. Metal-Enclosed Power Switchgear

1. Switchgear shall not be located within 7.6 m (25 ft) horizontally indoors or 3.0 m (10 ft) outdoors of storage containers, vessels, utilization equipment, or devices containing flammable liquids or gases.
   EXCEPTION: If an intervening barrier, designed to mitigate the potential effects of flammable liquids or gases, is installed, the distances listed above do not apply.
   The restrictions are not intended to apply to the power transformer(s) supplying the switchgear.
2. Enclosed switchgear rooms shall have at least two means of egress, one at each extreme of the area, not necessarily in opposite walls. Doors shall swing out and be equipped with panic bars, pressure plates, or other devices that are normally latched but open under simple pressure.
   EXCEPTION: One door may be used when required by physical limitations if means are provided for unhampered exit during emergencies.
3. Space shall be maintained in front of switchgear to allow breakers to be removed and turned without obstruction.
4. Space shall be maintained in the rear of the switchgear to allow for door opening to at least 90 degrees open, or a minimum of 900 mm (3 ft) without obstruction when removable panels are used.
5. Permanently mounted devices, panelboards, etc., located on the walls shall not encroach on the space requirements in 180B4.
6. Where columns extend into the room beyond the wall surface, the face of the column shall not encroach on the space requirements in 180B4.
7. Low-voltage cables or conductors, except those to be connected to equipment within the compartment, shall not be routed through the medium- or high-voltage divisions of switchgear unless installed in rigid metal conduit or isolated by rigid metal barriers.

8. Low-voltage conductors routed from medium- or high-voltage sections of switchgear shall terminate in a low-voltage section before being routed external to the switchgear.
9. Conductors entering switchgear shall be insulated for the higher operating voltage in that compartment or be separated from insulated conductors of other voltage ratings.
10. Switchgear enclosures shall be suitable for the environment in which they are installed.
11. A safety sign shall be placed in each cubicle containing more than one high-voltage source.

    NOTE: ANSI Z535.1-1998, ANSI Z535.2-1998, ANSI Z535.3-1998, ANSI Z535.4-1998, and ANSI Z535.5-1998 contain information regarding safety signs.
12. The location of control devices shall be readily accessible to personnel. Instruments, relays, and other devices requiring reading or adjustments should be so placed that work can readily be performed from the working space.

C. Dead-Front Power Switchboards

Dead-front power switchboards with uninsulated rear connections shall be installed in rooms or spaces that are capable of being locked, with access limited to qualified personnel.

D. Motor Control Centers

1. Motor control centers shall not be connected to systems having higher short-circuit capability than the bus bracing can withstand. Where current-limiting fuses are employed on the source side of the bus, the bus bracing and breaker-interrupting rating are determined by the peak let-through characteristic of the current-limiting fuse.
2. A safety sign shall be placed in each cubicle containing more than one voltage source.

    NOTE: ANSI Z535.1-1998, ANSI Z535.2-1998, ANSI Z535.3-1998, ANSI Z535.4-1998, and ANSI Z535.5-1998 contain information regarding safety signs.

E. Control Switchboards

1. Cabinets containing solid-state logic devices, electron tubes, or relay logic devices such as boiler analog, burner safety, annunciators, computers, inverters, precipitator logic, soot blower control, load control, telemetering, totalizing microwave radio, etc., are covered under these rules.
2. Where carpeting is installed in rooms containing control switchboards, it shall be of an antistatic type and shall minimize the release of noxious, corrosive, caustic, or toxic gas under any condition.
3. Layout of the installation shall provide adequate clearance in front of, or rear of, panels if applicable, to allow meters to be read without use of stools or auxiliary devices.
4. Where personnel access to control panels, such as benchboards, is required, cables shall be routed through openings separate from the personnel opening. Removable, sliding, or hinged panels are to be installed to close the personnel opening when not in use.

181. Metal-Enclosed Bus

A. General Requirements for All Types of Bus

1. Busways shall be installed only in accessible areas.
2. Busways, unless specifically approved for the purpose, shall not be installed: where subject to severe physical damage or corrosive vapors; in hoistways; in any classified hazardous location; outdoors or in damp locations.
3. Deadends of busway shall be closed.
4. Busways should be marked with the voltage and current rating for which they are designed, in such manner as to be visible after installation.

B. Isolated-Phase Bus

1. The minimum clearance between an isolated-phase bus and any magnetic material shall be the distance recommended by the manufacturer to avoid overheating of the magnetic material.
2. Nonmagnetic conduit should be used to protect the conductors for bus-alarm devices, thermocouples, space heaters, etc., if routed within the manufacturer's recommended minimum distance to magnetic material and parallel to isolated-phase bus enclosures.

3. When enclosure drains are provided for isolated-phase bus, necessary piping shall be provided to divert water away from electrical equipment.
4. Wall plates for isolated-phase bus shall be nonmagnetic, such as aluminum or stainless steel.
5. Grounding conductors for isolated-phase bus accessories should not be routed through ferrous conduit.

Section 19.
Surge Arresters

190. General Requirements

If arresters are required, they shall be located as close as practical to the equipment they protect.

NOTE: See IEEE Std C62.1-1989 [B37] and IEEE Std C62.11-1999 [B38] for additional information.

191. Indoor Locations

Arresters, if installed inside of buildings, shall be enclosed or shall be located well away from passageways and combustible parts.

192. Grounding Conductors

Grounding conductors shall be run as directly as practical between the arresters and ground and be of low impedance and ample current-carrying capacity and shall be grounded in accordance with the methods outlined in Section 9.

193. Installation

Arresters shall be installed in such a manner and location that neither the expulsion of gases nor the arrester disconnector is directed upon live parts in the vicinity.

Part 2.
Safety Rules for the Installation and Maintenance of Overhead Electric Supply and Communication Lines

Section 20.
Purpose, Scope, and Application of Rules

200. Purpose

The purpose of Part 2 of this code is the practical safeguarding of persons during the installation, operation, or maintenance of overhead supply and communication lines and their associated equipment.

201. Scope

Part 2 of this code covers supply and communication conductors and equipment in overhead lines. It covers the associated structural arrangements of such systems and the extension of such systems into buildings. The rules include requirements for spacing, clearances, and strength of construction. They do not cover installations in electric supply stations except as required by Rule 162A.

NOTE 1: Part 4 contains the approach distances and work rules required of supply and communciation employers and their employees working on or near supply and communication lines and equipment.

NOTE 2: The approach distances to energized parts, and other requirements applicable to the activities of utility or non-utility construction personnel, and others in close proximity to existing supply lines are governed by the Occupational Health and Safety Administration (OSHA), federal, state, or local statutes or regulations.

202. Application of Rules

The general requirements for application of these rules are contained in Rule 013. However, when a structure is replaced, the arrangement of equipment shall conform to the current edition of Rule 238C.

Section 21.
General Requirements

210. Referenced Sections

The Introduction (Section 1), Definitions (Section 2), References (Section 3), and Grounding Methods (Section 9) shall apply to the requirements of Part 2.

211. Number 211 not used in this edition.

212. Induced Voltages

Rules covering supply-line influence and communication-line susceptiveness have not been detailed in this code. Cooperative procedures are recommended in the control of voltages induced from proximate facilities. Therefore, reasonable advance notice should be given to owners or operators of other proximate facilities that may be adversely affected by new construction or changes in existing facilities.

213. Accessibility

All parts that must be examined or adjusted during operation shall be arranged so as to be accessible to authorized persons by the provision of adequate climbing spaces, working spaces, working facilities, and clearances between conductors.

214. Inspection and Tests of Lines and Equipment

A. When In Service

1. Initial Compliance With Rules

   Lines and equipment shall comply with these safety rules when placed in service.

2. Inspection

   Lines and equipment shall be inspected at such intervals as experience has shown to be necessary.

   NOTE: It is recognized that inspections may be performed in a separate operation or while performing other duties, as desired.

3. Tests

   When considered necessary, lines and equipment shall be subjected to practical tests to determine required maintenance.

4. Record of Defects

   Any defects affecting compliance with this code revealed by inspection or tests, if not promptly corrected, shall be recorded; such records shall be maintained until the defects are corrected.

5. Remedying Defects

   Lines and equipment with recorded defects that could reasonably be expected to endanger life or property shall be promptly repaired, disconnected, or isolated.

B. When Out of Service

1. Lines Infrequently Used

   Lines and equipment infrequently used shall be inspected or tested as necessary before being placed into service.

2. Lines Temporarily Out of Service

   Lines and equipment temporarily out of service shall be maintained in a safe condition.

3. Lines Permanently Abandoned

   Lines and equipment permanently abandoned shall be removed or maintained in a safe condition.

215. Grounding of Circuits, Supporting Structures, and Equipment

A. Methods

Grounding required by these rules shall be in accordance with the applicable methods given in Section 9.

B. Circuits

1. Common Neutral

A conductor used as a common neutral for primary and secondary circuits shall be effectively grounded as specified in Section 9.

2. Other Neutrals

Primary line, secondary line, and service neutral conductors shall be grounded as specified in Section 9.

EXCEPTION: Circuits designed for ground-fault detection and impedance-current-limiting devices.

3. Other Conductors

Line or service conductors, other than neutral conductors, that are intentionally grounded, shall be grounded as specified in Section 9.

4. Surge Arresters

Where the operation of surge arresters is dependent upon grounding, they shall be grounded in accordance with the methods outlined in Section 9.

5. Use of Earth as Part of Circuit

a. Supply circuits shall not be designed to use the earth normally as the sole conductor for any part of the circuit.

b. Monopolar operation of a bipolar HVDC system is permissible for emergencies and limited periods for maintenance.

C. Non-Current-Carrying Parts

1. General

Metal or metal-reinforced supporting structures, including lamp posts; metal conduits and raceways; cable sheaths; messengers; metal frames, cases, and hangers of equipment; and metal switch handles and operating rods shall be effectively grounded.

EXCEPTION 1: This rule does not apply to frames, cases, and hangers of equipment and switch handles and operating rods that are 2.45 m (8 ft) or more above readily accessible surfaces or are otherwise isolated or guarded and where the practice of not grounding such items has been a uniform practice over a well-defined area.

EXCEPTION 2: This rule does not apply to isolated or guarded equipment cases in certain specialized applications, such as series capacitors where it is necessary that equipment cases be either ungrounded or connected to the circuit. Such equipment cases shall be considered as energized and shall be suitably identified.

EXCEPTION 3: This rule does not apply to equipment cases, frames, equipment hangers, conduits, messengers, raceways, and cable sheaths enclosing or supporting only communication conductors, provided they are not exposed to contact with open supply conductors of over 300 V.

2. Guys

Guys shall be effectively grounded if attached to a supporting structure carrying any supply conductor of more than 300 V or if vulnerable to accidental energization by such conductors due to a slack conductor or guy.

EXCEPTION 1: This rule does not apply to guys containing an insulator or insulators installed in accordance with and meeting the requirements of Rule 279A.

EXCEPTION 2: This rule does not apply to guys attached to supporting structures if all supply conductors are in cable conforming to the requirements of Rules 230C1, 230C2, or 230C3.

EXCEPTION 3: This rule does not apply if the guy is attached to a supporting structure on private right-of-way if all the supply circuits exceeding 300 V meet the requirements of Rule 220B2.

3. Multiple Messengers on the Same Structure

Communication cable messengers exposed to power contacts, power induction, or lightning, shall be bonded together at intervals specified in Rule 92C.

216. Arrangement of Switches

A. Accessibility
Switches or their control mechanisms shall be installed so as to be accessible to authorized persons.

B. Indicating Open or Closed Position
Switch position shall be visible or clearly indicated.

C. Locking
Switch-operating mechanisms that are accessible to unauthorized persons shall have provisions for locking in each operational position.

D. Uniform Position
The handles or control mechanisms for all switches throughout any system should have consistent positions when opened and uniformly different positions when closed in order to minimize operating errors. Where this practice is not followed, the switches should be marked to minimize mistakes in operation.

E. Remotely controlled, automatic transmission, or distribution overhead line switching devices shall have local provisions to render remote or automatic controls inoperable.

217. General

A. Supporting Structures
1. Protection of Structures
a. Mechanical Damage
Appropriate physical protection shall be provided for supporting structures subject to vehicular traffic abrasion that would materially affect their strength.
NOTE: Nothing in this rule is intended to require protection or marking of structural components located outside of the traveled ways of roadways or established parking areas. Experience has shown that it is not practical to protect structures from contact by out-of-control vehicles operating outside of established traveled ways.
b. Fire
Supporting structures shall be placed and maintained so as to be exposed as little as is practical to brush, grass, rubbish, or building fires.
c. Attached to Bridges
Supporting structures attached to bridges for the purpose of carrying open supply conductors exceeding 600 V shall be posted with appropriate safety signs.
NOTE: ANSI Z535.1-1998, ANSI Z535.2-1998, ANSI Z535.3-1998, ANSI Z535.4-1998, and ANSI Z535.5-1998 contain information regarding safety signs.
2. Readily Climbable Supporting Structures
a. Readily climbable supporting structures, such as closely latticed poles, towers, or bridge attachments, carrying open supply conductors energized at more than 300 V, which are adjacent to roads, regularly traveled pedestrian thoroughfares, or places where persons frequently gather (such as schools or public playgrounds), shall be equipped with barriers to inhibit climbing by unqualified persons or posted with appropriate safety signs.
EXCEPTION: This rule does not apply where access to the supporting structure is limited by a fence meeting the height requirements of Rule 110A1.
NOTE: ANSI Z535.1-1998, ANSI Z535.2-1998, ANSI Z535.3-1998, ANSI Z535.4-1998, and ANSI Z535.5-1998 contain information regarding safety signs.
b. Steps
Steps permanently installed on supporting structures shall not be closer than 2.45 m (8 ft) from the ground or other accessible surface.
EXCEPTION 1: This rule does not apply where supporting structures are isolated.

EXCEPTION 2: This rule does not apply where access to the supporting structure is limited by a fence meeting the height requirements of Rule 110A1.

c. Standoff Brackets

Standoff brackets on supporting structures shall be arranged so that there is not less than 2.45 m (8 ft) between either:

(1) The lowest bracket and ground or other accessible surface, or

(2) The two lowest brackets.

EXCEPTION: This rule does not apply where supporting structures are isolated.

3. Identification

Supporting structures, including those on bridges, on which supply or communication conductors are maintained shall be so constructed, located, marked, or numbered so as to facilitate identification by employees authorized to work thereon.

4. Obstructions

Signs, posters, notices, and other attachments shall not be placed on supporting structures without concurrence of the owner. Supporting structures should be kept free from other climbing hazards such as tacks, nails, vines, and through bolts not properly trimmed.

5. Decorative Lighting

Attachment of decorative lighting on structures shall not be made without the concurrence of the owners and occupants.

B. Unusual Conductor Supports

Where line conductors are attached to structures other than those used solely or principally for their support, all rules shall be complied with as far as they apply. Such additional precautions as may be deemed necessary by the administrative authority shall be taken to avoid damage to the structures or injury to the persons using them. The supporting of conductors on trees and roofs should be avoided.

218. Tree Trimming

A. General

1. Trees that may interfere with ungrounded supply conductors should be trimmed or removed.

   NOTE: Normal tree growth, the combined movement of trees and conductors under adverse weather conditions, voltage, and sagging of conductors at elevated temperatures are among the factors to be considered in determining the extent of trimming required.

2. Where trimming or removal is not practical, the conductor should be separated from the tree with suitable materials or devices to avoid conductor damage by abrasion and grounding of the circuit through the tree.

B. At Line Crossings, Railroad Crossings, and Limited-Access Highway Crossings

The crossing span and the adjoining span on each side of the crossing should be kept free from overhanging or decayed trees or limbs that otherwise might fall into the line.

Section 22.
Relations Between Various Classes of Lines and Equipment

220. Relative Levels

A. Standardization of Levels

The levels at which different classes of conductors are to be located should be standardized by agreement of the utilities concerned.

B. Relative Levels: Supply and Communication Conductors

1. Preferred Levels

   Where supply and communication conductors cross each other or are located on the same structures, the supply conductors should be carried at the higher level.

   EXCEPTION: This rule does not apply to trolley feeders, which may be located for convenience approximately at the level of the trolley-contact conductor.

2. Special Construction for Supply Circuits, the Voltage of Which Is 600 V or Less and Carrying Power Not in Excess of 5 kW

   Where all circuits are owned or operated by one party or where cooperative consideration determines that the circumstances warrant and the necessary coordinating methods are employed, single-phase ac or two-wire dc circuits carrying a voltage of 600 V or less between conductors, with transmitted power not in excess of 5 kW, when involved in the joint use of structures with communication circuits, may be installed in accordance with footnote 1 of Table 235-5, under the following conditions:

   a. That such supply circuits are of covered conductor not smaller than AWG No. 8 medium hard-drawn copper or its equivalent in strength, and the construction otherwise conforms with the requirements for supply circuits of the same class.

   b. That the supply circuits be placed on the end and adjacent pins of the lowest through signal support arm and that a 750 mm (30 in) climbing space be maintained from the ground up to a point at least 600 mm (24 in) above the supply circuits. The supply circuits shall be rendered conspicuous by the use of insulators of different form or color from others on the pole line or by stenciling the voltage on each side of the support arm between the pins carrying each supply circuit, or by indicating the voltage by means of metal characters.

   c. That there shall be a vertical clearance of at least 600 mm (2 ft) between the support arm carrying these supply circuits and the next support arm above. The other pins on the support arm carrying the supply circuits may be occupied by communication circuits used in the operation or control of signal system or other supply system if owned, operated, and maintained by the same company operating the supply circuits.

   d. That such supply circuits shall be equipped with arresters and fuses installed in the supply end of the circuit and where the signal circuit is ac, the protection shall be installed on the secondary side of the supply transformer. The arresters shall be designed so as to break down at approximately twice the voltage between the wires of the circuit, but the breakdown voltage of the arrester need not be less than 1 kV. The fuses shall have a rating not in excess of approximately twice the maximum operating current of the circuit, but their rating need not be less than 10 A. The fuses likewise in all cases shall have a rating of at least 600 V, and where the supply transformer is a step-down transformer, shall be capable of opening the circuit successfully in the event the transformer primary voltage is impressed upon them.

   e. Such supply circuits in cable meeting the requirements of Rules 230C1, 230C2, or 230C3 may be installed below communication attachments, with not less than 400 mm (16 in) vertical separation between the supply cable and the lowest communication attachment. Communication circuits other than those used in connection with the operation of the supply circuits shall not be carried in the same cable with such supply circuits.

f. Where such supply conductors are carried below communication conductors, transformers and other apparatus associated therewith shall be attached only to the sides of the support arm in the space between and at no higher level than such supply wires.

g. Lateral runs of such supply circuits carried in a position below the communication space shall be protected through the climbing space by wood molding or equivalent covering, or shall be carried in insulated multiple-conductor cable, and such lateral runs shall be placed on the underside of the support arm.

C. Relative Levels: Supply Lines of Different Voltage Classifications (as classified in Table 235-5)

1. At Crossings or Conflicts

Where supply conductors of different voltage classifications cross each other or structure conflict exists, the higher-voltage lines should be carried at the higher level.

2. On Structures Used Only by Supply Conductors

Where supply conductors of different voltage classifications are on the same structures, relative levels should be as follows:

a. Where all circuits are owned by one utility, the conductors of higher voltage should be placed above those of lower voltage.

b. Where different circuits are owned by separate utilities, the circuits of each utility may be grouped together, and one group of circuits may be placed above the other group provided that the circuits in each group are located so that those of higher voltage are at the higher levels and that any of the following conditions is met:

(1) A vertical spacing of not less than that required by Table 235-5 is maintained between the nearest line conductors of the respective utilities.

(2) Conductors of a lower voltage classification placed at a higher level than those of a higher classification shall be placed on the opposite side of the structure.

(3) Ownership and voltage are prominently displayed.

D. Identification of Overhead Conductors

All conductors of electric supply and communication lines should, as far as is practical, be arranged to occupy uniform positions throughout, or shall be constructed, located, marked, numbered, or attached to distinctive insulators or crossarms, so as to facilitate identification by employees authorized to work thereon. This does not prohibit systematic transposition of conductors.

E. Identification of Equipment on Supporting Structures

All equipment of electric supply and communication lines should be arranged to occupy uniform positions throughout or shall be constructed, located, marked, or numbered so as to facilitate identification by employees authorized to work thereon.

221. Avoidance of Conflict

Two separate lines, either of which carries supply conductors, should be so separated from each other that neither conflicts with the other. If this is not practical, the conflicting line or lines should be separated as far as practical and shall be built to the grade of construction required by Section 24 for a conflicting line, or the two lines shall be combined on the same structures.

222. Joint Use of Structures

Joint use of structures should be considered for circuits along highways, roads, streets, and alleys. The choice between joint use of structures and separate lines shall be determined through cooperative consideration of all the factors involved, including the character of circuits, the total number and weight of conductors, tree conditions, number and location of branches and service drops, structure conflicts, availability of right-of-way, etc. Where such joint use is mutually agreed upon, it shall be subject to the appropriate grade of construction specified in Section 24.

223. Communications Protective Requirements

A. Where Required

Where communication apparatus is handled by other than qualified persons, it shall be protected by one or more of the means listed in Rule 223B if such apparatus is permanently connected to lines subject to any of the following:

1. Lightning
2. Contact with supply conductors whose voltage to ground exceeds 300 V
3. Transient rise in ground potential exceeding 300 V
4. Steady-state induced voltage of a hazardous level

   NOTE: When communication cables will be in the vicinity of supply stations where large ground currents may flow, the effect of these currents on communication circuits should be evaluated.

B. Means of Protection

Where communication apparatus is required to be protected under Rule 223A, protective means adequate to withstand the voltage expected to be impressed shall be provided by insulation, protected where necessary by surge arresters used in conjunction with fusible elements. Severe conditions may require the use of additional devices such as auxiliary arresters, drainage coils, neutralizing transformers, or isolating devices.

224. Communication Circuits Located Within the Supply Space and Supply Circuits Located Within the Communication Space

A. Communication Circuits Located in the Supply Space

1. Communication circuits located in the supply space shall be installed and maintained only by personnel authorized and qualified to work in the supply space in accordance with the applicable rules of Sections 42 and 44.
2. Communication circuits located in the supply space shall meet the following clearance requirements, as applicable:
   a. Insulated communication cables supported by an effectively grounded messenger shall have the same clearances as neutrals meeting Rule 230E1 from communication circuits located in the communication space and from supply conductors located in the supply space. See Rules 235 and 238.
   b. Fiber-optic cables located in the supply space shall meet the requirements of Rule 230F.
   c. Open-wire communication circuits permitted by other rules to be in the supply space shall have the same clearances from communication circuits located in the communication space and from other circuits located in the supply space as required by Rule 235 for open supply conductors of 0–750 V.

      EXCEPTION: Service drops meeting Rule 224A3a and 224A3b may originate in the supply space on a line structure or in the span and terminate in the communication space on the building or structure being served.
3. Communication circuits located in the supply space in one portion of the system may be located in the communication space in another portion of the system if the following requirements are met:
   a. Where the communication circuit is, at any point, located above an energized supply conductor or cable, the communication circuit shall be protected by fuseless surge arresters, drainage coils, or other suitable devices to limit the normal communication circuit voltage to 400 V or less to ground.

      NOTE: The grades of construction for communication conductors with inverted levels apply.
   b. Where the communication circuit is always located below the supply conductors, the communication protection shall meet the requirements of Rule 223.
   c. The transition(s) between the supply space and the communication space shall occur on a single structure; no transition shall occur between line structures.

      EXCEPTION: Service drops meeting Rule 224A3a and Rule 224A3b may originate in the supply space on a line structure or in the span and terminate in the communication space on the building or structure being served.

d. The construction and protection shall be consistently followed throughout the extent of such section of the communications system.

B. Supply Circuits Used Exclusively in the Operation of Communication Circuits

Circuits used for supplying power solely to apparatus forming part of a communications system shall be installed as follows:

1. Open-wire circuits shall have the grades of construction, clearances, insulation, etc., prescribed elsewhere in these rules for supply or communication circuits of the voltage concerned.
2. Special circuits operating at voltages in excess of 90 V ac or 150 V dc and used for supplying power solely to communications equipment may be included in communication cables under the following conditions:
   a. Such cables shall have a conductive sheath or shield that is effectively grounded, and each such circuit shall be carried on conductors that are individually enclosed with an effectively grounded shield.
   b. All circuits in such cables shall be owned or operated by one party and shall be maintained only by qualified personnel.
   c. Supply circuits included in such cables shall be terminated at points accessible only to qualified personnel.
   d. Communication circuits brought out of such cables, if they do not terminate in a repeater station or terminal office, shall be protected or arranged so that in the event of failure within the cable, the voltage on the communication circuit will not exceed 400 V to ground.
   e. Terminal apparatus for the power supply shall be so arranged that the live parts are inaccessible when such supply circuits are energized.

   EXCEPTION: The requirements of Rule 224B2 do not apply to communication circuits where the transmitted power does not exceed 150 W.

225. Electric Railway Construction

A. Trolley-Contact Conductor Fastenings

All overhead trolley-contact conductors shall be supported and arranged so that the breaking of a single contact conductor fastening will not allow the trolley conductor live span wire, or current-carrying connection, to come within 3.0 m (10 ft) (measured vertically) from the ground, or from any platform accessible to the general public.

Span-wire insulation for trolley-contact conductors shall comply with Rule 279B.

B. High-Voltage Contact Conductors

Trolley-contact conductors energized at more than 750 V shall be suspended in such a way that, if broken at one point, the conductor will not come within 3.6 m (12 ft) (measured vertically) of the ground, or any platform accessible to the public.

C. Third Rails

Third rails shall be protected by adequate guards composed of wood or other suitable insulating material.

EXCEPTION: This rule does not apply where third rails are on fenced right-of-way.

D. Prevention of Loss of Contact at Railroad Crossings at Grade

At crossings at grade with other railroads or other electrified railway systems, contact conductors shall be arranged as set forth in the following specifications, 1, 2, 3, 4, and 5, following, whichever apply:

1. Where the crossing span exceeds 30 m (100 ft), catenary construction shall be used for overhead trolley-contact conductors.
2. When pole trolleys, using either wheels or sliding shoes, are used:

a. The trolley-contact conductor shall be provided with live trolley guards of suitable construction; or

b. The trolley-contact conductor should be at a uniform height above its own track throughout the crossing span and the next adjoining spans. Where it is not practical to maintain a uniform height, the change in height shall be made in a gradual manner.

EXCEPTION: Rule 225D2 does not apply where the crossing is protected by signals or interlocking.

3. When collectors of the pantograph type are used, the contact conductor and track through the crossing should be maintained in a condition where the rocking of pantograph-equipped cars or locomotives will not de-wire the pantograph. If this cannot be done, auxiliary contact conductors shall be installed. Wire height shall conform with Rule 225D2.

4. Where two electrified tracks cross:

a. When the trolley-contact conductors are energized from different supply circuits, or from different phases of the same circuit, the trolley-conductor crossover shall be designed to insulate both conductors from each other. The design shall not permit either trolley collector to contact any conductor or part energized at a different voltage than at which it is designed to operate.

b. Trolley-contact crossovers used to insulate trolley conductors of the same voltage but of different circuit sections shall be designed to limit the likelihood of both sections being simultaneously contacted by the trolley collector.

5. When third rail construction is used, and the length of the third rail gap at the crossings is such that a car or locomotive stopping on the crossing can lose propulsion power, the crossing shall be protected by signals or interlocking.

E. Guards Under Bridges

Trolley guards of suitable construction shall be provided where the trolley-contact conductor is so located that a trolley pole leaving the conductor can make simultaneous contact between it and the bridge structure.

Section 23.
Clearances

230. General

A. Application

This section covers all clearances, including climbing spaces, involving overhead supply and communication lines.

NOTE: The more than 70 years of historical development and specification of clearances in Rules 232, 233, and 234 were reviewed for consistency among themselves and with modern practice and were appropriately revised in both concept and content for the 1990 Edition. See Appendix A.

1. Permanent and Temporary Installations

The clearances of Section 23 are required for permanent and temporary installations.

2. Emergency Installations

The clearances required in Section 23 may be decreased for emergency installations if the following conditions are met.

NOTE: See Rule 014.

a. Open supply conductors of 0 to 750 V and supply cables meeting Rule 230C; and communication conductors and cables, guys, messengers, and neutral conductors meeting Rule 230E1 shall be suspended not less than 4.8 m (15.5 ft) above areas where trucks are expected, or 2.70 m (9 ft) above areas limited to pedestrians or restricted traffic only where vehicles are not expected during the emergency, unless Section 23 permits lesser clearances.

For the purpose of this rule, trucks are defined as any vehicle exceeding 2.5 m (8 ft) in height. Areas not subject to truck traffic are areas where truck traffic is neither normally encountered nor reasonably anticipated or is otherwise limited.

Spaces and ways subject to pedestrians or restricted traffic only are those areas where riders on horseback, vehicles, or other mobile units exceeding 2.5 m (8 ft) in height are prohibited by regulation or permanent terrain configurations or are otherwise neither normally encountered nor reasonably anticipated or are otherwise limited.

b. Vertical clearances of open supply conductors above 750 V shall be increased above the applicable value of Rule 230A2a as appropriate for the voltage involved and the given local conditions.

c. Reductions in horizontal clearances permitted by this rule shall be in accordance with accepted good practice for the given local conditions during the term of the emergency.

d. Supply and communication cables may be laid directly on grade if they are guarded or otherwise located so that they do not unduly obstruct pedestrian or vehicular traffic and are appropriately marked. Supply cables operating above 600 V shall meet either Rule 230C or Rule 350B.

e. No clearance is specified for areas where access is limited to qualified personnel only.

B. Measurement of Clearance and Spacing

Unless otherwise stated, all clearances shall be measured from surface to surface and all spacings shall be measured center to center. For clearance measurement, live metallic hardware electrically connected to line conductors shall be considered a part of the line conductors. Metallic bases of potheads, surge arresters, and similar devices shall be considered a part of the supporting structure.

C. Supply Cables

For clearance purposes, supply cables, including splices and taps, conforming to any of the following requirements are permitted lesser clearances than open conductors of the same voltage. Cables should be capable of withstanding tests applied in accordance with an applicable standard.

1. Cables that are supported on or cabled together with an effectively grounded bare messenger or neutral, or with multiple concentric neutral conductors, where any associated neutral conductor(s) meet(s) the requirements of Rule 230E1 and where the cables also meet one of the following:

a. Cables of any voltage having an effectively grounded continuous metal sheath or shield, or

b. Cables designed to operate on a multi-grounded system at 22 kV or less and having semiconducting insulation shielding in combination with suitable metallic drainage.

2. Cables of any voltage, not included in Rule 230C1, covered with a continuous auxiliary semiconducting shield in combination with suitable metallic drainage and supported on and cabled together with an effectively grounded bare messenger.
3. Insulated, nonshielded cable operated at not over 5 kV phase to phase, or 2.9 kV phase to ground, supported on and cabled together with an effectively grounded bare messenger.

D. Covered Conductors

Covered conductors shall be considered bare conductors for all clearance requirements except that spacing between conductors of the same or different circuits, including grounded conductors, may be reduced below the requirements for open conductors when the conductors are owned, operated, or maintained by the same party and when the conductor covering provides sufficient dielectric strength to limit the likelihood of a short circuit in case of momentary contact between conductors or between conductors and the grounded conductor. Intermediate spacers may be used to maintain conductor spacing and to provide support.

E. Neutral Conductors

1. Neutral conductors that are effectively grounded throughout their length and associated with circuits of 0 to 22 kV to ground may have the same clearances as guys and messengers.
2. All other neutral conductors of supply circuits shall have the same clearances as the phase conductors of the circuit with which they are associated.

F. Fiber-Optic Cable

1. Fiber-optic—supply cable
   a. Cable defined as "fiber-optic—supply" supported on a messenger that is effectively grounded throughout its length shall have the same clearance from communications facilities as required for a neutral conductor meeting Rule 230E1.
   b. Cable defined as "fiber-optic—supply" that is entirely dielectric, or supported on a messenger that is entirely dielectric, shall have the same clearance from communications facilities as required for a neutral conductor meeting Rule 230E1.
   c. Fiber-optic—supply cables supported on or within messengers not meeting Rule 230F1a or 230F1b shall have the same clearances from communications facilities required for such messengers.
   d. Fiber-optic—supply cables supported on or within a conductor(s), or containing a conductor(s) or cable sheath(s) within the fiber-optic cable assembly shall have the same clearances from communications facilities required for such conductors. Such clearance shall be not less than that required under Rule 230F1a, 230F1b, or 230F1c, as applicable.
   e. Fiber-optic—supply cables meeting Rule 224A3 are considered to be communication cables when located in the communication space.
2. Fiber-optic—communication cable

   Cable defined as "fiber-optic—communication" shall have the same clearance from supply facilities as required for a communication messenger.

G. Alternating- and Direct-Current Circuits

The rules of this section are applicable to both ac and dc circuits. For dc circuits, the clearance requirements shall be the same as those for ac circuits having the same crest voltage to ground.

NOTE: Although the corresponding crest voltage for a common sinusoidal ac circuit may be calculated by multiplying its rms value by 1.414 (square root of 2), this may not be appropriate for other type ac circuits. An example of the latter is represented by non-sinusoidal power supplies such as used in some coaxial cable type communication systems.

H. Constant-Current Circuits

The clearances for constant-current circuits (such as series lighting circuits) shall be determined on the basis of their normal full-load voltage.

I. Maintenance of Clearances and Spacings

The clearances and spacing required shall be maintained at the values and under the conditions specified in Section 23 of the applicable edition.

NOTE: See Rule 013 to determine the applicable edition.

231. Clearances of Supporting Structures From Other Objects

Supporting structures, support arms and equipment attached thereto, and braces shall have the following clearances from other objects. The clearance shall be measured between the nearest parts of the objects concerned.

A. From Fire Hydrants

Not less than 1.2 m (4 ft).

EXCEPTION: Where conditions do not permit, a clearance of not less than 900 mm (3 ft) is allowed.

B. From Streets, Roads, and Highways

1. Where there are curbs: supporting structures, support arms, anchor guys, or equipment attached thereto, up to 4.6 m (15 ft) above the road surface shall be located a sufficient distance from the street side of the curbs to avoid contact by ordinary vehicles using and located on the traveled way. For a redirectional curb, such distance shall be not less than 150 mm (6 in). For paved or concrete swale-type curbs, such facilities shall be located behind the curb.
2. Where there are no curbs, supporting structures should be located a sufficient distance from the roadway to avoid contact by ordinary vehicles using and located on the traveled way.
3. Location of overhead utility installations on highways with narrow rights-of-way or on urban streets with closely abutting improvements are special cases that must be resolved in a manner consistent with the prevailing limitations and conditions.
4. Where a governmental authority exercising jurisdiction over structure location has issued a permit for, or otherwise approved, specific locations for supporting structures, that permit or approval shall govern.

C. From Railroad Tracks

Where railroad tracks are parallel to or crossed by overhead lines, all portions of the supporting structures, support arms, anchor guys, and equipment attached thereto less than 6.7 m (22 ft) above the nearest track rail shall have horizontal clearances not less than the values required by Rule 231C1 or Rule 231C2 for the situation concerned.

NOTE: See Rule 234I.

1. Not less than 3.6 m (12 ft) from the nearest track rail.

   EXCEPTION 1: A clearance of not less than 2.13 m (7 ft) may be allowed where the supporting structure is not the controlling obstruction, provided sufficient space for a driveway is left where cars are loaded or unloaded.

   EXCEPTION 2: Supports for overhead trolley-contact conductors may be located as near their own track rail as conditions require. If very close, however, permanent screens on cars will be necessary to protect passengers.

   EXCEPTION 3: Where necessary to provide safe operating conditions that require an uninterrupted view of signals, signs, etc., along tracks, the parties concerned shall cooperate in locating structures to provide the necessary clearance.

   EXCEPTION 4: At industrial sidings, a clearance of not less than 2.13 m (7 ft) shall be permitted, provided sufficient space is left where cars can be loaded or unloaded.
2. The clearances of Rule 231C1 may be reduced by agreement with the railroad(s).

232. Vertical Clearances of Wires, Conductors, Cables, and Equipment Above Ground, Roadway, Rail, or Water Surfaces

A. Application

The vertical clearances specified in Rule 232B1 apply under the following conductor temperature and loading conditions, whichever produces the largest final sag.

1. 50 °C (120 °F), no wind displacement.

2. The maximum conductor temperature for which the line is designed to operate, if greater than 50 °C (120 °F), with no wind displacement.
3. 0 °C (32 °F), no wind displacement, with radial thickness of ice, if any, specified in Rule 250B for the loading district concerned.

   EXCEPTION: The conductor temperature and loading condition for trolley and electrified railroad contact conductors shall be 15 °C (60 °F), no wind displacement, final unloaded sag, or initial unloaded sag in cases where these facilities are maintained approximately at initial unloaded sags.

   NOTE: The phase and neutral conductors of a supply line are normally considered separately when determining the sag of each due to temperature rise.

B. Clearance of Wires, Conductors, Cables, Equipment, and Support Arms Mounted on Supporting Structures

1. Clearance to Wires, Conductors, and Cables
   The vertical clearance of wires, conductors, and cables above ground in generally accessible places, roadway, rail, or water surfaces, shall be not less than that shown in Table 232-1.
2. Clearance to Unguarded Rigid Live Parts of Equipment
   The vertical clearance above ground, roadway, or water surfaces for unguarded rigid live parts such as potheads, transformer bushings, surge arresters, and short lengths of supply conductors connected thereto, which are not subject to variation in sag, shall be not less than that shown in Table 232-2. For clearances of drip loops of service drops, see Table 232-1.
3. Clearance to Support Arms and Equipment Cases
   The vertical clearance of equipment cases, support arms, platforms, and braces that extend beyond the surface of the structure shall be not less than that shown in Table 232-2. These clearances do not apply to internal structural braces for latticed towers, X-braces between poles, and pole-type push braces.
4. Street and Area Lighting
   a. The vertical clearance of street and area lighting luminaires shall be not less than that shown in Table 232-2. For this purpose, grounded luminaire cases and brackets shall be considered as effectively grounded equipment cases; ungrounded luminaire cases and brackets shall be considered as a rigid live part of the voltage contained.

      EXCEPTION: This rule does not apply to post-top mounted luminaires with grounded or entirely dielectric cases.
   b. Insulators, as specified in Rule 279A, should be inserted at least 2.45 m (8 ft) from the ground in metallic suspension ropes or chains supporting lighting units of series circuits.

C. Additional Clearances for Wires, Conductors, Cables, and Unguarded Rigid Live Parts of Equipment
Greater clearances than specified by Rule 232B shall be provided where required by Rule 232C1.

1. Voltages Exceeding 22 kV
   a. For voltages between 22 and 470 kV, the clearance specified in Rule 232B1 (Table 232-1) or Rule 232B2 (Table 232-2) shall be increased at the rate of 10 mm (0.4 in) per kilovolt in excess of 22 kV. For voltages exceeding 470 kV, the clearance shall be determined by the method given in Rule 232D. All clearances for lines over 50 kV shall be based on the maximum operating voltage.

      EXCEPTION: For voltages exceeding 98 kV ac to ground or 139 kV dc to ground, clearances less than those required above are permitted for systems with known maximum switching-surge factors (see Rule 232D).
   b. For voltages exceeding 50 kV, the additional clearance specified in Rule 232C1a shall be increased 3% for each 300 m (1000 ft) in excess of 1000 m (3300 ft) above mean sea level.

c. For voltages exceeding 98 kV ac to ground, either the clearances shall be increased or the electric field, or the effects thereof, shall be reduced by other means as required to limit the steady-state current due to electrostatic effects to 5 mA rms if the largest anticipated truck, vehicle, or equipment under the line were short-circuited to ground. The size of the anticipated truck, vehicle, or equipment used to determine these clearances may be less than but need not be greater than that limited by federal, state, or local regulations governing the area under the line. For this determination, the conductors shall be at a final unloaded sag at 50 °C (120 °F).

D. Alternate Clearances for Voltages Exceeding 98 kV AC to Ground or 139 kV DC to Ground

The clearances specified in Rules 232B and 232C may be reduced for circuits with known switching-surge factors, but shall be not less than the alternate clearance, which is computed by adding the reference height from Rule 232D2 to the electrical component of clearance from Rule 232D3.

1. Sag Conditions of Line Conductors

   The vertical clearance shall be maintained under the conductor temperature and loading condition given in Rule 232A.

2. Reference Heights

   The reference height shall be selected from Table 232-3.

3. Electrical Component of Clearance

   a. The electrical component (D) shall be computed using the following equations. Selected values of D are listed in Table 232-4.

$$D = 1.00\left[\frac{V \cdot (PU) \cdot a}{500K}\right]^{1.667} bc \quad \text{(m)}$$

$$D = 3.28\left[\frac{V \cdot (PU) \cdot a}{500K}\right]^{1.667} bc \quad \text{(ft)}$$

   where

   V = maximum ac crest operating voltage to ground or maximum dc operating voltage to ground in kilovolts,

   PU = maximum switching-surge factor expressed in per-unit peak voltage to ground and defined as a switching-surge level for circuit breakers corresponding to 98% probability that the maximum switching surge generated per breaker operation does not exceed this surge level, or the maximum anticipated switching-surge level generated by other means, whichever is greater,

   a = 1.15, the allowance for three standard deviations,

   b = 1.03, the allowance for nonstandard atmospheric conditions,

   c = 1.2, the margin of safety,

   K = 1.15, the configuration factor for conductor-to-plane gap.

   b. The value of D shall be increased 3% for each 300 m (1000 ft) in excess of 450 m (1500 ft) above mean sea level.

   c. For voltages exceeding 98 kV ac to ground, either the clearances shall be increased or the electric field, or the effects thereof, shall be reduced by other means as required to limit the steady state current due to electrostatic effects to 5 mA, rms, if the largest anticipated truck, vehicle, or equipment under the line were short-circuited to ground. The size of the anticipated truck, vehicle, or equipment used to determine these clearances may be less than but need not be greater than that limited by federal, state, or local regulations governing the area under the line. For this determination, the conductors shall be at a final unloaded sag at 50 °C (120 °F).

4. Limit

   The alternate clearance shall be not less than the clearance given in Tables 232-1 or 232-2 computed for 98 kV ac to ground in accordance with Rule 232C.

m

Table 232-1
Vertical Clearance of Wires, Conductors, and Cables Above Ground, Roadway, Rail, or Water Surfaces[25]

(Voltages are phase to ground for effectively grounded circuits and those other circuits where all ground faults are cleared by promptly de-energizing the faulted section, both initially and following subsequent breaker operations. See the definitions section for voltages of other systems. See Rules 232B1, 232C1a, and 232D4.)

Nature of surface underneath wires, conductors, or cables	Insulated communication conductors and cable; messengers; surge-protection wires; grounded guys; ungrounded guys exposed to 0 to 300 V[11, 15]; neutral conductors meeting Rule 230E1; supply cables meeting Rule 230C1 (m)	Noninsulated communication conductors; supply cables of 0 to 750 V meeting Rules 230C2 or 230C3 (m)	Supply cables over 750 V meeting Rules 230C2 or 230C3; open supply conductors, 0 to 750 V; ungrounded guys exposed to over 300 V to 750 V[14] (m)	Open supply conductors, over 750 V to 22 kV; ungrounded guys exposed to 750 V to 22 kV[14] (m)	Trolley and electrified railroad contact conductors and associated span or messenger wires	
					0 to 750 V to ground (m)	Over 750 V to 22 kV to ground (m)
Where wires, conductors, or cables cross over or overhang						
1. Track rails of railroads (except electrified railroads using overhead trolley conductors) [2, 16, 22]	7.2	7.3	7.5	8.1	6.7[4]	6.7[4]
2. Roads, streets, and other areas subject to truck traffic[23]	4.7	4.9	5.0	5.6	5.5[5]	6.1[5]
3. Driveways, parking lots, and alleys[23]	4.7[7, 13]	4.9[7, 13]	5.0[7]	5.6	5.5[5]	6.1[5]
4. Other land traversed by vehicles, such as cultivated, grazing, forest, orchard, etc.[26]	4.7	4.9	5.0	5.6	—	—
5. Spaces and ways subject to pedestrians or restricted traffic only [9]	2.9	3.6[8]	3.8[8]	4.4	4.9	5.5
6. Water areas not suitable for sailboating or where sailboating is prohibited[21]	4.0	4.4	4.6	5.2	—	—
7. Water areas suitable for sailboating including lakes, ponds, reservoirs, tidal waters, rivers, streams, and canals with an unobstructed surface area of[17, 18, 19, 20, 21]						
a. Less than 0.08 km^2	5.3	5.5	5.6	6.2	—	—
b. Over 0.08 to 0.8 km^2	7.8	7.9	8.1	8.7	—	—

Table 232-1 *(Continued)*
Vertical Clearance of Wires, Conductors, and Cables Above Ground, Roadway, Rail, or Water Surfaces[25]

(Voltages are phase to ground for effectively grounded circuits and those other circuits where all ground faults are cleared by promptly de-energizing the faulted section, both initially and following subsequent breaker operations. See the definitions section for voltages of other systems. See Rules 232B1, 232C1a, and 232D4.)

Nature of surface underneath wires, conductors, or cables	Insulated communication conductors and cable; messengers; surge-protection wires; grounded guys; ungrounded guys exposed to 0 to 300 V[11, 15]; neutral conductors meeting Rule 230E1; supply cables meeting Rule 230C1 (m)	Noninsulated communication conductors; supply cables of 0 to 750 V meeting Rules 230C2 or 230C3 (m)	Supply cables over 750 V meeting Rules 230C2 or 230C3; open supply conductors, 0 to 750 V; ungrounded guys exposed to over 300 V to 750 V[14] (m)	Open supply conductors, over 750 V to 22 kV; ungrounded guys exposed to 750 V to 22 kV[14] (m)	Trolley and electrified railroad contact conductors and associated span or messenger wires: 0 to 750 V to ground (m)	Trolley and electrified railroad contact conductors and associated span or messenger wires: Over 750 V to 22 kV to ground (m)
c. Over 0.8 to 8 km^2	9.6	9.8	9.9	10.5	—	—
d. Over 8 km^2	11.4	11.6	11.7	12.3	—	—
8. Established boat ramps and associated rigging areas; areas posted with sign(s) for rigging or launching sail boats	Clearance above ground shall be 1.5 m greater than in 7 above, for the type of water areas served by the launching sites					
Where wires, conductors, or cables run along and within the limits of highways or other road rights-of-way but do not overhang the roadway						
9. Roads, streets, or alleys	4.7[24]	4.9	5.0	5.6	5.5[5]	6.1[5]
10. Roads in rural districts where it is unlikely that vehicles will be crossing under the line	4.1[10, 12]	4.3[10]	4.4[10]	5.0	5.5[5]	6.1[5]

[1] Where subways, tunnels, or bridges require it, less clearance above ground or rails than required by Table 232-1 may be used locally. The trolley and electrified railroad contact conductor should be graded very gradually from the regular construction down to the reduced elevation.

[2] For wires, conductors, or cables crossing over mine, logging, and similar railways that handle only cars lower than standard freight cars, the clearance may be reduced by an amount equal to the difference in height between the highest loaded car handled and 6.1 m, but the clearance shall not be reduced below that required for street crossings.

[3] This footnote not used in this edition.

[4] In communities where 6.4 m has been established, this clearance may be continued if carefully maintained. The elevation of the contact conductor should be the same in the crossing and next adjacent spans. (See Rule 225D2 for conditions that must be met where uniform height above rail is impractical.)

[5] In communities where 4.9 m has been established for trolley and electrified railroad contact conductors 0 to 750 V to ground, or 5.5 m for trolley and electrified railroad contact conductors exceeding 750 V, or where local conditions make it impractical to obtain in the clearance given in the table, these reduced clearances may be used if carefully maintained.

[6] This footnote not used in this edition.

[7] Where the height of a building or other installation does not permit service drops to meet these values, the clearances over residential driveways only may be reduced to the following:

		(meters)
(a)	Insulated supply service drops limited to 300 V to ground	3.8
(b)	Insulated drip loops of supply service drops limited to 300 V to ground	3.2
(c)	Supply service drops limited to 150 V to ground and meeting Rules 230C1 or 230C3	3.6
(d)	Drip loops only of service drops limited to 150 V to ground and meeting Rules 230C1 or 230C3	3.0
(e)	Insulated communication service drops	3.5

[8] Where the height of a building or other installation does not permit service drops to meet these values, the clearances may be reduced to the following:

		(meters)
(a)	Insulated supply service drops limited to 300 V to ground	3.2
(b)	Insulated drip loops of supply service drops limited to 300 V to ground	3.2
(c)	Supply service drops limited to 150 V to ground and meeting Rules 230C1 or 230C3	3.0
(d)	Drip loops only of supply service drops limited to 150 V to ground and meeting Rules 230C1 or 230C3	3.0

[9]Spaces and ways subject to pedestrians or restricted traffic only are those areas where riders on horses or other large animals, vehicles, or other mobile units exceeding a total height of 2.45 m, are prohibited by regulation or permanent terrain configurations, or are otherwise not normally encountered nor reasonably anticipated.

[10] Where a supply or communication line along a road is located relative to fences, ditches, embankments, etc., so that the ground under the line would not be expected to be traveled except by pedestrians, the clearances may be reduced to the following values:

		(meters)
(a)	Insulated communication conductor and communication cables	2.9
(b)	Conductors of other communication circuits	2.9
(c)	Supply cables of any voltage meeting Rule 230C1, supply cables limited to 150 V to ground meeting Rules 230C2 or 230C3, and neutral conductors meeting Rule 230E1	2.9
(d)	Insulated supply conductors limited to 300 V to ground	3.8
(e)	Guys	2.9

[11] No clearance from ground is required for anchor guys not crossing tracks, rails, streets, driveways, roads, or pathways.

[12] This clearance may be reduced to 4.0 m for communication conductors and guys.

[13] Where this construction crosses over or runs along alleys, driveways, or parking lots not subject to truck traffic this clearance may be reduced to 4.6 m.

[14] Ungrounded guys and ungrounded portions of span guys between guy insulators shall have clearances based on the highest voltage to which they may be exposed due to a slack conductor or guy.

[15] Anchor guys insulated in accordance with Rule 279 may have the same clearance as grounded guys.

[16] Adjacent to tunnels and overhead bridges that restrict the height of loaded rail cars to less than 6.1 m, these clearances may be reduced by the difference between the highest loaded rail car handled and 6.1 m, if mutually agreed to by the parties at interest.

[17] For controlled impoundments, the surface area and corresponding clearances shall be based upon the design high-water level.

[18] For uncontrolled water flow areas, the surface area shall be that enclosed by its annual high-water mark. Clearances shall be based on the normal flood level; if available, the 10-year flood level may be assumed as the normal flood level.

[19] The clearance over rivers, streams, and canals shall be based upon the largest surface area of any 1.6 km-long segment that includes the crossing. The clearance over a canal, river, or stream normally used to provide access for sailboats to a larger body of water shall be the same as that required for the larger body of water.

[20] Where an overwater obstruction restricts vessel height to less than the applicable reference height given in Table 232-3, the required clearance may be reduced by the difference between the reference height and the overwater obstruction height, except that the reduced clearance shall be not less than that required for the surface area on the line-crossing side of the obstruction.

[21] Where the US Army Corps of Engineers, or the state, or surrogate thereof has issued a crossing permit, clearances of that permit shall govern.

[22] See Rule 234I for the required horizontal and diagonal clearances to rail cars.

[23] For the purpose of this rule, trucks are defined as any vehicle exceeding 2.45 m in height. Areas not subject to truck traffic are areas where truck traffic is not normally encountered nor reasonably anticipated.

[24] Communication cables and conductors may have a clearance of 4.6 m where poles are back of curbs or other deterrents to vehicular traffic.

[25] The clearance values shown in this table are computed by adding the applicable Mechanical and Electrical (M & E) value of Table A-1 to the applicable Reference Component of Table A-2a of Appendix A.

[26] When designing a line to accommodate oversized vehicles, these clearance values shall be increased by the difference between the known height of the oversized vehicle and 4.3 m.

---

ft

Table 232-1
Vertical Clearance of Wires, Conductors, and Cables Above Ground, Roadway, Rail, or Water Surfaces[25]

(Voltages are phase to ground for effectively grounded circuits and those other circuits where all ground faults are cleared by promptly de-energizing the faulted section, both initially and following subsequent breaker operations. See the definitions section for voltages of other systems. See Rules 232B1, 232C1a, and 232D4.)

Nature of surface underneath wires, conductors, or cables	Insulated communication conductors and cable; messengers; surge-protection wires; grounded guys; ungrounded guys exposed to 0 to 300 V [11,15]; neutral conductors meeting Rule 230E1; supply cables meeting Rule 230C1 (ft)	Noninsulated communication conductors; supply cables of 0 to 750 V meeting Rules 230C2 or 230C3 (ft)	Supply cables over 750 V meeting Rules 230C2 or 230C3; open supply conductors, 0 to 750 V; ungrounded guys exposed to over 300 V to 750 V[14] (ft)	Open supply conductors, over 750 V to 22 kV; ungrounded guys exposed to 750 V to 22 kV[14] (ft)	Trolley and electrified railroad contact conductors and associated span or messenger wires	
					0 to 750 V to ground (ft)	Over 750 V to 22 kV to ground (ft)
Where wires, conductors, or cables cross over or overhang						
1. Track rails of railroads (except electrified railroads using overhead trolley conductors)[2,16,]	23.5	24.0	24.5	26.5	22.0[4]	22.0[4]
2. Roads, streets, and other areas subject to truck traffic [23]	15.5	16.0	16.5	18.5	18.0[5]	20.0[5]

ft

Table 232-1 *(Continued)*
Vertical Clearance of Wires, Conductors, and Cables Above Ground, Roadway, Rail, or Water Surfaces[25]

(Voltages are phase to ground for effectively grounded circuits and those other circuits where all ground faults are cleared by promptly de-energizing the faulted section, both initially and following subsequent breaker operations. See the definitions section for voltages of other systems. See Rules 232B1, 232C1a, and 232D4.)

Nature of surface underneath wires, conductors, or cables	Insulated communication conductors and cable; messengers; surge-protection wires; grounded guys; ungrounded guys exposed to 0 to 300 V[11, 15]; neutral conductors meeting Rule 230E1; supply cables meeting Rule 230C1 (ft)	Noninsulated communication conductors; supply cables of 0 to 750 V meeting Rules 230C2 or 230C3 (ft)	Supply cables over 750 V meeting Rules 230C2 or 230C3; open supply conductors, 0 to 750 V; ungrounded guys exposed to over 300 V to 750 V[14] (ft)	Open supply conductors, over 750 V to 22 kV; ungrounded guys exposed to 750 V to 22 kV[14] (ft)	Trolley and electrified railroad contact conductors and associated span or messenger wires: 0 to 750 V to ground (ft)	Trolley and electrified railroad contact conductors and associated span or messenger wires: Over 750 V to 22 kV to ground (ft)
3. Driveways, parking lots, and alleys[23]	15.5[7, 13]	16.0[7, 13]	16.5[7]	18.5	18.0[5]	20.0[5]
4. Other land traversed by vehicles, such as cultivated, grazing, forest, orchards, etc.[26]	15.5	16.0	16.5	18.5	—	—
5. Spaces and ways subject to pedestrians or restricted traffic only[9]	9.5	12.0[8]	12.5[8]	14.5	16.0	18.0
6. Water areas not suitable for sailboating or where sailboating is prohibited[21]	14.0	14.5	15.0	17.0	—	—
7. Water areas suitable for sailboating including lakes, ponds, reservoirs, tidal waters, rivers, streams, and canals with an unobstructed surface area of[17, 18, 19, 20, 21]						
a. Less than 20 acres	17.5	18.0	18.5	20.5	—	—
b. Over 20 to 200 acres	25.5	26.0	26.5	28.5	—	—
c. Over 200 to 2000 acres	31.5	32.0	32.5	34.5	—	—
d. Over 2000 acres	37.5	38.0	38.5	40.5	—	—
8. Established boat ramps and associated rigging areas; areas posted with sign(s) for rigging or launching sail boats	Clearance above ground shall be 5 ft greater than in 7 above, for the type of water areas served by the launching site					
Where wires, conductors, or cables run along and within the limits of highways or other road rights-of-way but do not overhang the roadway						
9. Roads, streets, or alleys	15.5[24]	16.0	16.5	18.5	18.0[5]	20.0[5]
10. Roads in rural districts where it is unlikely that vehicles will be crossing under the line	13.5[10, 12]	14.0[10]	14.5[10]	16.5	18.0[5]	20.0[5]

[1] Where subways, tunnels, or bridges require it, less clearance above ground or rails than required by Table 232-1 may be used locally. The trolley and electrified railroad contact conductor should be graded very gradually from the regular construction down to the reduced elevation.

[2] For wires, conductors, or cables crossing over mine, logging, and similar railways that handle only cars lower than standard freight cars, the clearance may be reduced by an amount equal to the difference in height between the highest loaded car handled and 20 ft, but the clearance shall not be reduced below that required for street crossings.

[3] This footnote not used in this edition.

[4] In communities where 21 ft has been established, this clearance may be continued if carefully maintained. The elevation of the contact conductor should be the same in the crossing and next adjacent spans. (See Rule 225D2 for conditions that must be met where uniform height above rail is impractical.)

[5] In communities where 16 ft has been established for trolley and electrified railroad contact conductors 0 to 750 V to ground, or 18 ft for trolley and electrified railroad contact conductors exceeding 750 V, or where local conditions make it impractical to obtain the clearance given in the table, these reduced clearances may be used if carefully maintained.

[6] This footnote not used in this edition.

[7] Where the height of a building or other installation does not permit service drops to meet these values, the clearances over residential driveways only may be reduced to the following:

		(feet)
(a)	Insulated supply service drops limited to 300 V to ground	12.5
(b)	Insulated drip loops of supply service drops limited to 300 V to ground	10.5
(c)	Supply service drops limited to 150 V to ground and meeting Rules 230C1 or 230C3	12.0
(d)	Drip loops only of service drops limited to 150 V to ground and meeting Rules 230C1 or 230C3	10.0
(e)	Insulated communication service drops	11.5

[8] Where the height of a building or other installation does not permit service drops to meet these values, the clearances may be reduced to the following:

		(feet)
(a)	Insulated supply service drops limited to 300 V to ground	10.5
(b)	Insulated drip loops of supply service drops limited to 300 V to ground	10.5
(c)	Supply service drops limited to 150 V to ground and meeting Rules 230C1 or 230C3	10.0
(d)	Drip loops only of supply service drops limited to 150 V to ground and meeting Rules 230C1 or 230C3	10.0

[9] Spaces and ways subject to pedestrians or restricted traffic only are those areas where riders on horses or other large animals, vehicles, or other mobile units exceeding a total height of 8 ft are prohibited by regulation or permanent terrain configurations, or are otherwise not normally encountered nor reasonably anticipated.

[10] Where a supply or communication line along a road is located relative to fences, ditches, embankments, etc., so that the ground under the line would not be expected to be traveled except by pedestrians, the clearances may be reduced to the following values:

		(feet)
(a)	Insulated communication conductor and communication cables.	9.5
(b)	Conductors of other communication circuits	9.5
(c)	Supply cables of any voltage meeting Rule 230C1, supply cables limited to 150 V to ground meeting Rules 230C2 or 230C3, and neutral conductors meeting Rule 230E1	9.5
(d)	Insulated supply conductors limited to 300 V to ground	12.5
(e)	Guys	9.5

[11] No clearance from ground is required for anchor guys not crossing tracks, rails, streets, driveways, roads, or pathways.

[12] This clearance may be reduced to 13 ft for communication conductors and guys.

[13] Where this construction crosses over or runs along alleys, driveways, or parking lots not subject to truck traffic this clearance may be reduced to 15 ft.

[14]Ungrounded guys and ungrounded portions of span guys between guy insulators shall have clearances based on the highest voltage to which they may be exposed due to a slack conductor or guy.

[15]Anchor guys insulated in accordance with Rule 279 may have the same clearance as grounded guys.

[16] Adjacent to tunnels and overhead bridges that restrict the height of loaded rail cars to less than 20 ft, these clearances may be reduced by the difference between the highest loaded rail car handled and 20 ft, if mutually agreed to by the parties at interest.

[17] For controlled impoundments, the surface area and corresponding clearances shall be based upon the design high-water level.

[18] For uncontrolled water flow areas, the surface area shall be that enclosed by its annual high-water mark. Clearances shall be based on the normal flood level; if available, the 10-year flood level may be assumed as the normal flood level.

[19] The clearance over rivers, streams, and canals shall be based upon the largest surface area of any 1-mi-long segment that includes the crossing. The clearance over a canal, river, or stream normally used to provide access for sailboats to a larger body of water shall be the same as that required for the larger body of water.

[20] Where an overwater obstruction restricts vessel height to less than the applicable reference height given in Table 232-3, the required clearance may be reduced by the difference between the reference height and the overwater obstruction height, except that the reduced clearance shall be not less than that required for the surface area on the line-crossing side of the obstruction.

[21] Where the US Army Corps of Engineers, or the state, or surrogate thereof has issued a crossing permit, clearances of that permit shall govern.

[22] See Rule 234I for the required horizontal and diagonal clearances to rail cars.

[23] For the purpose of this Rule, trucks are defined as any vehicle exceeding 8 ft in height. Areas not subject to truck traffic are areas where truck traffic is not normally encountered nor reasonably anticipated.

[24] Communication cables and conductors may have a clearance of 15 ft where poles are back of curbs or other deterrents to vehicular traffic.

[25] The clearance values shown in this table are computed by adding the applicable Mechanical and Electrical (M & E) value of Table A-1 to the applicable Reference Component of Table A-2a of Appendix A.

[26] When designing a line to accommodate oversized vehicles, these clearance values shall be increased by the difference between the known height of the oversized vehicle and 14 ft.

m

Table 232-2
Vertical Clearance of Equipment Cases, Support Arms, Platforms, Braces and Unguarded Rigid Live Parts Above Ground, Roadway, or Water Surfaces [8]
(Voltages are phase to ground for effectively grounded circuits and those other circuits where all ground faults are cleared by promptly de-energizing the faulted section, both initially and following subsequent breaker operations. See the definitions section for voltages of other systems. See Rules 232B2, 232B3, 232C1a, and 232D4.)

Nature of surface below	Nonmetallic or effectively grounded support arms and equipment cases (m)	Unguarded rigid live parts of 0 to 750 V and ungrounded cases that contain equipment connected to circuits of more than 750 V (m)	Unguarded rigid live parts of over 750 V to 22 kV and ungrounded cases that contain equipment connected to circuits of over 750 V to 22 kV (m)
1. Where rigid parts overhang			
a. Roads, streets, and other areas subject to truck traffic[4]	4.6	4.9	5.5
b. Driveways, parking lots, and alleys	4.6	4.9[1]	5.5
c. Other land traversed by vehicles such as cultivated land, grazing land, forest, orchard, etc.	4.6[7]	4.9	5.5
d. Spaces and ways subject to pedestrians or restricted traffic only[5]	3.4[7]	3.6[1(b)]	4.3
2. Where rigid parts are along and within the limits of highways or other road rights-of-way but do not overhang the roadway			
a. Roads, streets, and alleys	4.6[7]	4.9	5.5
b. Roads in rural districts where it is unlikely that vehicles will be crossing under the line	4.0[7]	4.3[2]	4.9
3. Water areas not suitable for sailboating or where sailboating is prohibited[9]	4.3	4.4	4.6

[1] This clearance may be reduced to the following values: (meters)
(a) Insulated live parts limited to 300 V to ground 3.6
(b) Insulated live parts limited to 150 V to ground 3.0

[2] Where a supply line along a road is limited to 300 V to ground and is located relative to fences, ditches, embankments, etc., so that the ground under the line would not be expected to be traveled except by pedestrians, this clearance may be reduced to 3.6 m.

[3] This footnote not used in this edition.

[4] For the purpose of this rule, trucks are defined as any vehicle exceeding 2.45 m in height. Areas not subject to truck traffic are areas where truck traffic is not normally encountered nor reasonably anticipated.

[5] Spaces and ways subject to pedestrians or restricted traffic only are those areas where riders on horseback, vehicles, or other mobile units exceeding 2.45 m in height, are prohibited by regulation or permanent terrain configurations or are otherwise not normally encountered nor reasonably anticipated.

[6] This footnote not used in this edition.

[7] Effectively grounded supply or communication equipment cases (such as fire alarm boxes, control boxes, communication terminals, meters or similar equipment cases) may be mounted at a lower level for accessibility, provided such cases do not unduly obstruct a walkway.

NOTE: See also Rule 234J2c.

[8] The clearance values shown in this table are computed by adding the applicable Mechanical and Electrical (M & E) value of Table A-1 to the applicable Reference Component of Table A-2a of Appendix A.

[9] Where the US Army Corps of Engineers, or the state, or surrogate thereof has issued a crossing permit, clearances of that permit shall govern.

ft

Table 232-2
Vertical Clearance of Equipment Cases, Support Arms, Platforms, Braces and Unguarded Rigid Live Parts Above Ground, Roadway, or Water Surfaces[8]
(Voltages are phase to ground for effectively grounded circuits and those other circuits where all ground faults are cleared by promptly de-energizing the faulted section, both initially and following subsequent breaker operations. See the definitions section for voltages of other systems.
See Rules 232B2, 232B3, 232C1a, and 232D4.)

Nature of surface below	Nonmetallic or effectively grounded support arms and equipment cases (ft)	Unguarded rigid live parts of 0 to 750 V and ungrounded cases that contain equipment connected to circuits of not more than 750 V (ft)	Unguarded rigid live parts of over 750 V to 22 kV and ungrounded cases that contain equipment connected to circuits of over 750 V to 22 kV (ft)
1. Where rigid parts overhang			
a. Roads, streets, and other areas subject to truck traffic[4]	15.0	16.0	18.0
b. Driveways, parking lots, and alleys	15.0	16.0[1]	18.0
c. Other land traversed by vehicles such as cultivated land, grazing land, forest, orchard, etc.	15.0[7]	16.0	18.0
d. Spaces and ways subject to pedestrians or restricted traffic only[5]	11.0[7]	12.0[1(b)]	14.0
2. Where rigid parts are along and within the limits of highways or other road rights-of-way but do not overhang the roadway			
a. Roads, streets, and alleys	15.0[7]	16.0	18.0
b. Roads in rural districts where it is unlikely that vehicles will be crossing under the line	13.0[7]	14.0[2]	16.0
3. Water areas not suitable for sailboating or where sailboating is prohibited[9]	14.0	14.5	15.0

[1] This clearance may be reduced to the following values: (feet)
(a) Insulated live parts limited to 300 V to ground 12
(b) Insulated live parts limited to 150 V to ground 10

[2] Where a supply line along a road is limited to 300 V to ground and is located relative to fences, ditches, embankments, etc., so that the ground under the line would not be expected to be traveled except by pedestrians, this clearance may be reduced to 12 ft.

[3] This footnote not used in this edition.

[4] For the purpose of this rule, trucks are defined as any vehicle exceeding 8 ft in height. Areas not subject to truck traffic are areas where truck traffic is not normally encountered nor reasonably anticipated.

[5] Spaces and ways subject to pedestrians or restricted traffic only are those areas where riders on horseback, vehicles, or other mobile units exceeding 8 ft in height, are prohibited by regulation or permanent terrain configurations or are otherwise not normally encountered nor reasonably anticipated.

[6] This footnote not used in this edition.

[7] Effectively grounded supply or communication equipment cases (such as fire alarm boxes, control boxes, communication terminals, meters, or similar equipment cases) may be mounted at a lower level for accessibility, provided such cases do not unduly obstruct a walkway.

NOTE: See also Rule 234J2c.

[8] The clearance values shown in this table are computed by adding the applicable Mechanical and Electrical (M & E) value of Table A-1 to the applicable Reference Component of Table A-2a of Appendix A.

[9] Where the US Army Corps of Engineers, or the state, or surrogate thereof has issued a crossing permit, clearances of that permit shall govern.

Table 232-3
Reference Heights
(See Rule 232D2)

Nature of surface underneath lines	(m)	(ft)
a. Track rails of railroads (except electrified railroads using overhead trolley conductors) [1]	6.7	22
b. Streets, alleys, roads, driveways, and parking lots	4.3	14
c. Spaces and ways subject to pedestrians or restricted traffic only [2]	3.0	10
d. Other land, such as cultivated, grazing, forest, or orchard, that is traversed by vehicles	4.3	14
e. Water areas not suitable for sailboating or where sailboating is prohibited	3.8	12.5
f. Water areas suitable for sailboating including lakes, ponds, reservoirs, tidal waters, rivers, streams, and canals with unobstructed surface area [3, 4]		
(1) Less than 0.08 km^2 (20 acres)	4.9	16
(2) Over 0.08 to 0.8 km^2 (20 to 200 acres)	7.3	24
(3) Over 0.8 to 8 km^2 (200 to 2000 acres)	9.0	30
(4) Over 8 km^2 (2000 acres)	11.0	36
g. In public or private land and water areas posted for rigging or launching sailboats, the reference height shall be 1.5 m (5 ft) greater than in f above, for the type of water areas serviced by the launching site		

[1] See Rule 234I for the required horizontal and diagonal clearances to rail cars.

[2] Spaces and ways subject to pedestrians or restricted traffic only are those areas where riders on horseback, vehicles, or other mobile units exceeding 2.45 m (8 ft) in height, are prohibited by regulation or permanent terrain configurations or are otherwise not normally encountered nor reasonably anticipated.

[3] For controlled impoundments, the surface area and corresponding clearances shall be based upon the design high-water level. For other waters, the surface area shall be that enclosed by its annual high-water mark, and clearances shall be based on the normal flood level. The clearances over rivers, streams, and canals shall be based upon the largest surface area of any 1600 m (1 mi) long segment that includes the crossing. The clearance over a canal or similar waterway providing access for sailboats to a larger body of water shall be the same as that required for the larger body of water.

[4] Where an overwater obstruction restricts vessel height to less than the applicable reference height, the required clearance may be reduced by the difference between the reference height and the overwater obstruction height, except that the reduced clearance shall not be less than that required for the surface area on the line-crossing side of the obstruction.

Table 232-4
Electrical Component of Clearance in Rule 232D3a
(Add 3% for each 300 m (1000 ft) in excess of 450 m (1500 ft) above mean sea level.
Increase clearance to limit electrostatic effects in accordance with Rule 232D3c.)

Maximum operating voltage phase to phase (kV)	Switching-surge factor (per unit)	Switching surge (kV)	Electrical component of clearance	
			(m)	(ft)
242	3.54 or less	700 or less	2.15	7.1[1]
362	2.37 or less	700 or less	2.15	7.1[1]
550	1.56 or less	700 or less	2.15	7.1[1]
	1.90	853	3.0	9.9
	2.00	898	3.3	10.8
	2.20	988	3.9	12.7
	2.40	1079	4.4	14.6
	2.60	1168	5.1	16.7
800	1.60	1045	4.2	13.9
	1.80	1176	5.2	16.9
	2.00	1306	6.1	20.1
	2.10 or more	1372 or more	6.6	21.8[2]

[1] Limited by Rule 232D4.
[2] Limited by Rules 232A and 232B.

233. Clearances Between Wires, Conductors, and Cables Carried on Different Supporting Structures

A. General

Crossings should be made on a common supporting structure, where practical. In other cases, the clearance between any two crossing or adjacent wires, conductors, or cables carried on different supporting structures shall be not less than that required by Rules 233B and 233C at any location in the spans. The clearance shall be not less than that required by application of a clearance envelope developed under Rule 233A2 to the positions on or within conductor movement envelopes developed under Rule 233A1 at which the two wires, conductors, or cables would be closest together. For purposes of this determination, the relevant positions of the wires, conductors, or cables on or within their respective conductor movement envelopes are those that can occur when (1) both are simultaneously subjected to the same ambient air temperature and wind loading conditions, and (2) each is subjected individually to the full range of its icing conditions and applicable design electrical loading.

Figure 233-1 is a graphical illustration of the application of Rule 233A. Alternate methods that assure compliance with these rules may be used.

1. Conductor Movement Envelope

a. Development

The conductor movement envelope shall be developed from the locus of the most displaced conductor positions defined below and shown in Fig 233-2.

(1) 15 °C (60 °F), no wind displacement, at both initial unloaded and final unloaded sag (conductor positions A and C).

(2) With the wire, conductor, or cable displaced from rest by a 290 Pa (6 lb/ft^2) wind at both initial and final sag at 15 °C (60 °F). The displacement of the wire, conductor, or cable shall include deflection of suspension insulators and flexible structures (conductor positions B and D).

EXCEPTION: Where the entire span is so close to a building, terrain feature, or other obstacle as to be sheltered from the wind flowing across the line in either direction, the wind pressure may be reduced to a 190 Pa (4 lb/ft^2) wind. Trees are not considered to shelter a line.

(3) Final sag at one of the following loading conditions, whichever produces the largest sag (conductor position E):

(a) 50 °C (120 °F), no wind displacement,

(b) The maximum conductor temperature for which the line is designed to operate, if greater than 50 °C (120 °F), with no wind displacement, or

(c) 0 °C (32 °F), no wind displacement, with radial thickness of ice, if any, specified in Rule 250B for the loading district concerned.

b. Sag Increase

No sag increase for either high operating temperatures or ice loading is required for trolley and electrified railroad contact conductors. Rule 233A1a(3) does not apply to these conductors.

2. Clearance Envelope

The clearance envelope shown in Fig 233-3 shall be determined by the horizontal clearance (*H*) required by Rule 233B and the vertical clearance (*V*) required by Rule 233C.

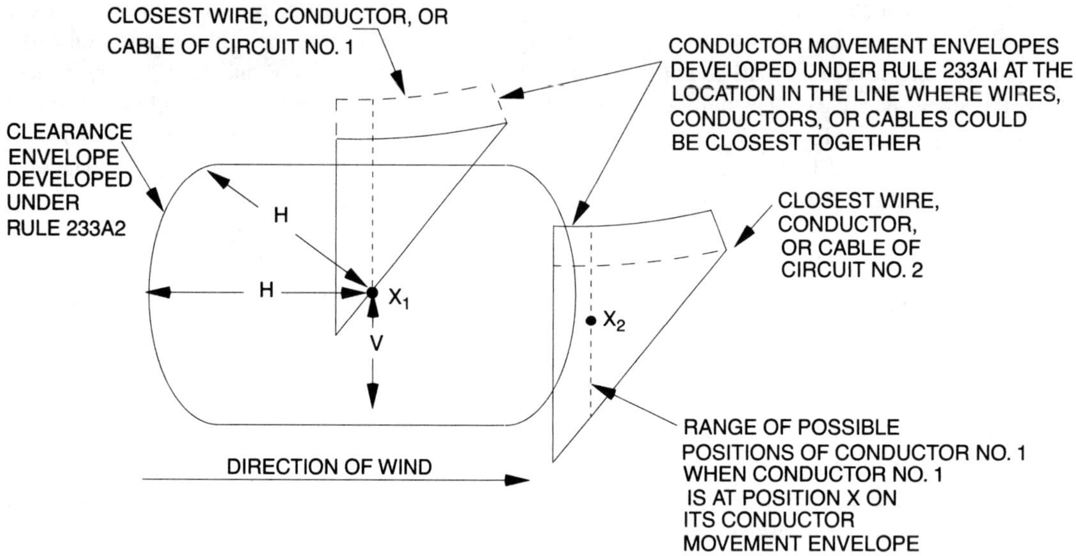

NOTE: In this illustration, Conductor No. 2 is closest at position X_2 to Conductor No. 1, where the latter is at position X_1.

Fig 233-1
Use of Clearance Envelope and Conductor Movement Envelopes to Determine Applicable Clearance

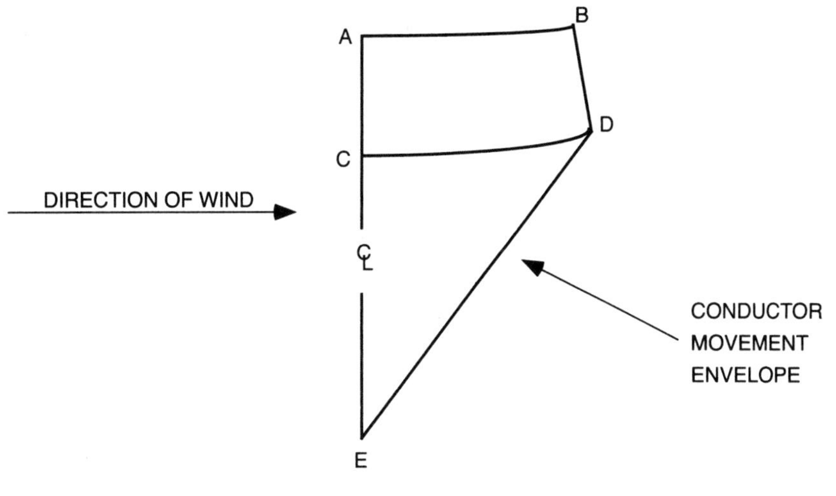

Point	Conductor Temperature	Sag	Ice Loading	Wind Displacement[1]
A	15 °C[5]	initial	none	none
B	15 °C[5]	initial	none	290 Pa (note 2)
C	15 °C[5]	final	none	none
D	15 °C[5]	final	none	290 Pa (note 2)
E_1 [3,4]	The greater of 50 °C or maximum operating temperature	final	none	none
E_2 [3,4]	0 °C	final	as applicable	none

[1] The direction of the wind shall be that which produces the minimum distance between conductors. The displacement of the wires, conductors, or cables includes the deflection of suspension insulators and flexible structures.

[2] Where the entire span is so close to a building, terrain feature, or other obstacle as to be sheltered from the wind flowing across the line in either direction, the wind pressure may be reduced to a 190 Pa wind. Trees are not considered to shelter a line.

[3] Point E shall be determined by whichever of the conditions described under E_1 and E_2 produces the greatest sag.

[4] Line D–E shall be considered to be straight unless the actual concavity characteristics are known.

[5] When one conductor movement envelope is lower than that of the other conductor, the lower envelope shall be developed with points A, B, C, and D at a conductor temperature equal to the ambient temperature used in determining E of the upper conductor movement envelope.

m

Fig 233-2
Conductor Movement Envelope

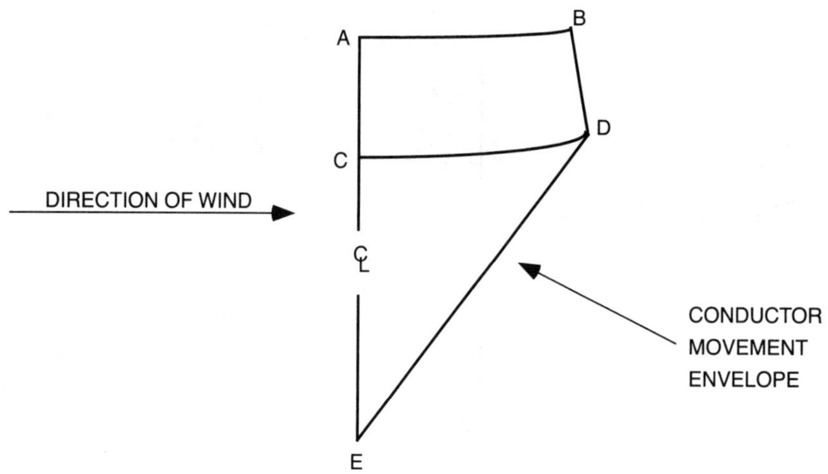

Point	Conductor Temperature	Sag	Ice Loading	Wind Displacement [1]
A	60 °F [5]	initial	none	none
B	60 °F [5]	initial	none	6 lb/ft^2 (note 2)
C	60 °F [5]	final	none	none
D	60 °F [5]	final	none	6 lb/ft^2 (note 2)
E_1 [3, 4]	The greater of 120 °F or maximum operating temperature	final	none	none
E_2 [3, 4]	32 °F	final	as applicable	none

[1] The direction of the wind shall be that which produces the minimum distance between conductors. The displacement of the wires, conductors, or cables includes the deflection of suspension insulators and flexible structures.

[2] Where the entire span is so close to a building, terrain feature, or other obstacle as to be sheltered from the wind flowing across the line in either direction, the wind pressure may be reduced to a 4 lb/ft^2 wind. Trees are not considered to shelter a line.

[3] Point E shall be determined by whichever of the conditions described under E_1 and E_2 produces the greatest sag.

[4] Line D–E shall be considered to be straight unless the actual concavity characteristics are known.

[5] When one conductor movement envelope is lower than that of the other conductor, the lower conductor envelope shall be developed with points A, B, C, and D at a conductor temperature equal to the ambient temperature used in determining E of the upper conductor movement envelope.

ft

Fig 233-2
Conductor Movement Envelope

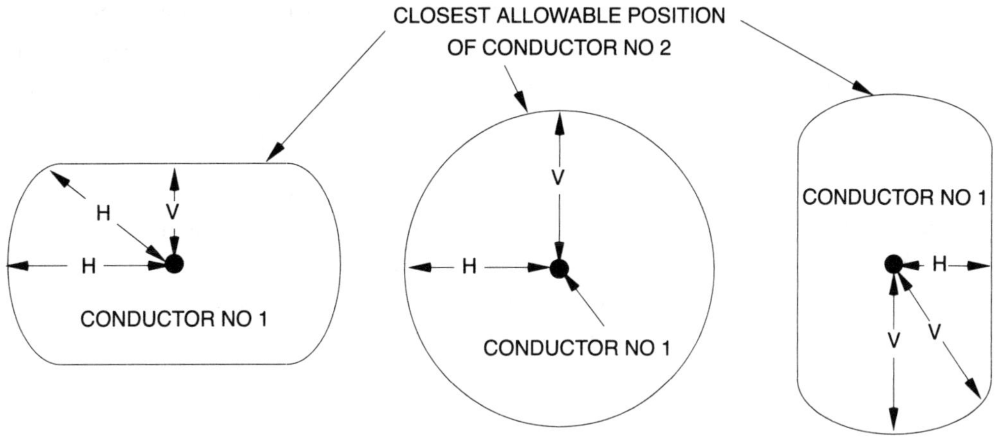

Fig 233-3
Clearance Envelope

B. Horizontal Clearance

1. Clearance Requirements

The horizontal clearance between crossing or adjacent wires, conductors, or cables carried on different supporting structures shall be not less than 1.50 m (5 ft). For voltages between the wires, conductors, or cables exceeding 129 kV, additional clearance of 10 mm (0.4 in) per kV over 129 kV shall be provided.

EXCEPTION: The horizontal clearance between anchor guys of different supporting structures may be reduced to 150 mm (6 in) and may be reduced to 600 mm (2 ft) between other guys, span wires, and neutral conductors meeting Rule 230E1.

2. Alternate Clearances for Voltages Exceeding 98 kV AC to Ground or 139 kV DC to Ground

The clearances specified in Rule 233B1 may be reduced for circuits with known switching-surge factors, but shall be not less than the alternate clearance derived from the computations required in Rules 235B3a and 235B3b.

C. Vertical Clearance

1. Clearance Requirements

The vertical clearance between any crossing or adjacent wires, conductors, or cables carried on different supporting structures shall be not less than that shown in Table 233-1.

EXCEPTION: No vertical clearance is required between wires, conductors, or cables that are electrically interconnected at the crossing.

2. Voltages Exceeding 22 kV

a. The clearance given in Table 233-1 shall be increased by the sum of the following: For the upper-level conductors between 22 and 470 kV, the clearance shall be increased at the rate of 10 mm (0.4 in) per kV in excess of 22 kV. For the lower-level conductors exceeding 22 kV, the additional clearance shall be computed at the same rate. For voltages exceeding 470 kV, the clearance shall be determined by the method given in Rule 233C3. The additional clearance shall be computed using the maximum operating voltage if above 50 kV and nominal voltage if below 50 kV.

EXCEPTION: For voltages exceeding 98 kV ac to ground or 139 kV dc to ground, clearances less than those required above are permitted for systems with known switching-surge factors. (See Rule 233C3.)

b. For voltages exceeding 50 kV, the additional clearance specified in Rule 233C2a shall be increased 3% for each 300 m (1000 ft) in excess of 1000 m (3300 ft) above mean sea level.

3. Alternate Clearances for Voltage Exceeding 98 kV AC to Ground or 139 kV DC to Ground

The clearances specified in Rules 233C1 and 233C2 may be reduced where the higher-voltage circuit has a known switching-surge factor, but shall be not less than the alternate clearance, which is computed by adding the reference height from Rule 233C3a to the electrical component of clearance from Rule 233C3b. For these computations, communication conductors and cables, guys, messengers, neutral conductors meeting Rule 230E1, and supply cables meeting Rule 230C1 shall be considered at zero voltage.

a. Reference Heights

The reference height shall be selected from Table 233-3.

b. Electrical Component of Clearance

(1) The electrical component (D) shall be computed using the following equations. Selected values of D are listed in Table 233-2.

$$D = 1.00\left[\frac{[V_H \cdot (PU) + V_L]a}{500K}\right]^{1.667} bc \quad \text{(m)}$$

$$D = 3.28\left[\frac{[V_H \cdot (PU) + V_L]a}{500K}\right]^{1.667} bc \quad \text{(ft)}$$

where

V_H = higher-voltage circuit maximum ac crest operating voltage to ground or maximum dc operating voltage to ground in kilovolts,

V_L = lower-voltage circuit maximum ac crest operating voltage to ground or maximum dc operating voltage to ground in kilovolts,

PU = higher-voltage circuit maximum switching-surge factor expressed in per-unit peak voltage to ground and defined as a switching-surge level for circuit breakers corresponding to 98% probability that the maximum switching surge generated per breaker operation does not exceed this surge level, or the maximum anticipated switching-surge level generated by other means, whichever is greater,

a = 1.15, the allowance for three standard deviations,

b = 1.03, the allowance for nonstandard atmospheric conditions,

c = 1.2, the margin of safety,

K = 1.4, the configuration factor for conductor-to-conductor gap.

(2) The value of D calculated by Rule 233C3b(1) shall be increased 3% for each 300 m (1000 ft) in excess of 450 m (1500 ft) above mean sea level.

c. Limit

The alternate clearance shall be not less than the clearance required by Rules 233C1 and 233C2 with the lower-voltage circuit at ground potential.

m

Table 233-1
Vertical Clearance Between Wires, Conductors, and Cables Carried on Different Supporting Structures

Voltages are phase to ground for effectively grounded circuits and those other circuits where all ground faults are cleared by promptly de-energizing the faulted section, both initially and following subsequent breaker operations. See the definitions section for voltages of other systems.
See Rules 233C1 and 233C2a.)

Lower level	Upper level				
	Effectively grounded communication guys, span wires and messengers, communication conductors and cables (m)	**Effectively grounded supply guys, span wires and messengers, neutral conductors meeting Rule 230E1, and surge-protection wires (m)**	**Supply cables meeting Rule 230C1, and supply cables of 0 to 750 V meeting Rule 230C2 or 230C3 (m)**	**Open supply conductors 0 to 750 V, and supply cables over 750 V meeting Rule 230C2 or 230C3 (m)**	**Open supply conductors over 750 V to 22 kV (m)**
1. Effectively grounded supply guys,[7] span wires and messengers, neutral conductors meeting Rule 230E1, and surge-protection wires	0.60 [1,2]	0.60 [1,2]	0.60 [2]	0.60	0.60
2. Effectively grounded communication guys,[7] span wires and messengers; communication conductors and cables	0.60[1]	0.60 [1,2]	0.60	1.20 [8]	1.50 [5]
3. Supply cables meeting Rule 230C1, and supply cables of 0 to 750 V meeting Rules 230C2 or 230C3	0.60	0.60	0.60	0.60	0.60
4. Open supply conductors, 0 to 750 V; supply cables over 750 V meeting Rule 230C2 or 230C3	1.20 [9]	0.60	0.60	0.60	0.60
5. Open supply conductors, 750 V to 22 kV	1.50[5,9]	0.60	0.60 [9]	0.60[9]	0.60
6. Trolley and electrified railroad contact conductors and associated span and messenger wires	1.20 [3]	1.20 [3]	1.20 [3]	1.20 [3,4]	1.80

[1] No clearance is specified between guys or span wires that are electrically interconnected.

[2] The clearance of communication conductors and their guy, span, and messenger wires from each other in locations where no other classes of conductors are involved may be reduced by mutual consent of the parties concerned, except for fire-alarm conductors and conductors used in the operation of railroads.

[3] Trolley and electrified railroad contact conductors of more than 750 V should have at least 1.80 m of clearance. This clearance should also be provided over lower-voltage trolley and electrified railroad contact conductors unless the crossover conductors are beyond reach of a trolley pole leaving the trolley-contact conductor or are suitably protected against damage from trolley poles leaving the trolley-contact conductor.

[4] Trolley and electrified railroad feeders are exempt from this clearance requirement for contact conductors if they are of the same nominal voltage and of the same system.

[5] This clearance may be reduced to 1.20 m where supply conductors of 750 V to 8.7 kV cross a communication line more than 6 ft horizontally from a communications structure.

[6] This footnote not used in this edition.

[7] These clearances may be reduced by not more than 25% to a guy insulator, provided that full clearance is maintained to its metallic end fittings and the guy wires. The clearance to an insulated section of a guy between two insulators may be reduced by not more than 25% provided that full clearance is maintained to the uninsulated portion of the guy.

[8] This clearance may be reduced to 0.60 m for supply service drops.

[9] In general, this type of crossing is not recommended.

ft

Table 233-1
Vertical Clearance Between Wires, Conductors, and Cables Carried on Different Supporting Structures
(Voltages are phase to ground for effectively grounded circuits and those other circuits where all ground faults are cleared by promptly de-energizing the faulted section, both initially and following subsequent breaker operations. See the definitions section for voltages of other systems. See Rules 233C1 and 233C2a.)

Lower level	Upper level				
	Effectively grounded communication guys, span wires and messengers, communication conductors and cables (ft)	Effectively grounded supply guys, span wires and messengers, neutral conductors meeting Rule 230E1, and surge-protection wires (ft)	Supply cables meeting Rule 230C1, and supply cables of 0 to 750 V meeting Rule 230C2 or 230C3 (ft)	Open supply conductors 0 to 750 V, and supply cables over 750 V meeting Rule 230C2 or 230C3 (ft)	Open supply conductors over 750 V to 22 kV (ft)
1. Effectively grounded supply guys,[7] span wires and messengers, neutral conductors meeting Rule 230E1, and surge-protection wires	2 [1, 2]	2 [1, 2]	2 [2]	2	2
2 Effectively grounded communication guys,[7] span wires and messengers; communication conductors and cables	2 [1,2]	2 [1]	2	4 [8]	5 [5]
3. Supply cables meeting Rule 230C1, and supply cables of 0 to 750 V meeting Rules 230C2 or 230C3	2	2	2	2	2
4. Open supply conductors, 0 to 750 V; supply cables over 750 V meeting Rule 230C2 or 230C3	4 [9]	2	2	2	2
5. Open supply conductors, 750 V to 22 kV	5 [5, 9]	2	2 [9]	2 [9]	2
6. Trolley and electrified railroad contact conductors and associated span and messenger wires	4 [3]	4 [3]	4 [3]	4 [3, 4]	6

[1] No clearance is specified between guys or span wires that are electrically interconnected.

[2] The clearance of communication conductors and their guy span, and messenger wires from each other in locations where no other classes of conductors are involved may be reduced by mutual consent of the parties concerned, except for fire-alarm conductors and conductors used in the operation of railroads.

[3] Trolley and electrified railroad contact conductors of more than 750 V should have at least 6 ft of clearance. This clearance should also be provided over lower-voltage trolley and electrified railroad contact conductors unless the crossover conductors are beyond reach of a trolley pole leaving the trolley-contact conductor or are suitably protected against damage from trolley poles leaving the trolley-contact conductor.

[4] Trolley and electrified railroad feeders are exempt from this clearance requirement for contact conductors if they are of the same nominal voltage and of the same system.

[5] This clearance may be reduced to 4 ft where supply conductors of 750 V to 8.7 kV cross a communication line more than 6 ft horizontally from a communications structure.

[6] This footnote not used in this edition.

[7] These clearances may be reduced by not more than 25% to a guy insulator, provided that full clearance is maintained to its metallic end fittings and the guy wires. The clearance to an insulated section of a guy between two insulators may be reduced by not more than 25% provided that full clearance is maintained to the uninsulated portion of the guy.

[8] This clearance may be reduced to 2 ft for supply service drops.

[9] In general, this type of crossing is not recommended.

m

Table 233-2
Clearance Between Supply Wires, Conductors, and Cables in Rule 233C3b(1)
(Add 3% for each 300 m in excess of 450 m above mean sea level.)

Higher-voltage circuit		Lower-voltage circuit						
Maximum operating voltage phase to phase (kV)	Switching-surge factor (per unit)	Maximum operating voltage phase to phase (kV)						
		121 (m)	145 (m)	169 (m)	242 (m)	362 (m)	550 (m)	800 (m)
242	3.3 or less	2.13 [1]	2.13 [1]	2.13 [1]	2.16 [1]			
362	2.4	2.80 [1]	2.80 [1]	2.80 [1]	2.80 [1]	2.90		
	2.6	2.80 [1]	2.80 [1]	2.80 [1]	2.80 [1]	3.1		
	2.8	2.80 [1]	2.80 [1]	2.80 [1]	3.0	3.4		
	3.0	2.80 [1]	2.90	3.0	3.3	3.7		
550	1.8	4.0 [1]	4.0 [1]	4.0 [1]	4.0 [1]	4.0 [1]	4.1	
	2.0	4.0 [1]	4.0 [1]	4.0 [1]	4.0 [1]	4.0 [1]	4.7	
	2.2	4.0 [1]	4.0 [1]	4.0 [1]	4.0 [1]	4.3	5.2	
	2.4	4.0 [1]	4.0 [1]	4.0 [1]	4.3	4.8	5.7	
	2.6	4.1 [2]	4.3 [2]	4.4	4.8	5.3	6.3	
800	1.6	5.4 [1]	5.4 [1]	5.4 [1]	5.4 [1]	5.4 [1]	5.6	6.9
	1.8	5.4 [1]	5.4 [1]	5.4 [1]	5.4 [1]	5.4 [1]	6.4	7.7
	2.0	5.4 [1]	5.4 [1]	5.4 [1]	5.6	6.2	7.0	8.4
	2.2	5.6 [2]	5.8 [2]	5.9 [2]	6.3 [2]	7.0 [2]	8.1 [2]	9.4 [2]

[1] Limited by Rule 233C3c.
[2] Need not be greater than the values specified in Rules 233C1 and 233C2.

Table 233-2
Clearance Between Supply Wires, Conductors, and Cables in Rule 233C3b(1)
(Add 3% for each 1000 ft in excess of 1500 ft above mean sea level.)

Higher-voltage circuit		Lower-voltage circuit						
Maximum operating voltage phase to phase (kV)	**Switching-surge factor (per unit)**	**Maximum operating voltage phase to phase (kV)**						
		121 (ft)	**145 (ft)**	**169 (ft)**	**242 (ft)**	**362 (ft)**	**550 (ft)**	**800 (ft)**
242	3.3 or less	7.0[1]	7.0[1]	7.0[1]	7.1[1]			
362	2.4	9.3[1]	9.3[1]	9.3[1]	9.3[1]	9.4		
	2.6	9.3[1]	9.3[1]	9.3[1]	9.3[1]	10.3		
	2.8	9.3[1]	9.3[1]	9.3[1]	9.7	11.3		
	3.0	9.3[1]	9.4[1]	9.7	10.7	12.3		
550	1.8	13.0[1]	13.0	13.0[1]	13.0[1]	13.0[1]	13.6	
	2.0	13.0[1]	13.0[1]	13.0[1]	13.0[1]	13.0[1]	15.3	
	2.2	13.0[1]	13.0[1]	13.0[1]	13.0[1]	14.1	17.0	
	2.4	13.0[1]	13.0[1]	13.0[1]	14.0	15.8	18.8	
	2.6	13.6[2]	14.1[2]	14.5	15.6	17.5	20.7	
800	1.6	17.8[1]	17.8[1]	17.8[1]	17.8[1]	17.8[1]	18.2	22.5
	1.8	17.8[1]	17.8[1]	17.8[1]	17.8[1]	17.8[1]	20.9	25.4
	2.0	17.8[1]	17.8[1]	17.8[1]	18.4	20.4	23.7	28.5
	2.2	18.4[2]	18.9[2]	19.4[2]	20.8[2]	23.1[2]	26.7[2]	31.5[2]

[1] Limited by Rule 233C3c.
[2] Need not be greater than the values specified in Rules 233C1 and 233C2.

Table 233-3
Reference Heights
(See Rule 233C3a.)

Reference height	(m)	(ft)
(1) Supply lines	0	0
(2) Communication lines	0.60	2

234. Clearance of Wires, Conductors, Cables, and Equipment From Buildings, Bridges, Rail Cars, Swimming Pools, and Other Installations

A. Application

1. Vertical and Horizontal Clearances (No Wind Displacement)

   The vertical and horizontal clearances specified in Rules 234B, 234C, 234D, 234E, 234F, and 234I apply under whichever conditions of the following conductor temperature and loading conditions produces the closest approach. Rules 234A1a, 234A1b, and 234A1c apply above and alongside subject installations; Rule 234A1d applies below and alongside subject installations.

   a. 50 °C (120 °F), no wind displacement, final sag.

   b. The maximum conductor temperature for which the line is designed to operate, if greater than 50 °C (120 °F), no wind displacement, final sag.

   c. 0 °C (32 °F), no wind displacement, final sag, with radial thickness of ice, if any, specified in Rule 250B for the applicable loading district.

   d. The minimum conductor temperature for which the line is designed, no wind displacement, initial sag.

   EXCEPTION: Vertical or lateral conductors or cables attached directly to the surface of a supporting structure in accordance with other rules are not subject to the provisions of this rule.

   NOTE: The phase and neutral conductors of a supply line are normally considered separately when determining the sag of each due to temperature rise.

2. Horizontal Clearances (With Wind Displacement)

   Where consideration of horizontal displacement under wind conditions is required, the wires, conductors, or cables shall be considered to be displaced from rest toward the installation by a 290 Pa (6 lb/ft^2) wind at final sag at 15 °C (60 °F). The displacement of a wire, conductor or cable shall include deflection of suspension insulators. The displacement of a wire, conductor, or cable shall also include deflection of a flexible structure if the highest wire, conductor, or cable attachment is 18 m (60 ft) or more above grade.

   EXCEPTION: Where the entire span is so close to a building, terrain feature, or other obstacle as to be sheltered from the wind flowing across the line in either direction, the wind pressure may be reduced to a 190 Pa (4 lb/ft^2) wind. Trees are not considered to shelter a line.

3. Transition Between Horizontal and Vertical Clearances

   The horizontal clearance governs above the level of the roof or top of an installation to the point where the diagonal equals the vertical clearance requirement. Similarly, the horizontal clearance governs above or below projections from buildings, signs, or other installations to the point where the diagonal equals the vertical clearance requirement. From this point the transitional clearance shall equal the vertical clearance as shown in Figs 234-1(a) and 234-1(b). This rule should not be interpreted as restricting the installation of a trolley-contact conductor over the approximate center line of the track it serves.

   EXCEPTION: When the horizontal clearance is greater than the vertical clearance, the vertical clearance governs beyond the roof or top of an installation, or projections from an installation, to the point where the diagonal equals the horizontal clearance requirement, as shown in Fig 234-1(c).

B. Clearances of Wires, Conductors, and Cables From Other Supporting Structures

Wires, conductors, or cables of one line passing near a lighting support, traffic signal support, or a supporting structure of a second line, without being attached thereto, shall have clearance from any part of such structure not less than the following:

1. Horizontal Clearances

   a. A horizontal clearance, without wind, of 1.50 m (5 ft) for voltages up to 50 kV.

   EXCEPTION: For effectively grounded guys and messengers, neutrals meeting Rule 230E1, and cables of 300 V or less to ground meeting the requirements of Rule 230C1, 230C2, or 230C3, the horizontal clearance may be reduced to 900 mm (3 ft).

Fig 234-1(a)
Clearance Diagram for Building

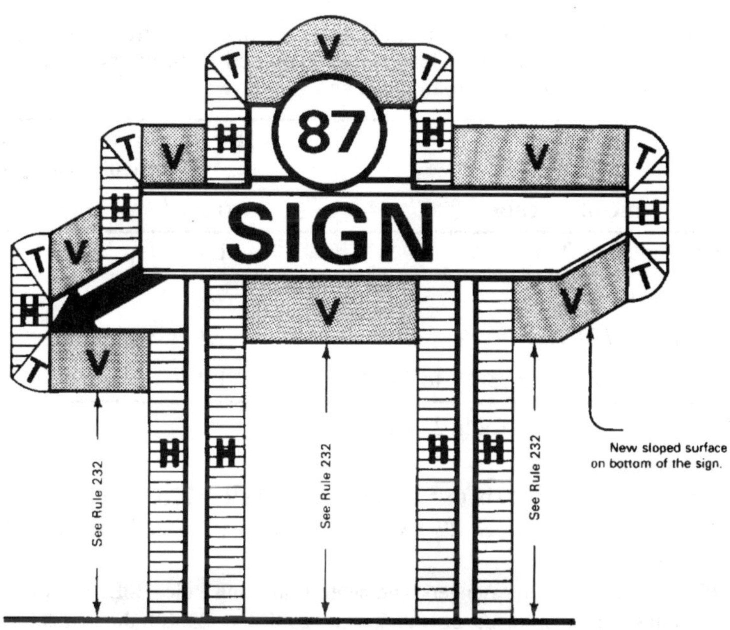

Fig 234-1(b)
Clearance Diagram for Other Structures

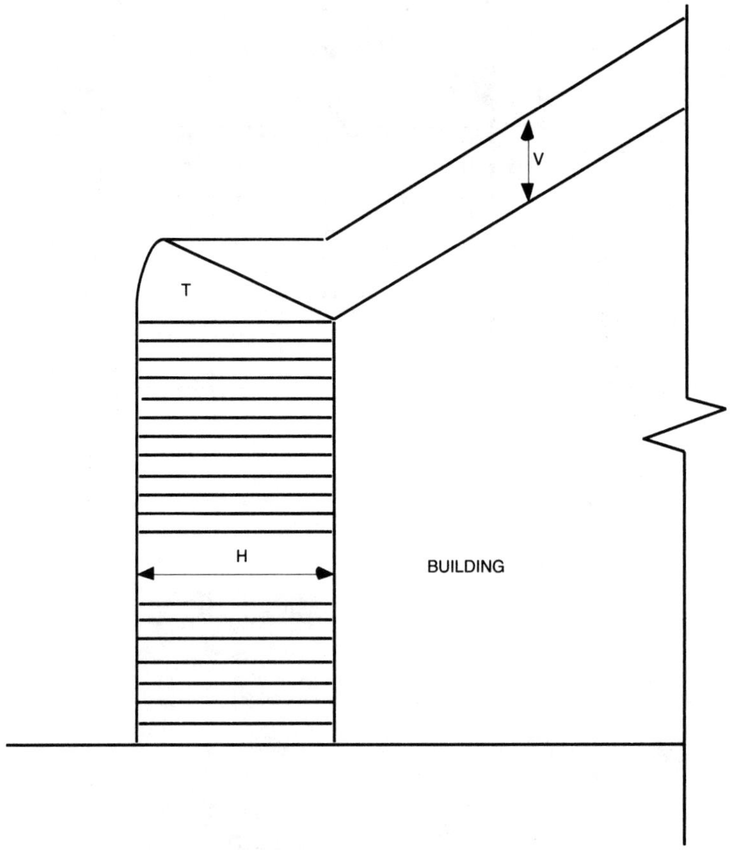

Fig 234-1(c)
Transitional Clearance When H is Greater Than V

b. When the following conductors and cables are displaced from rest under the wind conditions of Rule 234A2, horizontal clearances from such conductors or cables to other supporting structures shall be not less than those shown below:

Conductor or cable	Horizontal clearance required when displaced by wind	
	(m)	(ft)
Open supply conductor, 0 to 750 V	1.1	3.5
230C2 cable, above 750 V	1.1	3.5
230C3 cable, above 750 V	1.1	3.5
Open supply conductors, over 750 V to 22 kV	1.4	4.5

See footnotes 9 and 10 to Table 234-1.

2. Vertical Clearances

A vertical clearance of 1.40 m (4.5 ft) for voltages below 22 kV and a vertical clearance of 1.70 m (5.5 ft) for voltages between 22 kV and 50 kV. *EXCEPTIONS 1* and *2* shall not be applied cumulatively.

EXCEPTION 1: For guys, messengers, and neutrals meeting Rule 230E1 and for cables of 300 V or less to ground meeting the requirements of Rule 230C1, 230C2, or 230C3, the vertical clearance may be reduced to 600 mm (2 ft).

NOTE: Clearances of wires, conductors, and cables from adjacent line structure guy wires are given in Rule 233.

EXCEPTION 2: The vertical clearances may be reduced by 600 mm (2 ft) if both of the following conditions are met:

a. The wires, conductors, or cables above and the supporting structure of another line below are operated and maintained by the same utility.
b. Employees do not work above the top of the supporting structure unless:
   (1) The upper circuit is de-energized and grounded per Rule 444D or temporarily insulated or repositioned, or
   (2) Other equivalent measures are taken.

C. Clearances of Wires, Conductors, Cables, and Rigid Live Parts From Buildings, Signs, Billboards, Chimneys, Radio and Television Antennas, Tanks, and Other Installations Except Bridges

1. Vertical and Horizontal Clearances

a. Clearances

Unguarded or accessible wires, conductors, cables, or rigid live parts may be located adjacent to buildings, signs, billboards, chimneys, radio and television antennas, tanks, and other installations and any projections therefrom. The vertical and horizontal clearances of such rigid and nonrigid parts shall be not less than the values given in Table 234-1 when at rest under the conditions specified in Rule 234A1. These facilities may be installed beside, over or under buildings, building projections and other installations, as illustrated in Figs 234-1(a), 234-1(b), and 234-1(c).

b. Horizontal Clearances Under Wind Displacement Conditions

When the following conductors and cables are displaced from rest under the wind conditions of Rule 234A2, horizontal clearances from such conductors or cables to buildings, signs, billboards, chimneys, radio and television antennas, and other installations shall be not less than those shown below:

Conductor or cable	Horizontal clearance required when displaced by wind	
	(m)	**(ft)**
Open supply conductor, 0 to 750 V	1.1	3.5
230C2 cable, above 750 V	1.1	3.5
230C3 cable, above 750 V	1.1	3.5
Open supply conductors, over 750 V to 22 kV	1.4	4.5

See footnotes 9 and 10 to Table 234-1.

2. Guarding of Supply Conductors and Rigid Live Parts

Where the clearances set forth in Table 234-1 cannot be obtained, supply conductors and rigid live parts shall be guarded.

NOTE: Supply cables meeting Rule 230C1 are considered to be guarded within the meaning of this rule.

3. Supply Conductors Attached to Buildings or Other Installations

Where the permanent attachment of supply conductors of any class to a building or other installation is necessary for an entrance, such conductors shall meet the following requirements over or along the installation to which the conductors are attached:

a. Energized service drop conductors, including splices and taps, shall be insulated or covered in accordance with the following:
   (1) For 0 to 750V, Rules 230C or 230D
   (2) For over 750 V, Rule 230C1

This rule does not apply to neutral conductors meeting Rule 230E1.

b. Conductors of more than 300 V to ground shall not be carried along or near the surface of the installation unless they are guarded or made inaccessible.

c. Wires or cables attached to and run along side the installation shall have clearances from the surface of the installation not less than 75 mm (3 in).

d. Service-drop conductors, including drip loops, shall not be readily accessible, and they shall have a clearance of not less than the following:

(1) 2.45 m (8 ft) vertical clearance from the highest point of roofs or balconies over which they pass.

EXCEPTION 1: Where the voltage between conductors meeting Rule 230D does not exceed 300 V, or where the voltage of cables meeting Rule 230C2 or 230C3 does not exceed 750 V, or where the cable meets Rule 230C1, and the roof or balcony is not readily accessible, the clearance may be not less than 900 mm (3 ft). A roof or a balcony is considered readily accessible to pedestrians if it can be casually accessed through a doorway, window, ramp, stairway, or permanently mounted ladder by a person, on foot, who neither exerts extraordinary physical effort nor employs special tools or devices to gain entry. A permanently mounted ladder is not considered a means of access if its bottom rung is 2.45 m (8 ft) or more from the ground or other permanently installed accessible surface.

EXCEPTION 2: Where a roof or a balcony is not readily accessible, and a service drop meeting one of the following conditions passes over a roof to terminate at a (through-the-roof) raceway or approved support located not more than 1.20 m (4 ft), measured horizontally from the nearest edge of the roof, the clearance above the roof may be maintained at not less than 457 mm (18 in) for a horizontal distance of 1.8 m (6 ft) from the raceway or support, and may be maintained at not less than 900 mm (3 ft) for the remainder of the horizontal distance that the cable or conductor passes over the roof.

(a) Voltage between conductors meeting Rule 230D of 300 V or less,

(b) Cables of 750 V or less meeting Rules 230C2 or 230C3, or

(c) Cables meeting Rule 230C1.

NOTE: See Fig 234-2.

(2) 900 mm (3 ft) in any direction from windows, doors, porches, fire escapes, or similar locations.

EXCEPTION 1: This does not apply to service-drop conductors meeting Rule 230C3 above the top level of a window.

EXCEPTION 2: This does not apply to windows that are not designed to open.

4. Communication Conductors Attached to Buildings or Other Installations

Communication conductors and cables may be attached directly to buildings or other installations.

5. Ladder Space

Where buildings or other installations exceed three stories [or 15 m (50 ft)] in height, overhead lines should be arranged where practical so that a clear space or zone at least 1.8 m (6 ft) wide will be left either adjacent to the building or beginning not over 2.45 m (8 ft) from the building to facilitate the raising of ladders where necessary for fire fighting.

EXCEPTION: This requirement does not apply where it is the unvarying rule of the local fire departments to exclude the use of ladders in alleys or other restricted places that are generally occupied by supply conductors and cables.

m

Table 234-1
Clearance of Wires, Conductors, Cables, and Unguarded Rigid Live Parts Adjacent but Not Attached to Buildings and Other Installations Except Bridges [12]

(Voltages are phase to ground for effectively grounded circuits and those other circuits where all ground faults are cleared by promptly de-energizing the faulted section, both initially and following subsequent breaker operations. See the definitions section for voltages of other systems. Clearances are with no wind displacement except where stated in the footnotes below. See Rules 234C1a, 234C2, and 234H4.)

Clearance of	Insulated communication conductors and cables; messengers; surge-protection wires; grounded guys; ungrounded guys exposed to 0 to 300 V[13]; neutral conductors meeting Rule 230E1; supply cables meeting Rule 230C1[5] (m)	Supply cables of 0 to 750 V meeting Rules 230C2 or 230C3 (m)	Unguarded rigid live parts, 0 to 750 V; noninsulated communication conductors; ungrounded equipment cases, 0 to 750 V; and ungrounded guys exposed to open supply conductors of over 300 V to 750 V[5] (m)	Supply cables over 750 V meeting Rules 230C2 or 230C3; open supply conductors, 0 to 750 V (m)	Unguarded rigid live parts, over 750 V to 22 kV; ungrounded equipment cases, 750 V to 22 kV; ungrounded guys exposed to over 750V to 22 kV[5] (m)	Open supply conductors, over 750 V to 22 kV (m)
1. Buildings						
a. Horizontal						
(1) To walls, projections, and guarded windows	1.40 [1, 2, 7]	1.50[1, 2]	1.50[1, 2]	1.70[1, 2, 9]	2.00[1, 2]	2.30[1, 2, 10, 11]
(2) To unguarded windows[8]	1.40	1.50	1.50	1.70 [9]	2.00	2.30 [10, 11]
(3) To balconies and areas readily accessible to pedestrians [3]	1.40	1.50	1.50	1.70[9]	2.00	2.30 [10, 11]
b. Vertical[14]						
(1) Over or under roofs or projections not readily accessible to pedestrians [3]	0.90	1.07	3.0	3.2	3.6	3.8
(2) Over or under balconies and roofs readily accessible to pedestrians [3]	3.2	3.4	3.4	3.5	4.0	4.1
(3) Over roofs accessible to vehicles but not subject to truck traffic[6]	3.2	3.4	3.4	3.5	4.0	4.1
(4) Over roofs accessible to truck traffic[6]	4.7	4.9	4.9	5.0	5.5	5.6

m

Table 234-1 *(Continued)*
Clearance of Wires, Conductors, Cables, and Unguarded Rigid Live Parts Adjacent but Not Attached to Buildings and Other Installations Except Bridges [12]
(Voltages are phase to ground for effectively grounded circuits and those other circuits where all ground faults are cleared by promptly de-energizing the faulted section, both initially and following subsequent breaker operations. See the definitions section for voltages of other systems. Clearances are with no wind displacement except where stated in the footnotes below.
See Rules 234C1a, 234C2, and 234H4.)

Clearance of	Insulated communication conductors and cables; messengers; surge-protection wires; grounded guys; ungrounded guys exposed to 0 to 300 V[13]; neutral conductors meeting Rule 230E1; supply cables meeting Rule 230C1[5] (m)	Supply cables of 0 to 750 V meeting Rules 230C2 or 230C3 (m)	Unguarded rigid live parts, 0 to 750 V; noninsulated communication conductors; ungrounded equipment cases, 0 to 750 V; and ungrounded guys exposed to open supply conductors of over 300 V to 750 V[5] (m)	Supply cables over 750 V meeting Rules 230C2 or 230C3; open supply conductors, 0 to 750 V (m)	Unguarded rigid live parts, over 750 V to 22 kV; ungrounded equipment cases, 750 V to 22 kV; ungrounded guys exposed to over 750V to 22 kV[5] (m)	Open supply conductors, over 750 V to 22 kV (m)
2. Signs, chimneys, billboards, radio and television antennas, tanks, and other installations not classified as buildings or bridges						
a. Horizontal [4]						
(1) To portions that are readily accessible to pedestrians[3]	1.40	1.50	1.50	1.70[9]	2.00	2.30[10, 11]
(2) To portions that are not readily accessible to pedestrians[3]	0.90	1.07	1.50[1, 2]	1.70[1, 2, 9]	2.00[1, 2]	2.30[1, 2, 10, 11]
b. Vertical						
(1) Over or under catwalks and other surfaces upon which personnel walk	3.2	3.4	3.4	3.5	4.0	4.1
(2) Over or under other portions of such installations[4]	0.90	1.07	1.70	1.80[1]	2.45	2.30

[1] Where building, sign, chimney, antenna, tank, or other installation does not require maintenance such as painting, washing, changing of sign letters, or other operations that would require persons to work or pass between wires, conductors, cables, or unguarded rigid live parts and structure, the clearance may be reduced by 0.60 m.

[2] Where available space will not permit this value, the clearance may be reduced by 0.60 m wires, provided the wires, conductors, or cables, including splices and taps, and unguarded rigid live parts have a covering that provides sufficient dielectric strength to limit the likelihood of a short circuit in case of momentary contact with a structure or building.

[3] A roof, balcony, or area is considered readily accessible to pedestrians if it can be casually accessed through a doorway, ramp, window, stairway, or permanently mounted ladder by a person on foot who neither exerts extraordinary physical effort nor employs special tools or devices to gain entry. A permanently mounted ladder is not considered a means of access if its bottom rung is 2.45 m or more from the ground or other permanently installed accessible surface.

[4] The required clearances shall be to the closest approach of motorized signs or moving portions of installations covered by Rule 234C.

[5] Ungrounded guys and ungrounded portions of guys between guy insulators shall have clearances based on the highest voltage to which they may be exposed to a slack conductor or guy.

[6] For the purpose of this rule, trucks are defined as any vehicle exceeding 2.45 m in height.

[7] This clearance may be reduced to 75 mm for the grounded portions of guys.

[8] Windows not designed to open may have the clearances permitted for walls and projections.

[9] The clearance at rest shall be not less than the value shown in this table. Also, when the conductor or cable is displaced by wind, the clearance shall be not less than 1.07 m; see Rule 234C1b.

[10] The clearance at rest shall be not less than the value shown in this table. Also, when the conductor or cable is displaced by wind, the clearance shall be not less than 1.40 m; see Rule 234C1b.

[11] Where available space will not permit this value, the clearance may be reduced to 2.00 m for conductors limited to 8.7 kV to ground.

[12] The clearance values shown in this table are computed by adding the applicable Mechanical and Electrical (M & E) value of Table A-1 to the applicable Reference Component of Table A-2b of Appendix A.

[13]The anchor end of guys insulated in accordance with Rule 279 may have the same clearance as grounded guys.

[14]For clearances above railings, walls, or parapets around balconies or roofs, use the clearances required for row 1b(1). For such clearances where an outside stairway exists, use the clearances required for row 2b(2).

ft

Table 234-1
Clearance of Wires, Conductors, Cables, and Unguarded Rigid Live Parts Adjacent but Not Attached to Buildings and Other Installations Except Bridges[12]

(Voltages are phase to ground for effectively grounded circuits and those other circuits where all ground faults are cleared by promptly de-energizing the faulted section, both initially and following subsequent breaker operations. See the definitions section for voltages of other systems. Clearances are with no wind displacement except where stated in the footnotes below.
See Rules 234C1a, 234C2, and 234H4.)

Clearance of	Insulated communication conductors and cables; messengers; surge-protection wires; grounded guys; ungrounded guys exposed to 0 to 300 V[13]; neutral conductors meeting Rule 230E1; supply cables meeting Rule 230C1[5] (ft)	Supply cables of 0 to 750 V meeting Rules 230C2 or 230C3 (ft)	Unguarded rigid live parts, 0 to 750 V; non-insulated communication conductors; ungrounded equipment cases, 0 to 750 V; ungrounded guys exposed to open supply conductors of over 300 V to 750 V[5] (ft)	Supply cables over 750 V meeting Rules 230C2 or 230C3; open supply conductors, 0 to 750 V (ft)	Unguarded rigid live parts, over 750 V to 22 kV; ungrounded equipment cases, 750 V to 22 kV; ungrounded guys exposed to over 750 V to 22 kV[5] (ft)	Open supply conductors, over 750 V to 22 kV (ft)
1. Buildings						
a. Horizontal						
(1) To walls, projections, and guarded windows	4.5[1, 2, 7]	5.0[1, 2]	5.0[1, 2]	5.5[1, 2, 9]	7.0[1, 2]	7.5[1, 2, 10, 11]
(2) To unguarded windows[8]	4.5	5.0	5.0	5.5[9]	7.0	7.5[10, 11]
(3) To balconies and areas readily accessible to pedestrians [3]	4.5	5.0	5.0	5.5[9]	7.0	7.5[10, 11]
b. Vertical[14]						
(1) Over or under roofs or projections not readily accessible to pedestrians[3]	3.0	3.5	10.0	10.5	12.0	12.5
(2) Over or under balconies and roofs readily accessible to pedestrians [3]	10.5	11.0	11.0	11.5	13.0	13.5
(3) Over roofs accessible to vehicles but not subject to truck traffic [6]	10.5	11.0	11.0	11.5	13.0	13.5
(4) Over roofs accessible to truck traffic [6]	15.5	16.0	16.0	16.5	18.0	18.5

ft

Table 234-1 *(Continued)*
Clearance of Wires, Conductors, Cables, and Unguarded Rigid Live Parts Adjacent but Not Attached to Buildings and Other Installations Except Bridges[12]
(Voltages are phase to ground for effectively grounded circuits and those other circuits where all ground faults are cleared by promptly de-energizing the faulted section, both initially and following subsequent breaker operations. See the definitions section for voltages of other systems. Clearances are with no wind displacement except where stated in the footnotes below.
See Rules 234C1a, 234C2, and 234H4.)

Clearance of	Insulated communication conductors and cables; messengers; surge-protection wires; grounded guys; ungrounded guys exposed to 0 to 300 V[13]; neutral conductors meeting Rule 230E1; supply cables meeting Rule 230C1[5] (ft)	Supply cables of 0 to 750 V meeting Rules 230C2 or 230C3 (ft)	Unguarded rigid live parts, 0 to 750 V; non-insulated communication conductors; ungrounded equipment cases, 0 to 750 V; ungrounded guys exposed to open supply conductors of over 300 V to 750 V[5] (ft)	Supply cables over 750 V meeting Rules 230C2 or 230C3; open supply conductors, 0 to 750 V (ft)	Unguarded rigid live parts, over 750 V to 22 kV; ungrounded equipment cases, 750 V to 22 kV; ungrounded guys exposed to over 750V to 22 kV[5] (ft)	Open supply conductors, over 750 V to 22 kV (ft)
2. Signs, chimneys, billboards, radio and television antennas, tanks, and other installations not classified as buildings or bridges						
a. Horizontal [4]						
(1) To portions that are readily accessible to pedestrians [3]	4.5	5.0	5.0[1, 2]	5.5[9]	7.0[1,2]	7.5[10, 11]
(2) To portions that are not readily accessible to pedestrians [3]	3.0	3.5	5.0[1, 2]	5.5[1, 2, 9]	7.0[1,2]	7.5[1, 2, 10, 11]
b. Vertical						
(1) Over or under catwalks and other surfaces upon which personnel walk	10.5	11.0	11.0	11.5	13.0	13.5
(2) Over or under other portions of such installations [4]	3.0	3.5	5.5	6.0[1]	7.5	8.0

[1] Where building, sign, chimney, antenna, tank, or other installation does not require maintenance such as painting, washing, changing of sign letters, or other operations that would require persons to work or pass between wires, conductors, cables or unguarded rigid live parts and structure, the clearance may be reduced by 2 ft.

[2] Where available space will not permit this value, the clearance may be reduced by 2 ft provided the wires, conductors, or cables, including splices and taps, and unguarded rigid live parts have a covering that provides sufficient dielectric strength to limit the likelihood of a short circuit in case of momentary contact with a structure or building.

[3] A roof, balcony, or area is considered readily accessible to pedestrians if it can be casually accessed through a doorway, ramp, window, stairway, or permanently mounted ladder by a person on foot who neither exerts extraordinary physical effort nor employs special tools or devices to gain entry. A permanently mounted ladder is not considered a means of access if its bottom rung is 8 ft or more from the ground or other permanently installed accessible surface.

[4] The required clearances shall be to the closest approach of motorized signs or moving portions of installations covered by Rule 234C.

[5] Ungrounded guys and ungrounded portion of guys between guy insulators shall have clearances based on the highest voltage to which they may be exposed to a slack conductor or guy.

[6] For the purpose of this rule, trucks are defined as any vehicle exceeding 8 ft in height.

[7] This clearance may be reduced to 3 in for the grounded portions of guys.

[8] Windows not designed to open may have the clearances permitted for walls and projections.

[9] The clearance at rest shall be not less than the value shown in this table. Also, when the conductor or cable is displaced by wind, the clearance shall be not less than 3.5 ft; see Rule 234C1b.

[10] The clearance at rest shall be not less than the value shown in this table. Also, when the conductor or cable is displaced by wind, the clearance shall be not less than 4.5 ft; see Rule 234C1b.

[11] Where available space will not permit this value, the clearance may be reduced to 7.0 ft for conductors limited to 8.7 kV to ground.

[12] The clearance values shown in this table are computed by adding the applicable Mechanical and Electrical (M & E) value of Table A-1 to the applicable Reference Component of Table A-2b of Appendix A.

[13] The anchor end of guys insulated in accordance with Rule 279 may have the same clearance as grounded guys.

[14] For clearances above railings, walls, or parapets around balconies or roofs, use the clearances required for row 1b(1). For such clearances where an outside stairway exists, use the clearances required for row 2b(2).

D. Clearance of Wires, Conductors, Cables, and Unguarded Rigid Live Parts From Bridges

1. Vertical and Horizontal Clearances

a. Clearances

Unguarded or accessible wires, conductors, cables, or rigid live parts may be located adjacent to or within a bridge structure. The vertical and horizontal clearances of such rigid and non-rigid parts shall be not less than the values given in Table 234-2 when at rest under the conditions specified in Rule 234A1, as illustrated in Figs 234-1(a) and 234-1(b).

EXCEPTION: This rule does not apply to insulated communication cables, effectively grounded guys, span wires, and surge protection wires; neutrals meeting Rule 230E1; and supply cables meeting Rule 230C1.

b. Horizontal Clearances Under Wind Displacement Conditions

When the following conductors and cables are displaced from rest under the wind conditions of Rule 234A2, horizontal clearances from such conductors or cables to bridges shall be not less than those shown below:

Conductor or cable	**Horizontal clearance required when displaced by wind**	
	(m)	**(ft)**
Open supply conductor, 0 to 750 V	1.1	3.5
230C2 cable, above 750 V	1.1	3.5
230C3 cable, above 750 V	1.1	3.5
Open supply conductors, over 750 V to 22 kV	1.4	4.5

See footnotes 8 and 9 to Table 234-2.

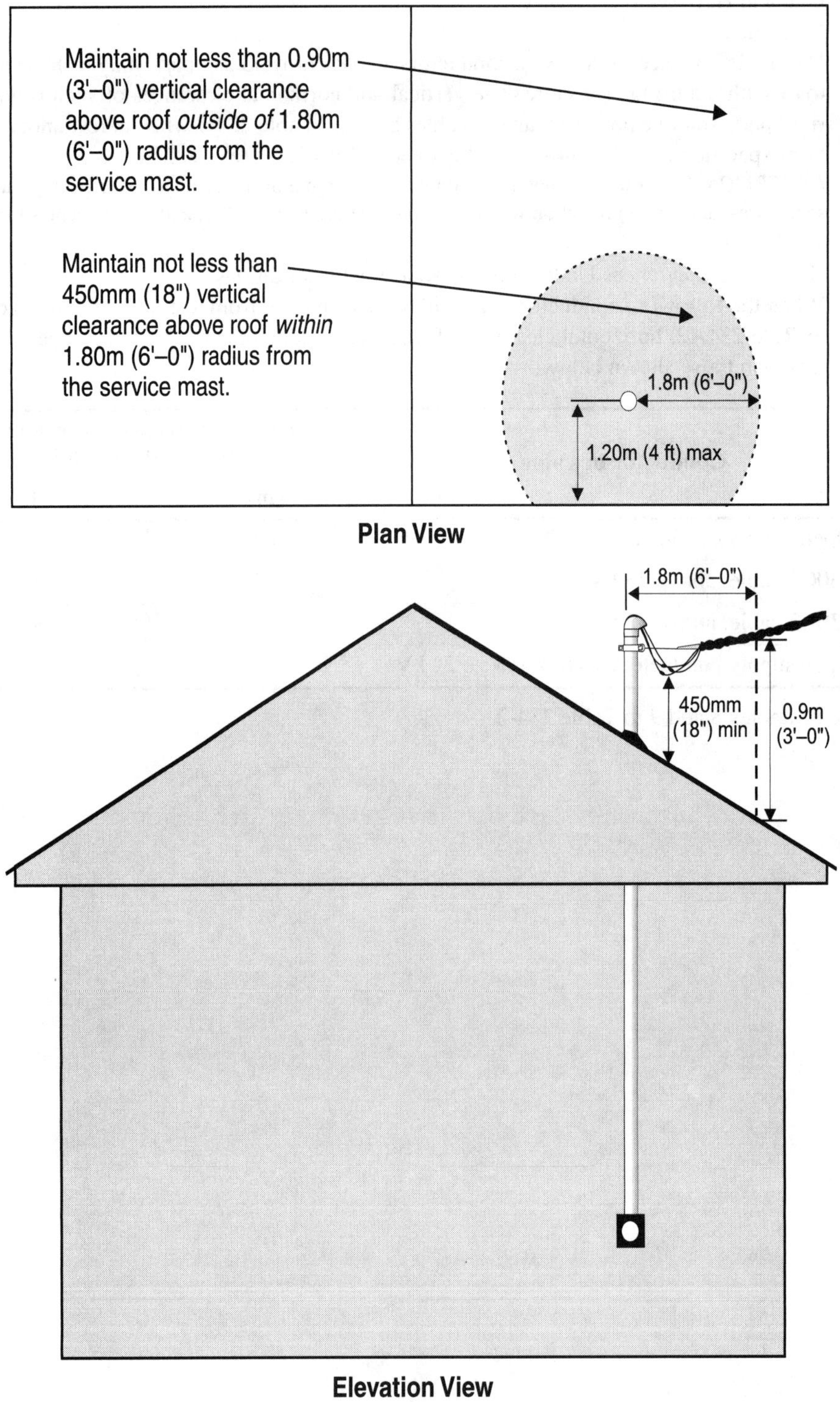

Fig 234-2
Clearances of Service Drop Terminating on Support Mast

2. Guarding Trolley-Contact Conductors Located Under Bridges
   a. Where Guarding Is Required
      Guarding is required where the trolley-contact conductor is located so that a trolley pole leaving the conductor can make simultaneous contact between it and the bridge structure.
   b. Nature of Guarding
      Guarding shall consist of a substantial inverted trough of nonconducting material located above the contact conductor, or of other suitable means of limiting the likelihood of contact between the trolley support and the bridge structure.

E. Clearance of Wires, Conductors, Cables, or Unguarded Rigid Live Parts Installed Over or Near Swimming Areas With No Wind Displacement
   1. Swimming Pools
      Where wires, conductors, cables, or unguarded rigid live parts are over a swimming pool or the surrounding area, the clearances in any direction shall be not less than those shown in Table 234-3 and illustrated in Fig 234-3.
      EXCEPTION 1: This rule does not apply to a pool fully enclosed by a solid or screened permanent structure.
      EXCEPTION 2: This rule does not apply to communication conductors and cables, effectively grounded surge-protection wires, neutral conductors meeting Rule 230E1, guys and messengers, supply cables meeting Rule 230C1, and supply cables of 0 to 750 V meeting Rules 230C2 or 230C3 when these facilities are 3 m (10 ft) or more horizontally from the edge of the pool, diving platform, or diving tower.
   2. Beaches and Waterways Restricted to Swimming
      Where rescue poles are used by lifeguards at supervised swimming beaches, the vertical and horizontal clearances shall be not less than those shown in Table 234-3. Where rescue poles are not used, the clearances shall be as specified in Rule 232.
   3. Waterways Subject to Water Skiing
      The vertical clearance shall be the same as that specified in Rule 232.

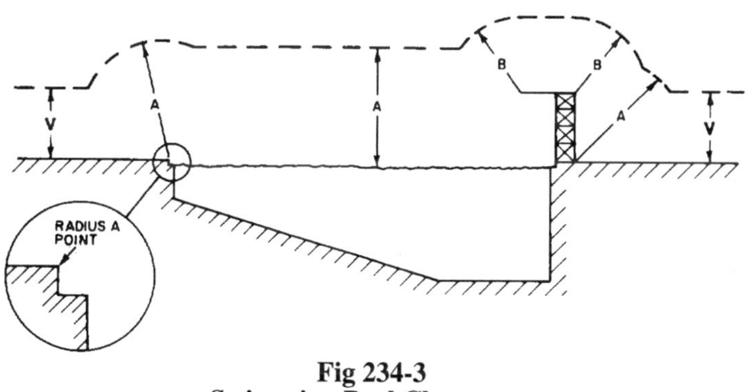

Fig 234-3
Swimming Pool Clearances

m

Table 234-2
Clearance of Wires, Conductors, Cables, and Unguarded Rigid Live Parts From Bridges

(Voltages are phase to ground for effectively grounded circuits and those other circuits where all ground faults are cleared by promptly de-energizing the faulted section, both initially and following subsequent breaker operations. See the definitions section for voltages of other systems. Clearances are with no wind displacement except where stated in the footnotes below. See Rules 234D1a and 234H4.)

	Unguarded rigid live parts, 0 to 750 V; noninsulated communication conductors; supply cables of 0 to 750 V meeting Rules 230C2 or 230C3[7]; ungrounded equipment cases, 0 to 750 V; ungrounded guys exposed to open supply conductors over 300 V to 750 V[4] (m)	Supply cables over 750 V meeting Rules 230C2 or 230C3[7]; open supply conductors, 0 to 750 V (m)	Open supply conductors, over 750 V to 22 kV (m)	Unguarded rigid live parts, over 750 V to 22 kV, ungrounded equipment cases, 750 V to 22 kV; ungrounded guys exposed to open supply conductors of over 750 V to 22 kV[4] (m)
1. Clearance over bridges [1]				
a. Attached [3]	0.90	1.07	1.70	1.50
b. Not attached	3.0	3.2	3.8	3.6
2. Clearance beside, under, or within bridge structure [6]				
a. Readily accessible portions of any bridge including wing, walls, and bridge attachments [1]				
(1) Attached [3]	0.90	1.07 [8]	1.70 [9]	1.50
(2) Not attached	1.50	1.70 [8]	2.30 [9]	2.00
b. Ordinarily inaccessible portions of bridges (other than brick, concrete, or masonry) and from abutments [2]				
(1) Attached [3, 5]	0.90	1.07 [8]	1.70 [9]	1.50
(2) Not attached [4, 5]	1.20	1.40 [8]	2.00 [9]	1.80

[1] Where over traveled ways on or near bridges, the clearances of Rule 232 apply also.

[2] Bridge seats of steel bridges carried on masonry, brick, or concrete abutments that require frequent access for inspection shall be considered as readily accessible portions.

[3] Clearance from supply conductors to supporting arms and brackets attached to bridges shall be the same as specified in Table 235-6 (Rule 235E1) if the supporting arms and brackets are owned, operated, or maintained by the same utility.

[4] Ungrounded guys and ungrounded portions of guys between guy insulators shall have clearances based on the highest voltage to which they may be exposed due to a slack conductor or guy.

[5] Where conductors passing under bridges are adequately guarded against contact by unauthorized persons and can be deenergized and grounded per Rule 444D for maintenance of the bridge, clearances of the conductors from the bridge, at any point, may have the clearances specified in Table 235-6 for clearance from surfaces of support arms plus one-half the final unloaded sag of the conductor at that point.

[6] Where the bridge has moving parts, such as a lift bridge, the required clearances shall be maintained throughout the full range of movement of the bridge or any attachment thereto.

[7] Where permitted by the bridge owner, supply cables may be run in rigid conduit attached directly to the bridge. Refer to Part 3 for installation rules.

[8] The clearance at rest shall be not less than the value shown in this table. Also, when the conductor or cable is displaced by wind, the clearance shall be not less than 1.07 m; see Rule 234D1b.

[9] The clearance at rest shall be not less than the value shown in this table. Also, when the conductor or cable is displaced by wind, the clearance shall be not less than 1.40 m; see Rule 234D1b.

ft

Table 234-2
Clearance of Wires, Conductors, Cables, and Unguarded Rigid Live Parts From Bridges

(Voltages are phase to ground for effectively grounded circuits and those other circuits where all ground faults are cleared by promptly de-energizing the faulted section, both initially and following subsequent breaker operations. See the definitions section for voltages of other systems. Clearances are with no wind displacement except where stated in the footnotes below. See Rules 234D1a and 234H4.)

	Unguarded rigid live parts, 0 to 750 V; noninsulated communication conductors; supply cables of 0 to 750 V meeting Rules 230C2 or 230C3[7], ungrounded equipment cases, 0 to 750 V; ungrounded guys exposed to open supply conductors over 300 V to 750 V[4] (ft)	Supply cables over 750 V meeting Rules 230C2 or 230C3[7]; open supply conductors, 0 to 750 V (ft)	Open supply conductors, over 750 V to 22 kV (ft)	Unguarded rigid live parts, over 750 V to 22 kV, ungrounded equipment cases, 750 V to 22 kV; ungrounded guys exposed to open supply conductors of over 750V to 22 kV[4] (ft)
1. Clearance over bridges[1]	3.0	3.5	5.5	5.0
a. Attached[3]				
b. Not attached	10.0	10.5	12.5	12.0
2. Clearance beside, under, or within bridge structure[6]				
a. Readily accessible portions of any bridge including wing, walls, and bridge attachments[1]				
(1) Attached[3]	3.0	3.5[8]	5.5[9]	5.0
(2) Not attached	5.0	5.5[8]	7.5[9]	7.0
b. Ordinarily inaccessible portions of bridges (other than brick, concrete, or masonry) and from abutments[2]				
(1) Attached[3, 5]	3.0	3.5[8]	5.5[9]	5.0
(2) Not attached[4, 5]	4.0	4.5[8]	6.5[9]	6.0

[1] Where over traveled ways on or near bridges, the clearances of Rule 232 apply also.

[2] Bridge seats of steel bridges carried on masonry, brick, or concrete abutments that require frequent access for inspection shall be considered as readily accessible portions.

[3] Clearance from supply conductors to supporting arms and brackets attached to bridges shall be the same as specified in Table 235-6 (Rule 235E1) if the supporting arms and brackets are owned, operated, or maintained by the same utility.

[4] Ungrounded guys and ungrounded portions of guys between guy insulators shall have clearances based on the highest voltage to which they may be exposed due to a slack conductor or guy.

[5] Where conductors passing under bridges are adequately guarded against contact by unauthorized persons and can be de-energized and grounded per Rule 444D for maintenance of the bridge, clearances of the conductors from the bridge, at any point, may have the clearances specified in Table 235-6 for clearance from surfaces of support arms plus one-half the final unloaded sag of the conductor at that point.

[6] Where the bridge has moving parts, such as a lift bridge, the required clearances shall be maintained throughout the full range of movement of the bridge or any attachment thereto.

[7] Where permitted by the bridge owner, supply cables may be run in rigid conduit attached directly to the bridge. Refer to Part 3 for installation rules.

[8] The clearance at rest shall be not less than the value shown in this table. Also, when the conductor or cable is displaced by wind, the clearance shall be not less than 3.5 ft; see Rule 234D1b.

[9] The clearance at rest shall be not less than the value shown in this table. Also, when the conductor or cable is displaced by wind, the clearance shall be not less than 4.5 ft; see Rule 234D1b.

m

Table 234-3
Clearance of Wires, Conductors, Cables, or Unguarded Rigid Live Parts Over or Near Swimming Pools [1]

(Voltages are phase to ground for effectively grounded circuits and those other circuits where all ground faults are cleared by promptly de-energizing the faulted section, both initially and following subsequent breaker operations. See the definitions section for voltages of other systems.
Clearances are with no wind displacement.
See Rules 234E1, 234E2, and 234H4.)

	Insulated communication conductors and cables; messengers; surge-protection wires; grounded guys; ungrounded guys exposed to 0 to 300 V [3]; neutral conductors meeting Rule 230E1; supply cables meeting Rule 230C1 (m)	Unguarded rigid live parts, 0 to 750 V; noninsulated communication conductors; supply cables of 0 to 750 V meeting Rules 230C2 or 230C3; ungrounded guys exposed to open supply conductors of over 300 V to 750 V [2] (m)	Supply cables over 750 V meeting Rules 230C2 or 230C3; open supply conductors, 0 to 750 V (m)	Unguarded rigid live parts over 750 V to 22 kV; ungrounded guys exposed to over 750 V to 22 kV [2] (m)	Open supply conductors, over 750 V to 22 kV (m)
A. Clearance in any direction from the water level, edge of pool, base of diving platform, or anchored raft	6.7	6.9	7.0	7.5	7.6
B. Clearance in any direction to the diving platform or tower	4.3	4.4	4.6	5.1	5.2
V. Vertical clearance over adjacent land	Clearance shall be as required by Rule 232.				

NOTE: A, B, and V are shown in Fig 234-3.

[1] The clearance values shown in this table are computed by adding the applicable Mechanical and Electrical (M & E) value of Table A-1 to the applicable Reference Component of Table A-2b of Appendix A.

[2] Ungrounded guys and ungrounded portions of guys between guy insulators shall have clearances based on the highest voltage to which they may be exposed due to a slack conductor or guy.

[3] Anchor guys insulated in accordance with Rule 279 may have the same clearance as grounded guys.

ft

Table 234-3
Clearance of Wires, Conductors, Cables, or Unguarded Rigid Live Parts Over or Near Swimming Pools [1]

(Voltages are phase to ground for effectively grounded circuits and those other circuits where all ground faults are cleared by promptly de-energizing the faulted section, both initially and following subsequent breaker operations. See the definitions section for voltages of other systems.
Clearances are with no wind displacement.
See Rules 234E1, 234E2, and 234H4.)

	Insulated communication conductors and cables; messengers; surge-protection wires; grounded guys; ungrounded guys exposed to 0 to 300 V [3]; neutral conductors meeting Rule 230E1; supply cables meeting Rule 230C1 (ft)	**Unguarded rigid live parts, 0 to 750 V; noninsulated communication conductors; supply cables of 0 to 750 V meeting Rules 230C2 or 230C3; ungrounded guys exposed to open supply conductors of over 300 V to 750 V [2] (ft)**	**Supply cables over 750 V meeting Rules 230C2 or 230C3; open supply conductors, 0 to 750 V (ft)**	**Unguarded rigid live parts over 750 V to 22 kV; ungrounded guys exposed to over 750 V to 22 kV [2] (ft)**	**Open supply conductors, over 750 V to 22 kV (ft)**
A. Clearance in any direction from the water level, edge of pool, base of diving platform, or anchored raft	22.0	22.5	23.0	24.5	25.0
B. Clearance in any direction to the diving platform or tower	14.0	14.5	15.0	16.5	17.0
V. Vertical clearance over adjacent land	Clearance shall be as required by Rule 232.				

NOTE: A, B, and V are shown in Fig 234-3.

[1] The clearance values shown in this table are computed by adding the applicable Mechanical and Electrical (M & E) value of Table A-1 to the applicable Reference Component of Table A-2b of Appendix A.

[2] Ungrounded guys and ungrounded portions of guys between insulators shall have clearances based on the highest voltage to which they may be exposed due to a slack conductor or guy.

[3] Anchor guys insulated in accordance with Rule 279 may have the same clearance as grounded guys.

F. Clearances of Wires, Conductors, Cables, and Rigid Live Parts From Grain Bins

1. Grain Bins Loaded by Permanently Installed Augers, Conveyers, or Elevator Systems

All portions of grain bins that are expected to be loaded by the use of a permanently installed auger, conveyer, or elevator system shall be considered as a building or other installation under Rule 234C for the purpose of determining appropriate clearances of wires, conductors, cables, and rigid live parts. In addition, the following clearances shall also apply without wind displacement.

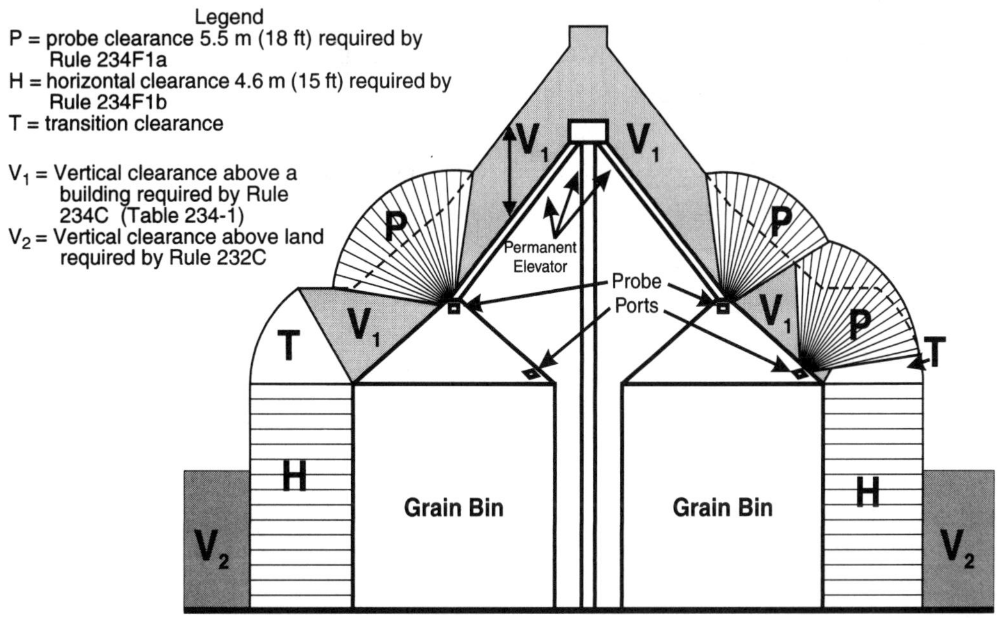

Dimension V is determined by Rule 234C and Table 234-1, Row 1b(2).

Fig 234-4(a)
Clearance Envelope for Grain Bins Filled by Permanently Installed Augers, Conveyors, or Elevators

a. A clearance of not less than 5.5 m (18 ft) in all directions above the grain bin shall be maintained from each probe port in the grain bin roof for all wires, conductors, and cables.

b. A horizontal clearance of not less than 4.6 m (15 ft) shall be maintained between grain bins and open supply conductors, 0 to 22 kV.

2. Grain Bins Loaded by Portable Augers, Conveyers, or Elevators (With No Wind Displacement)

a. The clearance of wires, conductors, cables, and rigid live parts from grain bins that are expected to be loaded by the use of a portable auger, conveyer, or elevator shall be not less than the values illustrated in Fig 234-4b.

EXCEPTION: Clearances of the following items on the nonloading side of grain bins shall be not less than those required by Rule 234C for clearances from buildings:

(1) Support arms; effectively grounded equipment cases

(2) Insulated communication conductors and cables, messengers, surge-protection wires, grounded guys, neutral conductors meeting Rule 230E1, and supply cables meeting Rule 230C1

(3) Supply cables of 0 to 750 V meeting Rules 230C2 or 230C3

b. Any side of a grain bin is considered to be a nonloading side if it is so designated, or if it is so closely abutting another structure or obstruction, or so close to a public road or other right-of-way that a portable auger, conveyor, or elevator is not reasonably anticipated to be used over that side or portion to fill the grain bin.

c. Where an agreement excludes the use of portable augers, conveyors, or elevators from a designated portion of a grain bin, such portion is considered to be a nonloading side.

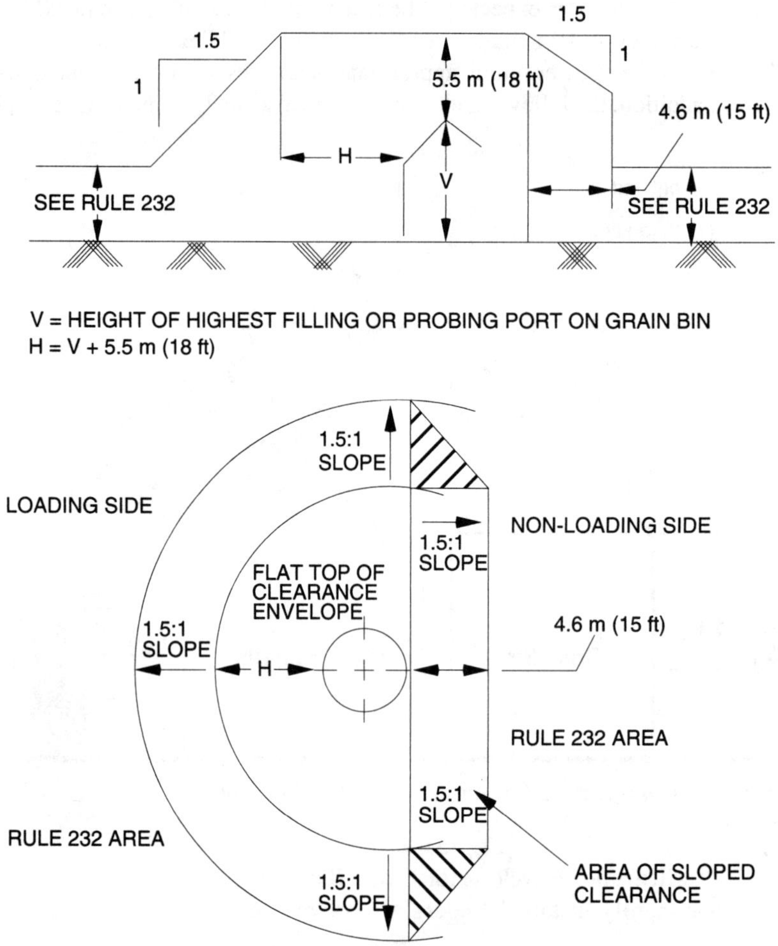

Fig 234-4(b)
Clearance Envelope for Grain Bins Filled by Portable Augers, Conveyors, or Elevators

G. Additional Clearances for Voltages Exceeding 22 kV for Wires, Conductors, Cables, and Unguarded Rigid Live Parts of Equipment
Greater clearances than specified in Rules 234B, 234C, 234D, 234E, 234F, and 234J shall be provided where required below.

1. For voltages between 22 and 470 kV, the clearance specified in Rules 234B, 234C, 234D, 234E, 234F, and 234J shall be increased at the rate of 10 mm (0.4 in) per kV in excess of 22 kV. For voltages exceeding 470 kV, the clearance shall be determined by the method given in Rule 234H. All clearances for lines over 50 kV shall be based on the maximum operating voltage.

   EXCEPTION 1: Where a clearance value is given for the 22 kV to 50 kV range, the voltage adder of Rule 234G1 applies to the voltage in excess of 50 kV.

   EXCEPTION 2: For voltages exceeding 98 kV ac to ground or 139 kV dc to ground, clearances less than those required above are permitted for systems with known maximum switching-surge factor. (See Rule 234H.)

2. For voltages exceeding 50 kV, the additional clearance specified in Rule 234G1 shall be increased 3% for each 300 m (1000 ft) in excess of 1000 m (3300 ft) above mean sea level.
3. For voltages exceeding 98 kV ac to ground, either the clearances shall be increased or the electric field, or the effects thereof, shall be reduced by other means, as required, to limit the steady-state current due to electrostatic effects to 5 mA, rms, if an ungrounded metal fence, building, sign, billboard, chimney, radio or television antenna, tank or other installation, or any ungrounded metal attachments thereto, were short-circuited to ground. For this determination, the conductor shall be at a final unloaded sag at 50 °C (120 °F).

H. Alternate Clearances for Voltages Exceeding 98 kV AC to Ground or 139 kV DC to Ground
The clearances specified in Rules 234B, 234C, 234D, 234E, 234F, 234G, and 234J may be reduced for circuits with known switching-surge factors, but shall be not less than the alternate clearance, which is computed by adding the reference distance from Rule 234H2 to the electrical component of clearance from Rule 234H3.

1. Sag Conditions of Line Conductors
   The vertical, horizontal, and diagonal clearances shall be maintained under the conductor temperature and loading conditions given in Rule 234A.
2. Reference Distances
   The reference distance shall be selected from Table 234-5.
3. Electrical Component of Clearance
   a. The electrical component (D) shall be computed using the following equations. Selected values of D are listed in Table 234-4.

$$D = 1.00\left[\frac{V \cdot (PU) \cdot a}{500K}\right]^{1.667} bc \quad \text{(m)}$$

$$D = 3.28\left[\frac{V \cdot (PU) \cdot a}{500K}\right]^{1.667} bc \quad \text{(ft)}$$

where

V = maximum ac crest operating voltage to ground or maximum dc operating voltage to ground in kilovolts,

PU = maximum switching-surge factor expressed in per-unit peak voltage to ground and defined as a switching-surge level for circuit breakers corresponding to 98% probability that the maximum switching surge generated per breaker operation does not exceed this surge level, or the maximum anticipated switching-surge level generated by other means, whichever is greater,

a = 1.15, the allowance for three standard deviations,

b = 1.03, the allowance for nonstandard atmospheric conditions,

c = the margin of safety:
1.2 for vertical clearances
1.0 for horizontal clearances,

K = 1.15, the configuration factor for conductor-to-plane gap.

b. The value of *D* shall be increased 3% for each 300 m (1000 ft) in excess of 450 m (1500 ft) above mean sea level.

4. Limit

The alternate clearance shall be not less than the clearance of Rule 234B, Table 234-1, Table 234-2, or Table 234-3, as applicable, computed for 98 kV ac rms to ground by Rule 234G1.

I. Clearance of Wires, Conductors, and Cables to Rail Cars

Where overhead wires, conductors, or cables run along railroad tracks, the clearance in any direction shall be not less than that shown in Fig 234-5. The values of *V* and *H* are as defined below:

V = vertical clearance from the wire, conductor, or cable above the top of the rail as specified in Rule 232 minus 6.1 m (20 ft), the assumed height of the rail car

H = horizontal clearance from the wire, conductor, or cable to the nearest rail, which is equal to the required vertical clearance above the rail minus 4.6 m (15 ft) as computed by the lesser of the following:

1. Rules 232B1 and 232C1
2. Rule 232D

These clearances are computed for railroads handling standard rail cars as common carriers in interchange service with other railroads. Where wires, conductors, or cables run along mine, logging, and similar railways that handle only cars smaller than standard freight cars, the value of *H* may be reduced by one-half the difference between the width of a standard rail car [3.3 m (10 ft, 8 in)] and the width of the narrower car.

J. Clearance of Equipment Mounted on Supporting Structures

1. Clearance to Unguarded Rigid Live Parts of Equipment

The horizontal and vertical clearances of unguarded rigid live parts such as potheads, transformer bushings, surge arresters, and short lengths of supply conductors connected thereto, which are not subject to variation in sag, shall be not less than those required by Rules 234C or 234D, as applicable.

2. Clearance to Equipment Cases

Equipment shall be mounted so that clearances are not less than that given by Rules 234J2a, 234J2b, and 234J2c.

a. Effectively grounded equipment cases may be located on or adjacent to buildings, bridges, or other structures provided that clearances for unguarded rigid live parts of such equipment, as specified in Rule 234J1, are maintained.

b. Equipment cases that are not effectively grounded shall be located so that the clearances of Rules 234C or 234D, as applicable, are maintained.

c. Equipment cases shall be located so as not to serve as a means of approach to unguarded rigid live parts by unqualified persons.

NOTE: Rule 234J is not subject to the loading conditions of Rule 234A.

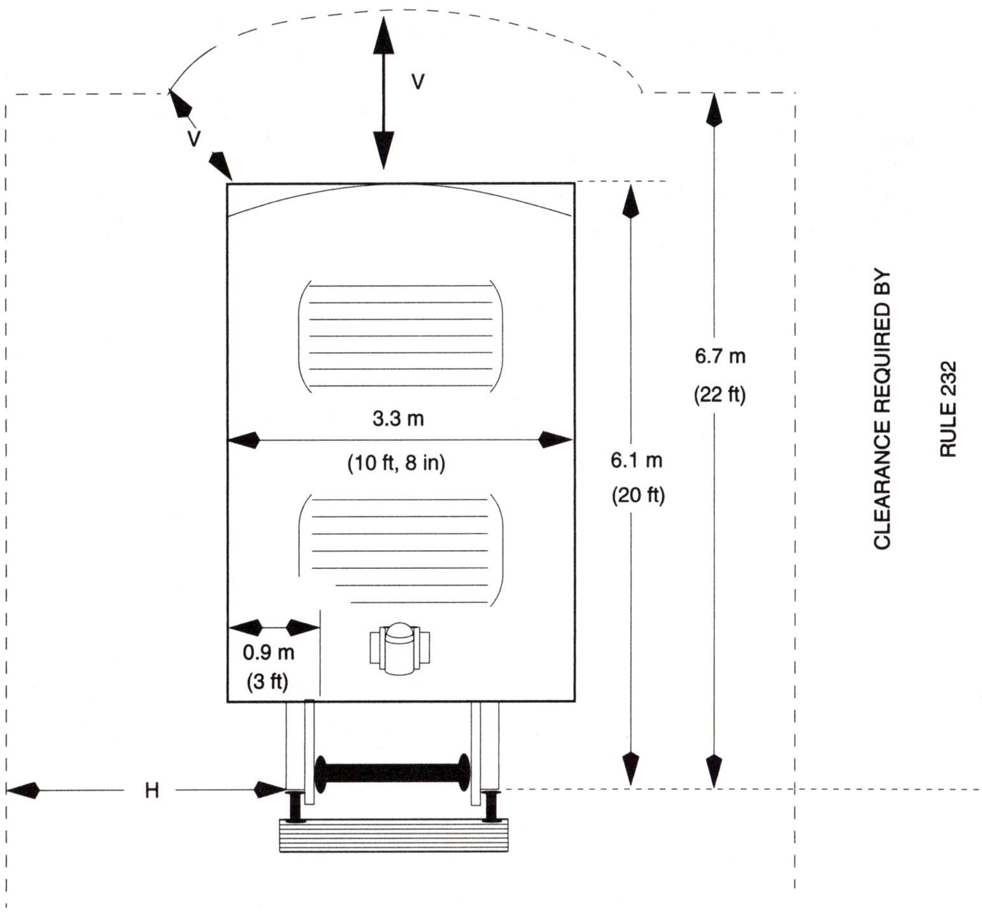

Fig 234-5
Rail Car Clearances

Table 234-4
Electrical Component of Clearance of Buildings, Bridges, and Other Installations
[Add 3% for each 300 m (1000 ft) in excess of 450 m (1500 ft) above mean sea level. See Rule 234H3a.]

Maximum operating voltage phase to phase (kV)	Switching-surge factor (per unit)	Switching surge (kV)	Electrical component of clearances			
			Vertical		Horizontal	
			(m)	(ft)	(m)	(ft)
242	2.0	395	0.82	2.7	0.70	2.3
	2.2	435	0.98	3.2	0.82	2.7
	2.4	474	1.13	3.7	0.94	3.1
	2.6	514	1.28	4.2	1.07	3.5
	2.8	553	1.40	4.8	1.20	4.0
	3.0	593	1.65	5.4	1.40	4.5
362	1.8	532	1.40	4.5	1.13	3.7
	2.0	591	1.65	5.4	1.40	4.5
	2.2	650	1.90	6.3	1.60	5.2
	2.4	709	2.20	7.3	1.85	6.1
	2.6	768	2.50	8.3	2.10	6.9
	2.8	828	2.90	9.4	2.40	7.8
	3.0	887	3.2	10.6	2.70	8.8
550	1.6	719	2.30	7.5	1.90	6.2
	1.8	808	2.80	9.1	2.30	7.6
	2.0	898	3.3	10.8	2.70	9.0
	2.2	988	3.9	12.7	3.2	10.6
	2.4	1079	4.4	14.6	3.7	12.2
	2.6	1168	5.1	16.7	4.2	13.9
800	1.6	1045	4.2	13.9	3.5	11.6
	1.8	1176	5.2	16.9	4.3	14.1
	2.0	1306	6.1	20.1	5.1	16.7
	2.2	1437	7.2	23.6	6.0	19.7
	2.4	1568	8.3	27.3	6.9	22.7

Table 234-5
Reference Distances
(See Rule 234H2.)

Reference distance	Vertical (m)	Vertical (ft)	Horizontal (m)	Horizontal (ft)
a. Buildings	2.70	9	0.90	3
b. Signs, chimneys, radio and television antennas, tanks, and other installations not classified as bridges or buildings	2.70	9	0.90	3
c. Superstructure of bridges [1, 2]	2.70	9	0.90	3
d. Supporting structures of another line	1.80	6	1.50	5
e. Dimension *A* of Fig 234-2	5.5	18	—	—
f. Dimension *B* of Fig 234-2	4.3	14	4.3	14

[1] Where over traveled ways on or near bridges, the clearances of Rule 232 apply also.

[2] Where the bridge has moving parts, such as a lift bridge, the required clearances shall be maintained throughout the full range of movement of the bridge or any attachment thereto.

235. Clearance for Wires, Conductors, or Cables Carried on the Same Supporting Structure

A. Application of Rule

1. Multiconductor Wires or Cables

Cables, and duplex, triple, or paired conductors supported on insulators or messengers meeting Rules 230C or 230D, whether single or grouped, for the purposes of this rule are considered single conductors even though they may contain individual conductors not of the same phase or polarity.

2. Conductors Supported by Messengers or Span Wires

Clearances between individual wires, conductors, or cables supported by the same messenger, or between any group and its supporting messenger, or between a trolley feeder, supply conductor, or communication conductor, and their respective supporting span wires, are not subject to the provisions of this rule.

3. Line Conductors of Different Circuits

a. Unless otherwise stated, the voltage between line conductors of different circuits shall be the greater of the following:

(1) The phasor difference between the conductors involved

NOTE: A phasor relationship of 180° is considered appropriate where the actual phasor relationship is unknown.

(2) The phase-to-ground voltage of the higher-voltage circuit.

b. When the circuits have the same nominal voltage, either circuit may be considered to be the higher-voltage circuit.

B. Horizontal Clearance Between Line Conductors

1. Fixed Supports

Line conductors attached to fixed supports shall have horizontal clearances from each other not less than the larger value required by either Rule 235B1a or Rule 235B1b for the situation concerned. Voltage is between the two conductors for which the clearance is being determined except for railway feeders, which are to ground.

EXCEPTION 1: The pin spacing at buckarm construction may be reduced as specified in Rule 236F to provide climbing space.

EXCEPTION 2: Grades D and N need meet only the requirements of Rule 235B1a.

EXCEPTION 3: These clearances do not apply to cables meeting Rule 230C or covered conductors of the same circuit meeting Rule 230D.

EXCEPTION 4: For voltages to ground exceeding 98 kV ac or 139 kV dc, clearances less than those required by a and b below are permitted for systems with known maximum switching-surge factors. (See Rule 235B3.)

a. Horizontal Clearance Between Line Conductors of the Same or Different Circuits
Clearances shall be not less than those given in Table 235-1.

b. Clearance According to Sags
The clearance at the supports of conductors of the same or different circuits of Grade B or C in no case shall be less than the values given by the following formulas, at a conductor temperature of 15 °C (60 °F), at final unloaded sag, no wind. The requirements of Rule 235B1a apply if they give a greater separation than this rule.

EXCEPTION: No requirement is specified for clearance between conductors of the same circuit when rated above 50 kV.

In the following, *S* is the apparent sag in inches of the conductor having the greater sag, and the clearance is in inches.

(1) For line conductors smaller than AWG No. 2: clearance = 7.6 mm per kilovolt (0.3 in) + $7\sqrt{S/(3-8)}$. (Table 235-2 shows selected values up to 46 kV.)

(2) For line conductors of AWG No. 2 or larger: clearance = 7.6 mm per kilovolt (0.3 in) per kilovolt + $8\sqrt{S/12}$. (Table 235-3 shows selected values up to 46 kV.)

(3) For voltages exceeding 814 kV, the clearance shall be determined by the alternate method given by Rule 235B3.

In the following, *S* is the apparent sag in millimeters of the conductor having the greater sag, and the clearance is in millimeters.

(1) For line conductors smaller than AWG No. 2: clearance = $7.6(kV) + 20.4\sqrt{S-610}$. (Table 235-2 shows selected values up to 46 kV.)

(2) For line conductors of AWG No. 2 or larger: clearance = $7.6(kV) + 8\sqrt{(2.12S)}$. (Table 235-3 shows selected values up to 46 kV.)

(3) For voltages exceeding 814 kV, the clearance shall be determined by the alternate method given by Rule 235B3.

(4) The clearance for voltages exceeding 50 kV specified in Rule 235B1b(1) and (2) shall be increased 3% for each 300 mm(1000 ft) in excess of 1000 m (3300 ft) above mean sea level. All clearances for lines over 50 kV shall be based on the maximum operating voltage.

Table 235-1
Horizontal Clearance Between Wires, Conductors, or Cables at Supports
(All voltages are between conductors involved except for railway feeders, which are to ground. See also Rules 235A, 235B3b, and 235B1a.)

Class of circuit	Clearance (mm)	Clearance (in)	Notes
Open communication conductors	150	6	Does not apply at conductor transposition points.
	75	3	Permitted where pin spacings less than 150 mm (6 in) have been in regular use. Does not apply at conductor transposition points.
Railway feeders: 0 to 750 V, AWG No. 4/0 or larger 0 to 750 V, smaller than AWG No. 4/0 Over 750 V to 8.7 kV	 150 300 300	 6 12 12	Where 250 to 300 mm (10 to 12 in) clearance has already been established by practice, it may be continued, subject to the provisions of Rule 235B1b, for conductors having apparent sags not over 900 mm (3 ft) and for voltages not exceeding 8.7 kV.
Supply conductors of the same circuit: 0 to 8.7 kV Over 8.7 to 50 kV Above 50 kV	 300 300 plus 10 per kV over 8.7 kV no value specified	 12 12 plus 0.4 per kV over 8.7 kV no value specified	
Supply conductors of different circuits: 0 to 8.7 kV Over 8.7 to 50 kV Over 50 kV to 814 kV	 300 300 plus 10 per kV over 8.7 kV 725 plus 10 per kV over 50 kV	 12 12 plus 0.4 per kV over 8.7 kV 28.5 plus 0.4 per kV over 50 kV	For all voltages above 50 kV, the additional clearance shall be increased 3% for each 300 m (1000 ft) in excess of 1000 m (3300 ft) above mean sea level. All clearances for voltages above 50 kV shall be based on the maximum operating voltage.

m

Table 235-2
Horizontal Clearances Between Line Conductors Smaller Than AWG No. 2 at Supports, Based on Sags
(See also Rule 235B1b.)

Voltage between conductors (kV)	Sag (mm)							
	915	1220	1830	2440	3050	4570	6095	But not less than[1]
	Horizontal clearance (mm)							
2.4	375	520	730	890	1025	1300	1525	300
4.16	390	535	745	905	1040	1315	1540	300
12.47	450	595	805	965	1100	1375	1605	345
13.2	455	605	810	970	1105	1385	1610	350
13.8	460	605	815	975	1110	1385	1615	355
14.4	465	610	820	980	1115	1390	1620	365
24.94	545	695	900	1060	1195	1475	1700	470
34.5	620	765	975	1135	1270	1545	1775	570
46	705	855	1060	1220	1355	1635	1860	685

[1] Clearance determined by Table 235-1, Rule 235B1a.

NOTE: Clearance = 7.6 (kV) + 20.4 $\sqrt{S - 610}$, where S is the sag in millimeters.

in

Table 235-2
Horizontal Clearances Between Line Conductors Smaller Than AWG No. 2 at Supports, Based on Sags
(See also Rule 235B1b.)

Voltage between conductors (kV)	Sag (in)							
	36	48	72	96	120	180	240	But not less than[1]
	Horizontal clearance (in)							
2.4	14.7	20.5	28.7	35.0	40.3	51.2	60.1	12.0
4.16	15.3	21.1	29.3	35.6	40.9	51.8	60.7	12.0
12.47	17.7	23.5	31.7	38.0	43.3	54.2	63.1	13.5
13.2	18.0	23.8	32.0	38.3	43.6	54.5	63.4	13.8
13.8	18.1	23.9	32.1	38.4	43.7	54.6	63.5	14.0
14.4	18.3	24.1	32.3	38.6	43.9	54.8	63.7	14.3
24.94	21.5	27.3	35.5	41.8	47.1	58.0	66.9	18.5
34.5	24.4	30.2	38.4	44.7	50.0	60.9	69.8	22.4
46	27.8	33.6	41.8	48.1	53.4	64.3	73.2	26.9

[1] Clearance determined by Table 235-1, Rule 235B1a.

NOTE: Clearance = 0.3 (kV) + 4.04 $\sqrt{S - 24}$, where S is the sag in inches.

m

Table 235-3
Horizontal Clearances Between Line Conductors AWG No. 2 or Larger at Supports, Based on Sags
(See also Rule 235B1b.)

Voltage between conductors (kV)	Sag (mm)							
	915	1220	1830	2240	3048	4572	6096	But not less than[1]
	Horizontal clearance (mm)							
2.4	370	425	515	590	660	805	925	300
4.16	385	440	530	605	675	820	940	300
12.47	445	500	600	670	735	880	1005	345
13.2	450	510	595	675	740	885	1010	350
13.8	455	510	600	680	745	890	1015	355
14.4	460	515	605	685	750	895	1020	365
24.94	540	595	685	760	835	975	1100	470
34.5	615	670	760	835	905	1050	1170	570
46	705	755	845	925	995	1140	1260	685

[1] Clearance determined by Table 235-1, Rule 235B1a.

NOTE: Clearance = 7.6 (kV) + $8\sqrt{2.12S}$, where S is the sag in millimeters.

in

Table 235-3
Horizontal Clearances Between Line Conductors AWG No. 2 or Larger at Supports, Based on Sags
(See also Rule 235B1b.)

Voltage between conductors (kV)	Sag (in)							
	36	48	72	96	120	180	240	But not less than[1]
	Horizontal clearance (in)							
2.4	14.6	16.7	20.2	23.3	26.0	31.7	36.5	12.0
4.16	15.1	17.3	20.8	23.8	26.5	32.2	37.0	12.0
12.47	17.6	19.7	23.3	26.3	29.0	34.7	39.5	13.5
13.2	17.8	20.0	23.5	26.5	29.2	34.9	39.7	13.8
13.8	18.0	20.1	23.7	26.7	29.4	35.1	39.9	14.0
14.4	18.2	20.3	23.8	26.9	29.6	35.3	40.1	14.3
24.94	21.3	23.5	27.0	30.0	32.8	38.4	43.2	18.5
34.5	24.2	26.4	29.9	32.9	35.6	41.3	46.1	22.4
46	27.7	29.8	33.3	36.4	39.1	44.8	49.6	26.9

[1] Clearance determined by Table 235-1, Rule 235B1a.

NOTE: Clearance = 0.3 (kV) + $8\sqrt{S/12}$, where S is the sag in inches.

2. Suspension Insulators
Where suspension insulators are used and are not restrained from movement, the clearance between conductors shall be increased so that one string of insulators may swing transversely throughout a range of insulator swing up to its maximum design swing angle without reducing the values given in Rule 235B1. The maximum design swing angle shall be based on a 290 Pa (6 lb/ft^2) wind on the conductor at final sag at 15 °C (60 °F). This may be reduced to a 190 Pa (4 lb/ft^2) wind in areas sheltered by buildings, terrains, or other obstacles. The displacement of the wires, conductors, and cables shall include deflection of flexible structures and fittings, where such deflection would reduce the horizontal clearance between two wires, conductors, or cables.
3. Alternate Clearances for Different Circuits Where One or Both Circuits Exceed 98 kV AC to Ground or 139 kV DC to Ground
The clearances specified in Rules 235B1 and 235B2 may be reduced for circuits with known switching-surge factors but shall be not less than the clearances derived from the following computations. For these computations, communication conductors and cables, guys, messengers, neutral conductors meeting Rule 230E1, and supply cables meeting Rule 230C1 shall be considered line conductors at zero voltage.
   a. Clearance
      (1) The alternate clearance shall be maintained under the expected loading conditions and shall be not less than the electrical clearance between conductors of different circuits computed from the following equation. For convenience, clearances for typical system voltages are shown in Table 235-4.

$$D = 1.00\left[\frac{V_{L-L} \cdot (PU) \cdot a}{500K}\right]^{1.667} b \quad \text{(m)}$$

$$= 3.28\left[\frac{V_{L-L} \cdot (PU) \cdot a}{500K}\right]^{1.667} b \quad \text{(ft)}$$

where

V_{L-L} = maximum ac crest operating voltage in kilovolts between phases of different circuits or maximum dc operating voltage between poles of different circuits. If the phases are of the same phase and voltage magnitude, one phase conductor shall be considered grounded,

PU = maximum switching-surge factor expressed in per-unit peak operating voltage between phases of different circuits and defined as a switching-surge level between phases for circuit breakers corresponding to 98% probability that the maximum switching surge generated per breaker operation does not exceed this surge level, or the maximum anticipated switching-surge level generated by other means, whichever is greater,

a = 1.15, the allowance for three standard deviations,

b = 1.03, the allowance for nonstandard atmospheric conditions,

K = 1.4, the configuration factor for a conductor-to-conductor gap.

      (2) The value of D shall be increased 3% for each 300 m (1000 ft) in excess of 450 m (1500 ft) above mean sea level.
   b. Limit
The clearance derived from Rule 235B3a shall not be less than the basic clearances given in Table 235-1 computed for 169 kV ac.

C. Vertical Clearance Between Line Conductors
All line wires, conductors, and cables located at different levels on the same supporting structure shall have vertical clearances not less than the following:

1. Basic Clearance for Conductors of Same or Different Circuits

The clearances given in Table 235-5 shall apply to line wires, conductors, or cables of 0 to 50 kV attached to supports. No value is specified for clearances between conductors of the same circuit exceeding 50 kV.

EXCEPTION 1: Line wires, conductors, or cables on vertical racks or separate brackets placed vertically and meeting the requirements of Rule 235G may have spacings as specified in that rule.

EXCEPTION 2: Where communication service drops cross under supply conductors on a common crossing structure, the clearance between the communication conductor and an effectively grounded supply conductor may be reduced to 100 mm (4 in) provided the clearance between the communication conductor and supply conductors not effectively grounded meets the requirements of Rule 235C as appropriate.

EXCEPTION 3: Supply service drops of 0 to 750 V running above and parallel to communication service drops may have a spacing of not less than 300 mm (12 in) at any point in the span including the point of their attachment to the building or structure being served provided that the nongrounded conductors are insulated and that the clearance as otherwise required by this rule is maintained between the two service drops at the pole.

EXCEPTION 4: This rule does not apply to conductors of the same circuit meeting Rule 230D.

2. Additional Clearances

Greater clearances than those required (by Rule 235C1) and given in Table 235-5 shall be provided under the following conditions. The increases are cumulative where more than one is applicable.

a. Voltage Related Clearances

(1) For voltages between 50 and 814 kV, the clearance between line wires, conductors, or cables of different circuits shall be increased 10 mm (0.4 in) per kilovolt in excess of 50 kV.

EXCEPTION: For voltages to ground exceeding 98 kV ac or 139 kV dc, clearances less than those required above are permitted for systems with known switching-surge factors. (See Rule 235C3.)

(2) The increase in clearance for voltages in excess of 50 kV specified in Rule 235C2a(1) shall be increased 3% for each 300 m (1000 ft) in excess of 1000 m (3300 ft) above mean sea level.

(3) All clearances for lines over 50 kV shall be based on the maximum operating voltage.

(4) No value is specified for clearances between conductors of the same circuit.

b. Sag-Related Clearances

(1) Line wires, conductors, and cables supported at different levels on the same structures shall have vertical clearances at the supporting structures so adjusted that the clearance at any point in the span shall be not less than any of the following:

(a) For voltages less than 50 kV between conductors, 75% of that required at the supports by Table 235-5.

EXCEPTION 1: Neutral conductors meeting Rule 230E1 and supply cables meeting Rule 230C1 (including their support brackets) running above and parallel to communication cables where the supply neutral or messenger is bonded to the communication messenger at intervals specified in Rule 92C, may have a clearance of 300 mm (12 in) at any point in the span provided that a clearance of 0.75 m (30 in) is maintained between the supply conductors and cables and the communication cables at the supporting poles.

EXCEPTION 2: When all parties involved are in agreement, for supply conductors of different utilities, vertical clearance at any point in the span need not exceed 75% of the values required at the supports for the same utility by Table 235-5.

(b) For voltages more than 50 kV between conductors, the value specified in Rule 235C2b(1)(a) increased in accordance with Rule 235C2a.

Table 235-4
Electrical Clearances in Rule 235B3a(1)
[Add 3% for each 300 m (1000 ft) in excess of 450 m (1500 ft) above mean sea level.]

Maximum operating voltage phase to phase (kV)	Switching surge factor (per unit)	Switching surge (kV)	Electrical component of clearance	
			(m)	(ft)
242	2.6 or less	890 or less	1.90	6.3 [1]
	2.8	958	2.20	7.2
	3.0	1027	2.50	8.1
	3.2 or more	1095 or more	2.70	8.8 [2]
362	1.8	893 or less	1.95	6.4 [1]
	2.0	1024	2.45	8.0
	2.2	1126	2.90	9.5
	2.4	1228	3.3	10.9
	2.6	1330	3.8	12.5
	2.7 or more	1382 or more	3.9	12.8 [2]
550	1.6	1245	3.4	11.2
	1.8	1399	4.1	13.6
	2.0	1555	4.9	16.2
	2.2	1711	5.8	19.0
	2.3	1789 or more	5.8	19.1 [2]
800	1.6	1810	6.3	20.8
	1.8	2037	7.7	25.3
	1.9 or more	2149 or more	8.4	27.4 [2]

[1] Limited by Rule 235B3b.
[2] Need not be greater than specified in Rules 235B1 and 235B2.

m

Table 235-5
Vertical Clearance Between Conductors at Supports
(When using column and row headings, voltages are phase to ground for effectively grounded circuits and those other circuits where all ground faults are cleared by promptly de-energizing the faulted section, both initially and following subsequent breaker operations. See the definitions section for voltages of other systems. See also Rules 235C1, 235C2, and 235F.)

Conductors and cables usually at lower levels	Conductors and cables usually at upper levels			
	Supply cables meeting Rule 230C1, 2, or 3; neutral conductors meeting Rule 230E1; communications cables meeting Rule 224A2a (m)	Open supply conductors		
		0 to 8.7 kV (m)	Over 8.7 to 50 kV	
			Same utility [7] (m)	Different utilities [8] (m)
1. Communication conductors and cables				
a. Located in the communication space	1.00 [1, 5]	1.00	1.00	1.00 plus 0.01 per kV [6] over 8.7 kV
b. Located in the supply space	0.41 [9, 10]	0.41 [10]	1.00 [10]	1.00 plus 0.01 per kV [6] over 8.7 kV
2. Supply conductors and cables				
a. Open conductors 0 to 750 V; supply cables meeting Rule 230C1, 2, or 3; neutral conductors meeting Rule 230E1	0.41 [9]	0.41 [2]	0.41 plus 0.01 per kV [6] over 8.7 kV	1.00 plus 0.01 per kV [6] over 8.7 kV
b. Open conductors over 750 V to 8.7 kV		0.41 [2]	0.41 plus 0.01 per kV [4, 6] over 8.7 kV	1.00 plus 0.01 per kV A[6] over 8.7 kV
c. Open conductors over 8.7 to 22 kV				
(1) If worked on energized with live-line tools and adjacent circuits are neither de-energized nor covered with shields or protectors			0.41 plus 0.01 per kV [6] over 8.7 kV	1.00 plus 0.01 per kV [6] over 8.7 kV
(2) If not worked on energized except when adjacent circuits (either above or below) are de-energized or covered by shields or protectors, or by the use of live-line tools not requiring line workers to go between live wires			0.41 plus 0.01 per kV [3, 6] over 8.7 kV	0.41 plus 0.01 per kV [3, 6] over 8.7 kV
d. Open conductors exceeding 22 kV, but not exceeding 50 kV			0.41 plus 0.01 per kV [3, 6] over 8.7 kV	0.41 plus 0.01 per kV [3, 6] over 8.7 kV

[1] Where supply circuits of 600 V or less, with transmitted power of 5000 W or less, are run below communication circuits in accordance with Rule 220B2, the clearance may be reduced to 0.41 m.

[2] Where conductors are operated by different utilities, a vertical clearance of not less than 1.00 m is recommended.

[3] These values do not apply to conductors of the same circuit or circuits being carried on adjacent conductor supports.

[4] May be reduced to 0.41 m where conductors are not worked on energized except when adjacent circuits (either above or below) are de-energized or covered by shields or protectors, or by the use of live-line tools not requiring line workers to go between live wires.

[5] May be reduced to 0.75 m for supply neutrals meeting Rule 230E1, entirely dielectric fiber-optic supply cables meeting Rule 230F1b and cables meeting Rule 230C1 where the supply neutral or messenger is bonded to the communication messenger.

[6] The greater of phasor difference or phase-to-ground voltage; see Rule 235A3.

[7] Example: For a 50 kV-to-ground conductor above a 22 kV-to-ground conductor, the required clearance is 0.41 m + 0.64 m = 1.05 m when the conductors are 180° out of phase.

[8] Example: For a 50 kV-to-ground conductor above a 22 kV-to-ground conductor, the required clearance is 1.00 m + 0.64 m = 1.64 m when the conductors are 180° out of phase.

[9] No clearance is specified between neutral conductors meeting Rule 230E1 and insulated communication cables located in the supply space and supported by an effectively grounded messenger.

[10] No clearance is specified between fiber-optic—supply cables meeting Rule 230F1b and supply cables and conductors.

in

Table 235-5
Vertical Clearance Between Conductors at Supports
(When using column and row headings, voltages are phase to ground for effectively grounded circuits and those other circuits where all ground faults are cleared by promptly de-energizing the faulted section, both initially and following subsequent breaker operations.
See the definitions section for voltages of other systems.
See also Rules 235C1, 235C2, and 235F.)

Conductors and cables usually at lower levels	Conductors and cables usually at upper levels			
	Supply cables meeting Rule 230C1, 2, or 3; neutral conductors meeting Rule 230E1; communications cables meeting Rule 224A2a (in)	Open supply conductors		
		0 to 8.7 kV (in)	Over 8.7 to 50 kV: Same utility [7] (in)	Over 8.7 to 50 kV: Different utilities [8] (in)
1. Communication conductors and cables				
a. Located in the communication space	40 [1, 5]	40	40	40 plus 0.4 per kV [6] over 8.7 kV
b. Located in the supply space	16 [9, 10]	16 [10]	40 [10]	40 plus 0.4 per kV [6] over 8.7 kV
2. Supply conductors and cables				
a. Open conductors 0 to 750 V; supply cables meeting Rule 230C1, 2, or 3; neutral conductors meeting Rule 230E1	16 [9]	16 [2]	16 plus 0.4 per kV [6] over 8.7 kV	40 plus 0.4 per kV [6] over 8.7 kV
b. Open conductors over 750 V to 8.7 kV		16 [2]	16 plus 0.4 per kV [4, 6] over 8.7 kV	40 plus 0.4 per kV [6] over 8.7 kV
c. Open conductors over 8.7 to 22 kV				
(1) If worked on energized with live-line tools and adjacent circuits are neither de-energized nor covered with shields or protectors			16 plus 0.4 per kV [6] over 8.7 kV	40 plus 0.4 per kV [6] over 8.7 kV
(2) If not worked on energized except when adjacent circuits (either above or below) are de-energized or covered by shields or protectors, or by the use of live-line tools not requiring line workers to go between live wires			16 plus 0.4 per kV [3, 6] over 8.7 kV	16 plus 0.4 per kV [3, 6] over 8.7 kV
d. Open conductors exceeding 22 kV, but not exceeding 50 kV			16 plus 0.4 per kV [3, 6] over 8.7 kV	16 plus 0.4 per kV [3, 6] over 8.7 kV

[1] Where supply circuits of 600 V or less, with transmitted power of 5000 W or less, are run below communication circuits in accordance with Rule 220B2, the clearance may be reduced to 16 in.

[2] Where conductors are operated by different utilities, a vertical clearance of not less than 40 in is recommended.

[3] These values do not apply to conductors of the same circuit or circuits being carried on adjacent conductor supports.

[4] May be reduced to 16 in where conductors are not worked on energized except when adjacent circuits (either above or below) are de-energized or covered by shields or protectors, or by the use of live line tools not requiring line workers to go between live wires.

[5] May be reduced to 30 in for supply neutrals meeting Rule 230E1, entirely dielectric fiber-optic supply cables meeting Rule 230F1b, and cables meeting Rule 230C1 where the supply neutral or messenger is bonded to the communication messenger.

[6] The greater of phasor difference or phase-to-ground voltage; see Rule 235A3.

[7] Example: For a 50 kV-to-ground conductor above a 22 kV-to-ground conductor, the required clearance is 16 in [+] 25 in [=] 41 in when the conductors are 180° out of phase.

[8] Example: For a 50 kV-to-ground conductor above a 22 kV-to-ground conductor, the required clearance is 40 in [+] 25 in [=] 65 in when the conductors are 180° out of phase.

[9] No clearance is specified between neutral conductors meeting Rule 230E1 and insulated communication cables located in the supply space and supported by an effectively grounded messenger.

[10] No clearance is specified between fiber-optic—supply cables meeting Rule 230F1b and supply cables and conductors.

(c) For purposes of this determination the vertical clearances required in Rule 235C2b(1)(a) and (b) apply to the following conductor temperature and loading conditions whichever produces the greater vertical clearance at the structure when:

i. The upper conductor is at final sag at the maximum operating temperature for which the line is designed to operate and the lower conductor is at final sag at the same ambient conditions as the upper conductor without electrical loading, and without ice loading, or

ii. The upper conductor is at final sag at 32 °F with the radial thickness of ice, if any, specified in Rule 250B for the loading district concerned and the lower conductor is at final sag at the same ambient conditions as the upper conductor without electrical loading, and without ice loading.

EXCEPTION: Rule 235C2b(1)(c) does not apply to conductors of the same utility when the conductors are the same size and type, and are installed at the same sag and tension. However, where experience in an area has shown that different ice conditions do occur between the upper and lower conductors, then Rule 235C2b(1)(c) shall apply.

(2) Sags should be readjusted when necessary to accomplish the foregoing, but not reduced sufficiently to conflict with the requirements of Rule 261H2. In cases where conductors of different sizes are strung to the same sag for the sake of appearance or to maintain unreduced clearance throughout storms, the chosen sag should be such as will keep the smallest conductor involved in compliance with the sag requirements of Rule 261H2.

(3) For span lengths in excess of 45 m (150 ft), vertical clearance at the structure between open supply conductors and communication cables or conductors shall be adjusted so that under conditions of conductor temperature of 15 °C (60 °F), no wind displacement and final unloaded sag, no open supply conductor of over 750 V but less than 50 kV shall be lower in the span than a straight line joining the points of support of the highest communication cable or conductor.

EXCEPTION: Effectively grounded supply conductors associated with systems of 50 kV or less need meet only the provisions of Rule 235C2b(1).

3. Alternate Clearances for Different Circuits Where One or Both Exceed 98 kV AC, or 139 kV DC to Ground

The clearances specified in Rules 235C1 and 235C2 may be reduced for circuits with known switching-surge factors, but shall not be less than the crossing clearances required by Rule 233C3.

4. Communication Worker Safety Zone

The clearance specifications of Rules 235C and 238 create a *communication worker safety zone* between the facilities located in the supply space and facilities located in the communication space, both at the structure and in the span between structures. Except as allowed by Rules 238C, 238D, and 239, no supply or communication facility shall be located in the communication worker safety zone.

D. Diagonal Clearance Between Line Wires, Conductors, and Cables Located at Different Levels on the Same Supporting Structure

No wire, conductor, or cable may be closer to any other wire, conductor, or cable than defined by the dashed line in Fig 235-1, where *V* and *H* are determined in accordance with other parts of Rule 235.

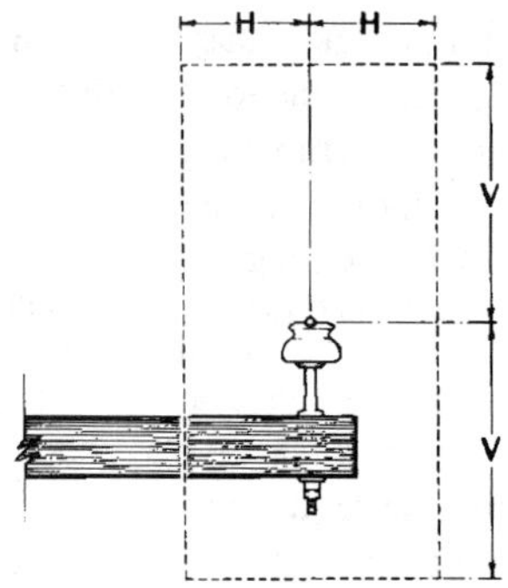

V = Vertical clearance

H = Horizontal clearance

Fig 235-1
Clearance Diagram for Energized Conductor

m

Table 235-6
Clearance in Any Direction From Line Conductors to Supports and to Vertical or Lateral Conductors, Span, or Guy Wires Attached to the Same Support
[See also Rule 235E1, 235E3b(2), and 235I.]

Clearance of line conductors from	Communication lines: In general (mm)	Communication lines: On jointly used structures (mm)	Supply lines, Circuit phase-to-phase voltage: 0 to 8.7 kV (mm)	Supply lines: Over 8.7 to 50 kV (mm)	Supply lines: Over 50 to 814 kV [4, 9] (mm)
1. Vertical and lateral conductors:					
a. Of the same circuit	75	75	75	75 plus 6.5 per kV over 8.7 kV	no value specified
b. Of other circuits[12, 13]	75	75	150[5]	150 plus 10 per kV over 8.7 kV	580 plus 10 per kV over 50 kV
2. Span or guy wires,[11] or messengers attached to same structure:					
a. When parallel to line	75[7]	150[1, 7]	300[1]	300 plus 10 per kV over 8.7 kV	740 plus 10 per kV over 50 kV
b. Anchor guys	75[7]	150[1, 7]	150[1]	150 plus 6.5 per kV over 8.7 kV	410 plus 6.5 per kV over 50 kV
c. All other	75[7]	150[1, 7]	150	150 plus 10 per kV over 8.7 kV	580 plus 10 per kV over 50 kV
3. Surface of support arms	75[2]	75[2]	75[6, 8]	75 plus 5 per kV over 8.7 kV[6, 8, 10]	280 plus 5 per kV over 50 kV
4. Surface of structures:					
a. On jointly used structures	—	125[2]	125[3, 6, 8]	125 plus 5 per kV over 8.7 kV[6, 8, 10]	330 plus 5 per kV over 50 kV
b. All other	75[2]	—	75[6, 8]	75 plus 5 per kV over 8.7 kV[6, 8, 10]	280 plus 5 per kV over 50 kV

[1] For guy wires, if practical. For clearances between span wires and communication conductors, see Rule 238C.

On jointly used structures, guys that pass within 300 mm of supply conductors, and also pass within 300 mm of communication cables, shall be protected with a suitable insulating covering where the guy passes the supply conductors, unless the guy is effectively grounded or insulated with a strain insulator at a point below the lowest supply conductor and above the highest communication cable.

The clearance from an insulated or effectively grounded guy to a communication cable may be reduced to 75 mm when abrasion protection is provided on the guy or communication cable.

[2] Communication conductors may be attached to supports on the sides or bottom of crossarms or surfaces of poles with less clearance.

[3] This clearance applies only to supply conductors at the support below communication conductors, on jointly used structures.

Where supply conductors are above communication conductors, this clearance may be reduced to 75 mm.

[4] All clearances for line over 50 kV shall be based on the maximum operating voltage. For voltages exceeding 814 kV, the clearance shall be determined by the alternate method given by Rule 235E3.

[5] For supply circuits of 0 to 750 V, this clearance may be reduced to 75 mm.

[6] A neutral conductor meeting Rule 230E1 may be attached directly to the structure surface.

[7] Guys and messengers may be attached to the same strain plates or to the same through bolts.

[8] For open supply circuits of 0 to 750 V and supply cables of all voltages meeting Rule 230C1, 2, or 3, this clearance may be reduced to 25 mm. No clearance is specified for phase conductors of such cables where they are physically restrained by a suitable bracket from abrasion against the pole.

[9] The additional clearance for voltages in excess of 50 kV specified in Table 235-6 shall be increased 3% for each 300 m in excess of 1000 m above mean sea level.

[10] Where the circuit is effectively grounded and the neutral conductor meets Rule 230E1, phase-to-neutral voltage shall be used to determine the clearance from the surface of support arms and structures.

[11] These clearances may be reduced by not more than 25% to a guy insulator, provided that full clearance is maintained to its metallic end fittings and the guy wires. The clearance to an insulated section of a guy between two insulators may be reduced by not more than 25% provided that full clearance is maintained to the uninsulated portion of the guy.

[12] Phase-to-phase voltages shall be determined according to Rule 235A3.

[13] These clearances apply to communication antennas operated at a radio frequency of 0 to 750 V. Also see Rules 235I4 and Rule 239.

ft

Table 235-6
Clearance in Any Direction From Line Conductors to Supports and to Vertical or Lateral Conductors, Span, or Guy Wires Attached to the Same Support

[See also Rule 235E1, 235E3b(2), and 235I.]

Clearance of line conductors from	Communication lines		Supply lines Circuit phase-to-phase voltage		
	In general (in)	On jointly used structures (in)	0 to 8.7 kV (in)	Over 8.7 to 50 kV (in)	Over 50 to 814 kV[4, 9] (in)
1. Vertical and lateral conductors:					
a. Of the same circuit	3	3	3	3 plus 0.25 per kV over 8.7 kV	no value specified
b. Of other circuits[12, 13]	3	3	6[5]	6 plus 0.4 per kV over 8.7 kV	23 plus 0.4 per kV over 50 kV
2. Span or guy wires,[11] or messengers attached to same structure:					
a. When parallel to line	3[7]	6[1, 7]	12[1]	12 plus 0.4 per kV over 8.7 kV	29 plus 0.4 per kV over 50 kV
b. Anchor guys	3[7]	6[1, 7]	6[1]	6 plus 0.25 per kV over 8.7 kV	16 plus 0.25 per kV over 50 kV
c. All other	3[7]	6[1, 7]	6	6 plus 0.4 per kV over 8.7 kV	23 plus 0.4 per kV over 50 kV
3. Surface of support arms	3[2]	3[2]	3[6, 8]	3 plus 0.2 per kV over 8.7 kV[6, 8, 10]	11 plus 0.2 per kV over 50 kV
4. Surface of structures:					
a. On jointly used structures	—	5[2]	5[3, 6, 8]	5 plus 0.2 per kV over 8.7 kV[6, 8, 10]	13 plus 0.2 per kV over 50 kV
b. All other	3[2]	—	3[6, 8]	3 plus 0.2 per kV over 8.7 kV[6, 8, 10]	11 plus 0.2 per kV over 50 kV

[1] For guy wires, if practical. For clearances between span wires and communication conductors, see Rule 238C.

On jointly used structures, guys that pass within 12 in of supply conductors, and also pass within 12 in of communication cables, shall be protected with a suitable insulating covering where the guy passes the supply conductors, unless the guy is effectively grounded or insulated with a strain insulator at a point below the lowest supply conductor and above the highest communication cable.

The clearance from an insulated or effectively grounded guy to a communication cable may be reduced to 3 in when abrasion protection is provided on the guy or communication cable.

[2] Communication conductors may be attached to supports on the sides or bottom of crossarms or surfaces of poles with less clearance.

[3] This clearance applies only to supply conductors at the support below communication conductors, on jointly used structures.

Where supply conductors are above communication conductors, this clearance may be reduced to 3 in.

[4] All clearances for line over 50 kV shall be based on the maximum operating voltage. For voltages exceeding 814 kV, the clearance shall be determined by the alternate method given by Rule 235E3.

[5] For supply circuits of 0 to 750 V, this clearance may be reduced to 3 in.

[6] A neutral conductor meeting Rule 230E1 may be attached directly to the structure surface.

[7] Guys and messengers may be attached to the same strain plates or to the same through bolts.

[8] For open supply circuits of 0 to 750 V and supply cables of all voltages meeting Rule 230C1, 2 or 3, this clearance may be reduced to 1 in. No clearance is specified for phase conductors of such cables where they are physically restrained by a suitable bracket from abrasion against the pole.

[9] The additional clearance for voltages in excess of 50 kV specified in Table 235-6 shall be increased 3% for each 1000 ft in excess of 3300 ft above mean sea level.

[10] Where the circuit is effectively grounded and the neutral conductor meets Rule 230E1, phase-to-neutral voltage shall be used to determine the clearance from the surface of support arms and structures.

[11] These clearances may be reduced by not more than 25% to a guy insulator, provided that full clearance is maintained to its metallic end fittings and the guy wires. The clearance to an insulated section of a guy between two insulators may be reduced by not more than 25% provided that full clearance is maintained to the uninsulated portion of the guy.

[12] Phase-to-phase voltages shall be determined according to Rule 235A3.

[13] These clearances apply to communication antennas operated at a radio frequency of 0 to 750 V. Also see Rules 235I4 and Rule 239.

E. Clearances in Any Direction From Line Conductors to Supports, and to Vertical or Lateral Conductors, Span, or Guy Wires Attached to the Same Support

1. Fixed Supports

   Clearances shall be not less than those given in Table 235-6.

   EXCEPTION: For voltages exceeding 98 kV ac to ground or 139 kV dc to ground, clearances less than those required by Table 235-6 are permitted for systems with known switching-surge factor. (See Rule 235E3.)

2. Suspension Insulators

   Where suspension insulators are used and are not restrained from movement, the clearance shall be increased so that the string of insulators may swing transversely throughout a range of insulator swing up to its maximum design swing angle without reducing the values given in Rule 235E1. The maximum design swing angle shall be based on a 290 Pa (6 lb/ft^2) wind on the conductor at final sag at 15 °C (60 °F). This may be reduced to a 190 Pa (4 lb/ft^2) wind in areas sheltered by buildings, terrain, or other obstacles. The displacement of the wires, conductors, and cables shall include deflection of flexible structures and fittings, where such deflection would reduce the clearance.

3. Alternate Clearances for Voltages Exceeding 98 kV AC to Ground or 139 kV DC to Ground

   The clearances specified in Rules 235E1 and 235E2 may be reduced for circuits with known switching-surge factors but shall not be less than the following:

   a. Alternate Clearances to Anchor Guys, Surge-Protection Wires, and Vertical or Lateral Conductors

      The alternate clearances shall be not less than the crossing clearances required by Rule 233B2 and Rules 233C3a and 233C3b for the conductor voltages concerned. For the purpose of this rule, anchor guys and surge-protection wires shall be assumed to be at ground potential. The limits of Rule 235E3b(2) shall apply to the clearance derived from Rules 233C3a and 233C3b.

b. Alternate Clearance to Surface of Support Arms and Structures

(1) Alternate Clearance

(a) Basic Computation

The alternate clearances shall be maintained under the expected loading conditions and shall be not less than the electrical clearances computed from the following equation. For convenience, clearances for typical system voltages are shown in Table 235-7.

$$D = 1.00\left[\frac{V \cdot (PU) \cdot a}{500K}\right]^{1.667} b \qquad \text{(m)}$$

$$D = 39.37\left[\frac{V \cdot (PU) \cdot a}{500K}\right]^{1.667} b \qquad \text{(in)}$$

where

V = maximum ac crest operating voltage to ground or maximum dc operating voltage to ground in kilovolts,

PU = maximum switching-surge factor expressed in per-unit peak voltage to ground and defined as a switching-surge level for circuit breakers corresponding to 98% probability that the maximum switching surge generated per breaker operation does not exceed this surge level, or the maximum anticipated switching-surge level generated by other means, whichever is greater,

a = 1.15, the allowance for three standard deviations with fixed insulator supports;
= 1.05, the allowance for one standard deviation with free-swinging manipulators,

b = 1.03, the allowance for nonstandard atmospheric conditions,

K = 1.2, the configuration factor for conductor-to-tower window.

(b) Atmospheric Correction

The value of D shall be increased 3% for each 300 m (1000 ft) in excess of 450 m (1500 ft) above mean sea level.

(2) Limits

The alternate clearance shall not be less than the clearance of Table 235-6 for 169 kV ac. The alternate clearance shall be checked for adequacy of clearance to workers and increased, if necessary, where work is to be done on the structure while the circuit is energized. (Also see Part 4.)

F. Clearances Between Circuits of Different Voltage Classifications Located in the Supply Space on the Same Support Arm

Circuits of any one voltage classification as given in Table 235-5 may be maintained in the supply space on the same support arm with supply circuits of the next consecutive voltage classification only under one or more of the following conditions:

1. If they occupy positions on opposite sides of the structure.
2. If in bridge-arm or sidearm construction, the clearance is not less than the climbing space required for the higher voltage concerned and provided for in Rule 236.
3. If the higher-voltage conductors occupy the outer positions and the lower-voltage conductors occupy the inner positions.
4. If series lighting or similar supply circuits are ordinarily dead during periods of work on or above the support arm concerned.
5. If the two circuits concerned are communication circuits (located in the supply space in accordance with Rule 224A), or one circuit is such a communication circuit and the other is a supply circuit of less than 8.7 kV, and both are owned by the same utility, provided they are installed as specified in Rule 235F1 or 235F2.

Table 235-7
Clearance in Any Direction From Line Conductors to Supports
(See also Rule 235E3b and 235E3b(1)(a).)

Maximum operating voltage phase to phase (kV)	Switching-surge factor (per unit)	Switching surge (kV)	Computed clearance to supports			
			Fixed		Free swinging at maximum angle	
			(m)	(in)	(m)	(in)
242	2.4	474	0.89	35 [1]	0.89	35 [1]
	2.6	514	1.00	40	0.89	35
	2.8	553	1.14	45	0.97	38
	3.0	593	1.27	50 [2]	1.10	43
	3.2	632	1.27	50 [2]	1.10	48
362	1.6	473	0.89	35 [1]	0.89	35 [1]
	1.8	532	1.07	42	0.89	36
	2.0	591	1.27	50	1.20	48
	2.2	650	1.50	59	1.20	51
	2.4	709	1.73	68	1.20	59
	2.5	739	1.85	73 [2]	1.20	63
550	1.6	719	1.80	70	1.50	60
	1.8	808	2.15	85	1.85	73
	2.0	898	2.60	101	2.20	87
	2.2	988	2.80	111 [2]	2.20	101
800	1.6	1045	3.3	130	2.80	111
	1.8	1176	4.0	158	2.80	135
	1.9	1241	4.1	161 [2]	3.8	148
	2.0	1306	4.1	161 [2]	4.1	161 [2]

[1] Limited by Rule 235E3(b)(2).
[2] Need not be greater than specified in Rules 235E1 and 235E2.

G. Conductor Spacing: Vertical Racks

Conductors or cables may be carried on vertical racks or separate brackets other than wood placed vertically on one side of the structure and securely attached thereto with less clearance between the wires, conductors, or cables than specified in Rule 235C if all the following conditions are met:

1. The voltage shall not be more than 750 V, except supply cables and conductors meeting Rules 230C1 or 230C2, which may carry any voltage.
2. Conductors shall be of the same material or materials, except that different materials may be used if their sag tension characteristics and arrangement are such that the spacing specified in Rule 235G3 is maintained under all service conditions.
3. Vertical spacing between conductors shall be not less than the following:

Span length		Vertical spacing between conductors	
(m)	(ft)	(mm)	(in)
0 to 45	0 to 150	100	4
45 to 60	150 to 200	150	6
60 to 75	200 to 250	200	8
75 to 90	250 to 300	300	12

EXCEPTION: The vertical spacing may be reduced where the conductors are held apart by intermediate spacers, but may not be less than 100 mm (4 in).

H. Clearance and Spacing Between Communication Conductors, Cables, and Equipment
   1. The spacing between messengers supporting communication cables should be not less than 300 mm (12 in) except by agreement between the parties involved.
   2. The clearances between the conductors, cables, and equipment of one communiation utility to those of another, anywhere in the span, shall be not less than 100 mm (4 in), except by agreement between the parties involved.

I. Clearances in Any Direction From Supply Line Conductors to Communication Antennas in the Supply Space Attached to the Same Supporting Structure
   1. General
      Communication antennas located in the supply space shall be installed and maintained only by personnel authorized and qualified to work in the supply space in accordance with the applicable rules of Sections 42 and 44. See also Rule 224A.
   2. Communication Antenna
      The clearance between a communication antenna operated at a radio frequency of 0 to 750 V and a supply line conductor shall be not less than the value given in Table 235-6, row 1b.
      NOTE: The antenna functions as a rigid, vertical or lateral open wire communication conductor.
   3. Equipment Case That Supports a Communication Antenna
      The clearance between an equipment case that supports a communication antenna and a supply line conductor shall be not less than the value given in Table 235-6, Row 4a.
   4. Vertical or Lateral Communication Conductors and Cables Attached to a Communication Antenna
      The clearance between a supply line conductor and the vertical or lateral communciation conductor and cable attached to a communication antenna shall be not less than the value given in Rule 239.

236. Climbing Space

The following requirements apply only to portions of structures that workers ascend.

A. Location and Dimensions
   1. A climbing space having the horizontal dimensions specified in Rule 236E shall be provided past any conductors, support arms, or other parts.
   2. The climbing space need be provided on one side or corner of the support only.
   3. The climbing space shall extend vertically past any conductor or other part between levels above and below the conductor as specified in Rules 236E, F, G, and I, but may otherwise be shifted from any side or corner of the support to any other side or corner.

B. Portions of Supporting Structures in Climbing Space
   Portions of the supporting structure, when included in one side or corner of the climbing space, are not considered to obstruct the climbing space.

C. Support Arm Location Relative to Climbing Space
   RECOMMENDATION: Support arms should be located on the same side of the pole.
   EXCEPTION: This recommendation does not apply where double crossarms are used on any pole or where crossarms on any pole are not all parallel.

D. Location of Equipment Relative to Climbing Space

1. All supply and communication equipment such as transformers, regulators, capacitors, cable terminals (potheads), amplifiers, loading coils, surge arresters, switches, etc., when located below conductors or other attachments, shall be mounted outside of the climbing space.
2. All exposed ungrounded conductive parts of luminaires and their supports that are not insulated from current-carrying parts shall be maintained at not less than 500 mm (20 in) from the surface of their supporting structure.

   EXCEPTION 1: This may be reduced to 125 mm (5 in) if located on the side of the structure opposite the designated climbing space.

   EXCEPTION 2: This does not apply where the equipment is located at the top or other vertical portion of the structure that is not subject to climbing.

E. Climbing Space Between Conductors

Climbing space between conductors shall be not less than the horizontal dimensions specified in Table 236-1. These dimensions are intended to provide a clear climbing space of 600 mm (24 in) while the conductors bounding the climbing space are covered with temporarily installed protective covering rated for the voltage involved. The climbing space shall be provided both along and across the line and shall be projected vertically not less than 1.0 m (40 in) above and below the limiting conductors. Where communication conductors are above supply conductors of more than 8.7 kV to ground or 15 kV line to line, the climbing space shall be projected vertically at least 1.50 m (60 in) above the highest supply conductors.

EXCEPTION 1: This rule does not apply if it is the unvarying practice of the employers concerned to prohibit employees from ascending beyond the conductors or equipment of a given line or structure unless the conductors or equipment are de-energized and grounded per Rule 444D.

EXCEPTION 2: For supply conductors carried on a structure in a position below communications facilities in the manner permitted in Rule 220B2, the climbing space need not extend more than 600 mm (2 ft) above such supply space.

EXCEPTION 3: If the conductors are owned, operated, or maintained by the same utility, the climbing space may be provided by temporarily moving the line conductors using live-line tools.

Table 236-1
Horizontal Clearance Between Conductors Bounding the Climbing Space

(All voltages are between the two conductors bounding the climbing space except for communication conductors, which are voltage to ground. Where the two conductors are in different circuits, the voltage between conductors shall be the arithmetic sum of the voltages of each conductor to ground for a grounded circuit, or phase to phase for an ungrounded conductor. See also Rule 236E.)

Character of conductors adjacent to climbing space	Voltage of conductors	Horizontal clearance between conductors bounding the climbing space [3]							
		On structures used solely by				On jointly used structures			
		Communication conductors		Supply conductors		Supply conductors above communication conductors		Communication conductors above supply conductors [1]	
		(m)	(in)	(m)	(in)	(m)	(in)	(m)	(in)
1. Communication conductors	0 to 150 V	0.60	no requirements		—		(note 2)	0.60	no requirements
	exceeding 150 V		24 recommended		—		(note 2)		24 recommended
2. Supply cables meeting Rule 230C1	all voltages				—		(note 2)		no requirements
3. Supply cables meeting Rule 230C2 or 3	all voltages	—	—	0.60	24	0.60	24	0.75	30
4. Open supply line conductors and supply cables meeting Rule 230D	0 to 750 V	—	—	0.60	24	0.60	24	0.75	30
	750 V to 15 kV	—	—	0.75	30	0.75	30	0.75	30
	15 kV to 28 kV	—	—	0.90	36	0.90	36	0.90	36
	28 kV to 38 kV	—	—	1.00	40	1.00	40		
	38 kV to 50 kV	—	—	1.17	46	1.17	46		
	50 kV to 73 kV	—	—	1.40	54	1.40	54		
	exceeding 73 kV	—	—	>1.40	>54				

[1] This relation of levels in general is not desirable and should be avoided.

[2] Climbing space shall be the same as required for the supply conductors immediately above, with a maximum of 0.75 m (30 in) except that a climbing space of 0.41 m (16 in) across the line may be employed for communication cables or conductors where the only supply conductors at a higher level are secondaries (0 to 750 V) supplying airport or airway marker lights or crossing over the communication line and attached to the pole top or to a pole-top extension fixture.

[3] Attention is called to the operating requirements of Rules 441A and 446C, Part 4, of this code.

F. Climbing Space on Buckarm Construction

Method of Providing Climbing Space on Buckarm Construction

The full width of climbing space shall be maintained on buckarm construction and shall extend vertically in the same position at least 1.0 m (40 in) [or 1.50 m (60 in) where required by Rule 236E] above and below any limiting conductor.

A six-pin crossarm having pin spacing of 370 mm (14.5 in) may be used to provide a 750 mm (30 in) climbing space on one corner of a junction pole by omitting the pole pins on all arms, and inserting pins midway between the remaining pins so as to give a spacing of 185 mm (7.25 in), provided that all of the following conditions are met:

(1) Circuits are less than 8.7 kV to ground or 15 kV line to line

(2) Span lengths do not exceed 45 m (150 ft)

(3) Sags do not exceed 380 mm (15 in) for wires of AWG No. 2 and larger sizes, or 750 mm (30 in) for wires smaller than AWG No. 2

(4) Each conductor on the end of every arm is tied to the same side of its insulator

(5) The spacing on the next pole is not less than 370 mm (14.5 in)

G. Climbing Space Past Longitudinal Runs Not on Support Arms

The full width of climbing space shall be provided past longitudinal runs and shall extend vertically in the same position from 1.0 m (40 in) below the run to a point 1.0 m (40 in) above [or 1.50 m (60 in) where required by Rule 236E]. The width of climbing space shall be measured from the longitudinal run concerned. Longitudinal runs on racks, or cables on messengers, are not considered as obstructing the climbing space if all wires concerned are covered by rubber protective equipment or otherwise guarded as an unvarying practice before workers climb past them. This does not apply where communication conductors are above the longitudinal runs concerned.

EXCEPTION 1: If a supply longitudinal run is placed on the side or corner of the supporting structure where climbing space is provided, the width of climbing space shall be measured horizontally from the center of the structure to the nearest supply conductors on support arms, under both of the following conditions:

(1) Where the longitudinal run consists of open supply conductors carrying not more than 750 V, or supply cables and conductors meeting Rule 230C, all voltages; and is supported close to the structure as by brackets, racks, or pins close to the structure.

(2) Where the nearest supply conductors on support arms are parallel to and on the same side of the structure as the longitudinal run and within 1.20 m (4 ft) above or below the run.

EXCEPTION 2: For supply conductors carried on a structure in a position below communications facilities in the manner permitted in Rule 220B2, the climbing space need not extend more than 600 mm (2 ft) above such supply space.

EXCEPTION 3: A service drop less than 750 V and meeting Rule 230C is not considered to obstruct the climbing space if all conductors concerned are covered by rubber protective equipment or otherwise guarded as an unvarying practice before workers climb past them, provided that such a service drop is (1) not closer to the longitudinal run at the point of attachment than the diameter of the pole plus 125 mm (5 in) measured horizontally, and (2) not closer than 950 mm (38 in) measured horizontally to the longitudinal run at a point 750 mm (30 in) on the run measured from the point of attachment at the pole. See Fig 236-1.

H. Climbing Space Past Vertical Conductors

Vertical runs physically protected by suitable conduit or other protective covering and securely attached without spacers to the surface of the line structure are not considered to obstruct the climbing space.

I. Climbing Space Near Ridge-Pin Conductors

The climbing space specified in Table 236-1 shall be provided above the top support arm to the ridge-pin conductor but need not be carried past it.

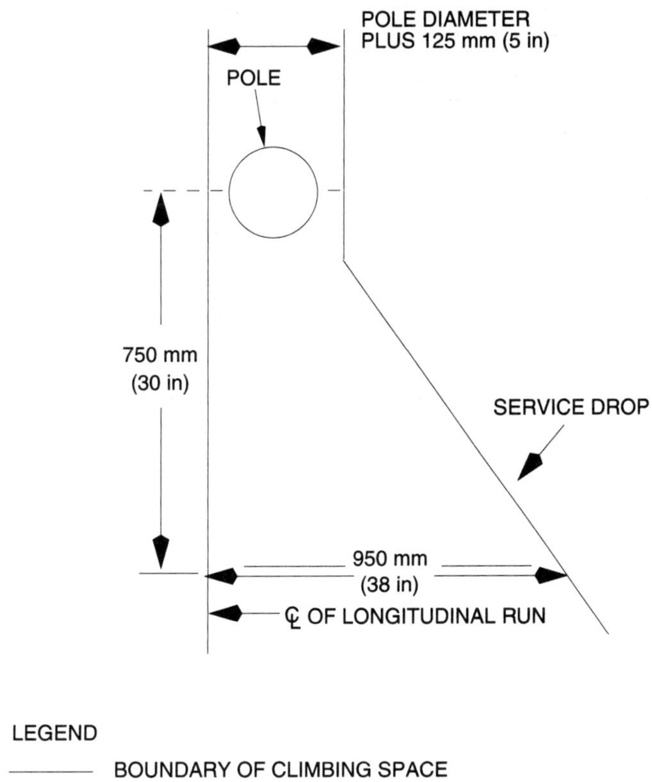

Fig 236-1
Rule 236G, Exception 3

237. Working Space

A. Location of Working Spaces

Working spaces shall be provided on the climbing face of the structure at each side of the climbing space.

B. Dimensions of Working Spaces

1. Along the Support Arm

The working space shall extend from the climbing space to the outmost conductor position on the support arm.

2. At Right Angles to the Support Arm

The working space shall have the same dimension as the climbing space (see Rule 236E). This dimension shall be measured horizontally from the face of the support arm.

3. Vertically

The working space shall have a height not less than that required by Rule 235 for the vertical separation of line conductors carried at different levels on the same support.

C. Location of Vertical and Lateral Conductors Relative to Working Spaces

The working spaces shall not be obstructed by vertical or lateral conductors. Such conductors shall be located on the opposite side of the pole from the climbing side or on the climbing side of the pole at a distance from the support arm at least as great as the width of climbing space required for the highest voltage conductors concerned. Vertical conductors enclosed in suitable conduit may be attached on the climbing side of the structure.

D. Location of Buckarms Relative to Working Spaces

Buckarms may be used under any of the following conditions, provided the climbing space is maintained. Climbing space may be obtained as in Rule 236F.

1. Standard Height of Working Space

   Lateral working space of the height required by Table 235-5 shall be provided between the crossing or tap line conductors attached to the buckarm and the main line conductors. This may be accomplished by increasing the spacing between the line support arms, as shown in Fig 237-1.

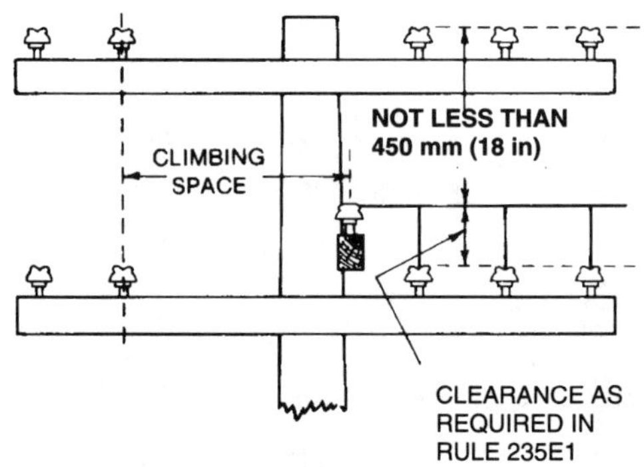

Fig 237-1
Obstruction of Working Space by Buckarm

2. Reduced Height of Working Space

   Where no circuits exceeding 8.7 kV to ground or 15 kV line to line are involved and the clearances of Rules 235B1a and 235B1b are maintained, conductors supported on buckarms may be placed between line conductors having normal vertical spacing, even though such buckarms obstruct the normal working space, provided that a working space of not less than 450 mm (18 in) in height is maintained either above or below line conductors and buckarm conductors.

   EXCEPTION: The above working space may be reduced to 300 mm (12 in) if both of the following conditions exist:

   (a) Not more than two sets of the line arms and buckarms are involved.

   (b) Working conditions are rendered safe by providing rubber protective equipment or other suitable devices to insulate and cover line conductors and equipment that are not being worked upon.

E. Guarding of Energized Equipment

Exposed energized parts of equipment such as switches, circuit breakers, surge arresters, etc., shall be enclosed or guarded if all of the following conditions apply:

1. The equipment is located below the top conductor support.
2. The equipment is located on the climbing side of the structure.
3. The requirements of Rule 441, Part 4, of this code cannot be met.

F. Working Clearances From Energized Equipment

All parts of equipment such as switches, fuses, transformers, surge arresters, luminaires and their support brackets, etc., or other connections that may require operation or adjustment while energized and exposed at such times, shall be so arranged with respect to each other, other equipment, vertical and lateral conductors, and portions of the supporting structure, including supporting platforms or structural members, that in adjustment or operation no portion of the body, including the hands, need be brought closer to any exposed energized parts or conductors than permitted in Part 4, Rules 441 or 446 of this code.

238. Vertical Clearance Between Certain Communications and Supply Facilities Located on the Same Structure

A. Equipment
For the purpose of measuring clearances under this rule, *equipment* shall be taken to mean non-current-carrying metal parts of equipment, including metal supports for cables or conductors, and metal support braces that are attached to metal supports or are less than 25 mm (1 in) from transformer cases or hangers that are not effectively grounded.

B. Clearances in General
Vertical clearances between supply conductors and communications equipment, between communication conductors and supply equipment, and between supply and communications equipment shall be as specified in Table 238-1, except as provided in Rule 238C.

C. Clearances for Span Wires or Brackets
Span wires or brackets carrying luminaires, traffic signals, or trolley conductors shall have at least the vertical clearances in millimeters or inches from communications equipment set forth in Table 238-2.

D. Clearance of Drip Loops of Luminaire or Traffic Signal Brackets
If a drip loop of conductors entering a luminaire bracket or traffic signal bracket from the surface of the structure is above a communication cable, the lowest point of the loop shall be at least 300 mm (12 in) above communication cable or through bolt.
EXCEPTION: The above clearance may be reduced to 75 mm (3 in) if the loop is covered by a suitable nonmetallic covering that extends at least 50 mm (2 in) beyond the loop.

E. Communication Worker Safety Zone
The clearance specifications of Rules 235C and 238 create a communication worker safety zone between the facilities located in the supply space and facilities located in the communication space, both at the structure and in the span between structures. Except as allowed by Rules 238C, 238D, and 239, no supply or communication facility shall be located in the communication worker safety zone.

Table 238-1
Vertical Clearance Between Supply Conductors and Communications Equipment, Between Communication Conductors and Supply Equipment, and Between Supply and Communications Equipment

(Voltages are phase to ground for effectively grounded circuits and those other circuits where all ground faults are cleared by promptly de-energizing the faulted section, both initially and following subsequent breaker operations. See the definitions section for voltages of other systems. See also Rule 238B.)

Supply voltage (kV)	Vertical clearance (m)	Vertical clearance (in)
1. Grounded conductor and messenger hardware and supports	0.75	30
2. 0 to 8.7	1.00	40 [1]
3. Over 8.7	1.00 plus 0.01 per kV over 8.7 kV	40 plus 0.4 per kV [1] over 8.7 kV

[1] Where non-current-carrying parts of supply equipment are effectively grounded and the associated neutral meeting Rule 230E1 or supply cables meeting Rule 230C1 (including the support brackets) are bonded to communication messengers at intervals meeting Rule 92C through out well-defined areas and where communication is at lower levels, clearances may be reduced to 0.75 m (30 in).

Table 238-2
Vertical Clearance of Span Wires and Brackets From Communication Lines

(See also Rule 238C.)

	Carrying luminaires or traffic signals				Carrying trolley conductors			
	Not effectively grounded		Effectively grounded		Not effectively grounded		Effectively grounded	
	(mm)	(in)	(mm)	(in)	(mm)	(in)	(mm)	(in)
Above communication support arms	500	20 [1]	500	20 [1]	500	20 [1]	500	20 [1]
Below communication support arms	1000	40 [3]	600	24	600	24	600	24
Above messengers carrying communication cables	500	20 [1]	100	4	300	12	100	4
Below messengers carrying communication cables	1000	40 [4]	100	4	300	12	100	4
From terminal box of communication cable	500	20 [1]	100	4	300	12 [2]	100	4
From communication brackets, bridle wire rings, or drive hooks	410	16 [1]	100	4	100	4	100	4

[1] This may be reduced to 300 mm (12 in) for either span wires or metal parts of brackets at points 1.0 m (40 in) or more from the structure surface.

[2] Where it is not practical to obtain a clearance of 300 mm (1 ft) from terminal boxes of communication cables, all metal parts of terminals shall have the greatest possible separation from fixtures or span wires including all supporting screws and bolts of both attachments.

[3] This may be reduced to 600 mm (24 in) for luminaires and traffic signals operating at less than 150 V to ground.

[4] This may be reduced to 500 mm (20 in) for luminaires and traffic signals operating at less than 150 V to ground.

239. Clearance of Vertical and Lateral Facilities From Other Facilities and Surfaces on the Same Supporting Structure

Vertical and lateral conductors shall have the clearances and separations required by this rule from other facilities or surfaces on the same supporting structure.

A. General

1. Grounding conductors, surge-protection wires, neutral conductors meeting Rule 230E1, insulated communication conductors and cables, supply cables meeting Rule 230C1, insulated supply cables of 0 to 750 V, or conduits may be placed directly on the supporting structure. These conductors, wires, cables, and conduits shall be securely attached to the surface of the structure. Cables not in conduit shall be installed in such a manner as to avoid abrasion at the point of attachment.
2. Supply circuits of the same or next voltage classification may be placed in the same duct, if each circuit or set of wires is enclosed in a metal sheath.
3. Paired communication conductors in rings may be attached directly to a structure or messenger.
4. Insulated supply circuits of 600 V or less and not exceeding 5000 W may be placed in the same cable with control circuits with which they are associated.
5. The term nonmetallic covering as used in Rule 239 refers to material other than a cable jacket that provides an additional barrier against physical contact.
6. Where guarding and protection are required by other rules, either conduit or U-guards may be used. A backing plate shall be used with a U-guard unless the U-guard fits tightly to the supporting structure surface.

B. Location of Vertical or Lateral Conductors Relative to Climbing Spaces, Working Spaces, and Pole Steps

Vertical or lateral conductors shall be located so that they do not obstruct climbing spaces, or lateral working spaces between line conductors at different levels, or interfere with the safe use of pole steps.

EXCEPTION: This rule does not apply to portions of the structure that workers do not ascend while the conductors in question are energized.

NOTE: See Rule 236H for vertical runs in conduit or other protective covering.

C. Conductors Not in Conduit

Conductors not encased in conduit shall have the same clearances from conduits as from other surfaces of structures.

D. Guarding and Protection Near Ground

1. Where within 2.45 m (8 ft) of the ground, or other areas readily accessible to the public, all vertical conductors and cables shall be guarded.

   EXCEPTION: This guarding may be omitted from grounding conductors used to ground multi-grounded circuits or equipment (communications or supply); communication cables or conductors; armored cables; or conductors used solely to protect structures from lightning.
2. Where guarding is required, conductors and cables shall be protected by covering or conduit that gives suitable mechanical protection.
3. When guarding is not required, conductors and cables shall be securely attached to the surface of the structure or to standoff brackets and located, where practical, on the portion of the structure having the least exposure to mechanical damage.
4. Guards that completely enclose grounding conductors of lightning-protection equipment shall be of nonmetallic materials or shall be bonded at both ends to the grounding conductor.

E. Requirements for Vertical and Lateral Supply Conductors on Supply Line Structures or Within Supply Space on Jointly Used Structures

1. General Clearances

   In general, clearances shall be not less than the values specified in Table 239-1 or Rule 235E.
2. Special Cases

   The following requirements apply only to portions of a structure that workers ascend while the conductors in question are energized.

a. General

If open-line conductors are within 1.20 m (4 ft) of the pole, vertical conductors shall be run in one of the following ways:

(1) The clearance between open vertical conductors and pole-surface shall be not less than that given in Table 239-2 within the zone specified in the table.

(2) Within the zone above and below open supply conductors as given in Table 239-2, vertical and lateral conductors or cables attached to the surface of the structure shall be enclosed in nonmetallic conduit or protected by non-metallic covering.

EXCEPTION: This conduit or covering may be omitted from grounding conductors, surge-protection wires, neutral conductors meeting Rule 230E1, supply cables meeting Rule 230C1, and jacketed multiple-conductor supply cables of 0 to 750V, where such conductors or cable are not in the climbing space.

For the purpose of this exception, a jacketed multiple-conductor cable is a cable with a jacket enclosing the entire cable assembly.

b. Conductors to Luminaires

On structures used only for supply lines or on jointly used structures where the luminaire bracket is 1.0 m (40 in) or more above all communication attachments, open wires may be run from the supply line arm directly to the head of a luminaire, provided the clearances of Table 239-1 are obtained and the open wires are securely supported at both ends.

F. Requirements for Vertical and Lateral Communication Conductors on Communication Line Structures or Within the Communication Space on Jointly Used Structures

1. Clearances From Communication Conductors

The clearances of uninsulated vertical and lateral communication conductors from other communication conductors (except those in the same ring run) and from guy, span, or messenger wires shall be not less than those given in Rule 235E1, Table 235-6.

2. Clearances From Supply Conductors

The vertical clearance of vertical and lateral insulated communication conductors shall be not less than 1.0 m (40 in) from any supply conductors (other than vertical runs or luminaire leads) of 8.7 kV or less, or 1.0 m (40 in) plus 10 mm (0.4 in) per kV over 8.7 to 50 kV. The additional clearance of Rule 235C2 is applicable when the voltage exceeds 50 kV.

EXCEPTION 1: May be reduced to 0.75 m (30 in) from supply neutrals meeting Rule 230E1, cables meeting Rule 230C1, and fiber optic-supply cables where the supply neutral or messenger is bonded to the communication messenger.

EXCEPTION 2: These clearances do not apply where the supply circuits involved are those carried in the manner specified in Rule 220B2.

G. Requirements for Vertical Supply Conductors and Cables Passing Through Communication Space on Jointly Used Line Structures

1. Guarding—General

Vertical supply conductors or cables attached to the structure shall be guarded with suitable conduit or covering from 1.0 m (40 in) above the highest communication attachment to 1.80 m (6 ft) below the lowest communication attachment.

EXCEPTION 1: This conduit or covering may be omitted from neutral conductors meeting Rule 230E1, supply cables meeting Rule 230C1, and jacketed multiple-conductor supply cables of 0 to 750 V, where such conductors or cable are not in the climbing space.

For the purpose of this exception, a jacketed multiple-conductor cable is a cable with a jacket enclosing the entire cable assembly.

EXCEPTION 2: This conduit or covering may be omitted from supply grounding conductors where there are no trolley or ungrounded traffic signal attachments, or ungrounded street lighting fixtures located below the communication attachment, provided:

(a) The grounding conductor is directly (metallically) connected to a conductor which forms part of an effective grounding system,

(b) The grounding conductor has no connection to supply equipment between the grounding electrode and the effectively grounded conductor unless the supply equipment has additional connections to the effectively grounded conductor, and

(c) The grounding conductor is bonded to grounded communication facilities at that structure.

2. Cables and Conductors in Conduit or Covering
   Cables and conductors of all voltages may be run in a nonmetallic conduit or covering or in a grounded metallic conduit or covering in accordance with Rule 239A1. Where a metallic conduit or covering is not bonded to grounded communications facilities at that structure, such metal conduit or covering shall have a nonmetallic covering from 1.0 m (40 in) above the highest communication attachment to 1.80 m (6 ft) below the lowest communication attachment.
3. Protection Near Trolley, Ungrounded Traffic Signal, or Ungrounded Luminaire Attachments
   Vertical supply conductors or cables attached to the structure shall be guarded with suitable nonmetallic conduit or covering on structures that carry a trolley or ungrounded traffic signal attachment or an ungrounded luminaire that is attached below the communication cable. The cable shall be protected with nonmetallic covering from 1.0 m (40 in) above the highest communication wire to 1.80 m (6 ft) below the lowest trolley attachment or ungrounded luminaire fixture or ungrounded traffic signal attachment.
4. Aerial Services
   Where supply cables are used as aerial services, the point where such cables leave the structure shall be at least 1.0 m (40 in) above the highest or 1.0 m (40 in) below the lowest communication attachment. Within the communication space, all splices and connections in the energized phase conductors shall be insulated.
5. Clearance From Through Bolts and Other Metal Objects
   Vertical runs of supply conductors or cables shall have a clearance of not less than 50 mm (2 in) from exposed through bolts and other exposed metal objects attached thereto that are associated with communication line equipment.

   EXCEPTION: Vertical runs of effectively grounded supply conductors may have a clearance of 25 mm (1 in).

H. Requirements for Vertical Communication Conductors Passing Through Supply Space on Jointly Used Structures
   All vertical runs of communication conductors passing through supply space shall be installed as follows:
   1. Metal-Sheathed Communication Cables
      Vertical runs of metal-sheathed communication cables shall be covered with suitable nonmetallic material, where they pass trolley feeders or other supply line conductors. This nonmetallic covering shall extend from a point 1.0 m (40 in) above the highest trolley feeders or other supply conductors, to a point 1.80 m (6 ft) below the lowest trolley feeders or other supply conductors, but need not extend below the top of any mechanical protection that may be provided near the ground.

      EXCEPTION 1: Communication cables may be run vertically on the pole through space occupied by railroad signal supply circuits in the lower position, as permitted in Rule 220B2, without covering within the supply space.

      EXCEPTION 2: Covering is not required in the supply space on metallic or concrete supporting structures.
   2. Communication Conductors
      Vertical runs of insulated communication conductors shall be covered with suitable nonmetallic material, to the extent required for metal-sheathed communication cables in Rule 239H1, where such conductors pass trolley feeders or supply conductors.

      EXCEPTION 1: Communication conductors may be run vertically on the structure through space occupied by railroad-signal supply circuits in the lower position, as permitted in Rule 220B2, without covering within the supply space.

      EXCEPTION 2: Covering is not required in the supply space on metallic or concrete supporting structures.
   3. Communication Grounding Conductors
      Vertical communication grounding conductors shall be covered with suitable nonmetallic material between points at least 1.80 m (6 ft) below and 1.0 m (40 in) above any trolley feeders or other supply line conductors by which they pass.

      EXCEPTION 1: Communication grounding conductors may be run vertically on the structure though space occupied by railroad-signal supply circuits in the lower position, as permitted in Rule 220B2, without covering within the supply space.

      EXCEPTION 2: Covering is not required in the supply space on metallic or concrete supporting structures.

4. Clearance From Through Bolts and Other Metal Objects
   Vertical runs of communication conductors or cables shall have a clearance of one-eighth of the pole circumference but not less than 50 mm (2 in) from exposed through bolts and other exposed metal objects attached thereto that are associated with supply line equipment.
   EXCEPTION: Vertical runs of effectively grounded communication cables may have a clearance of 25 mm (1 in).

I. Operating Rods
Effectively grounded or insulated operating rods of switches are permitted to pass through the communication space, but shall be located outside of the climbing space.

J. Additional Rules for Standoff Brackets
1. Standoff brackets may be used to support the conduit(s). Cable insulation appropriate for the intended service is required; non-metallic conduit shall not be used to meet basic insulation requirements.
   NOTE: See Rule 217A2.
2. Standoff brackets may be used to support the following types of cable enclosed within a single outer jacket or sheath (cable only without conduit):
   a. Communication
   b. 230C1a supply (any voltage)
   c. Supply less than 750 V
   NOTE: See Rule 217A2.

m

Table 239-1
Clearance of Open Vertical and Lateral Conductors
(Circuit Phase-to-Phase Voltage. See also Rules 239E1 and 239E2b.)

Clearance of open vertical and lateral conductors	0 to 8.7 kV (mm)	Over 8.7 to 50 kV (mm)	Over 50 kV [4] (mm)
From surfaces of supports	75 [1, 2]	75 plus 5 per kV over 8.7 kV	280 plus 5 per kV over 50 kV
From span, guy, and messenger wires [5]	150	150 plus 10 per kV over 8.7 kV [3]	580 plus 10 per kV over 50 kV [3]

[1] A neutral conductor meeting Rule 230E1 may be attached directly to the structure surface.

[2] For supply circuits of 0 to 750 V, this clearance may be reduced to 25 mm.

[3] Multiplier may be reduced to 6.5 mm/kV for anchor guys.

[4] The additional clearance for voltages in excess of 50 kV specified in Table 239-1 shall be increased 3% for each 300 m in excess of 1000 m above mean sea level.

[5] These clearances may be reduced by not more than 25% to a guy insulator, provided that full clearance is maintained to its metallic end fittings and the guy wires. The clearance to an insulated section of a guy between two insulators may be reduced by not more than 25% provided that full clearance is maintained to the uninsulated portion of the guy.

in

Table 239-1
Clearance of Open Vertical and Lateral Conductors
(Circuit Phase-to-Phase Voltage. See also Rules 239E1 and 239E2b.)

Clearance of open vertical and lateral conductors	0 to 8.7 kV (in)	Over 8.7 to 50 kV (in)	Over 50 kV [4] (in)
From surfaces of supports	3 [1, 2]	3 plus 0.2 per kV over 8.7 kV	11 plus 0.2 per kV over 50 kV
From span, guy, and messenger wires [5]	6	6 plus 0.4 per kV over 8.7 kV [3]	23 plus 0.4 per kV over 50 kV [3]

[1] A neutral conductor meeting Rule 230E1 may be attached directly to the structure surface.

[2] For supply circuits of 0 to 750 V, this clearance may be reduced to 1 in.

[3] Multiplier may be reduced to 0.25 in/kV for anchor guys.

[4] The additional clearance for voltages in excess of 50 kV specified in Table 239-1 shall be increased 3% for each 1000 ft in excess of 3300 ft above mean sea level.

[5] These clearances may be reduced by not more than 25% to a guy insulator, provided that full clearance is maintained to its metallic end fittings and the guy wires. The clearance to an insulated section of a guy between two insulators may be reduced by not more than 25% provided that full clearance is maintained to the uninsulated portion of the guy.

m

Table 239-2
Clearance Between Open Vertical Conductors and Pole Surface

[Voltages are phase to ground for effectively grounded circuits and those other circuits where all ground faults are cleared by promptly de-energizing the faulted section, both initially and following subsequent breaker operations. See the definitions section for voltages of other systems. See also Rule 239E2a(1) and 239E2a(2).]

Voltage (kV)	Distance above and below open supply conductors where clearances apply (m)	Clearance between vertical conductor and pole surface (mm)
0 to 22	1.80	480
22 to 30	1.80	560
30 to 50	1.80	760

in

Table 239-2
Clearance Between Open Vertical Conductors and Pole Surface

[Voltages are phase to ground for effectively grounded circuits and those other circuits where all ground faults are cleared by promptly de-energizing the faulted section, both initially and following subsequent breaker operations. See the definitions section for voltages of other systems. See also Rule 239E2a(1) and 239E2a(2).]

Voltage (kV)	Distance above and below open supply conductors where clearances apply (ft)	Clearance between vertical conductor and pole surface (in)
0 to 22	6	19
22 to 30	6	22
30 to 50	6	30

Section 24.
Grades of Construction

240. General

A. The grades of construction are specified in this section on the basis of the required strengths for safety. Where two or more conditions define the grade of construction required, the grade used shall be the highest one required by any of the conditions.

B. For the purposes of this section, the voltage values for direct-current circuits shall be considered equivalent to the rms values for alternating-current circuits.

241. Application of Grades of Construction to Different Situations

A. Supply Cables
For the purposes of these rules, supply cables are classified by two types as follows:
Type 1—Supply cables conforming to Rules 230C1, 230C2, or 230C3 shall be installed in accordance with Rule 261I.
Type 2—All other supply cables are required to have the same grade of construction as open-wire conductors of the same voltage.

B. Order of Grades
The relative order of grades for supply and communication conductors and supporting structures is B, C, and N, with Grade B being the highest.

C. At Crossings
Wires, conductors, or other cables of one line are considered to be at crossings when they cross over another line, whether or not on a common supporting structure, or when they cross over or overhang a railroad track or the traveled way of a limited access highway. Joint-use or collinear construction in itself is not considered to be at crossings.
1. Grade of Upper Line
Conductors and supporting structures of a line crossing over another line shall have the grade of construction specified in Rules 241C3, 242, and 243.
2. Grade of Lower Line
Conductors and supporting structures of a line crossing under another line need only have the grades of construction that would be required if the line at the higher level were not there.
3. Multiple Crossings
   a. Where a line crosses in one span over two or more other lines, or where one line crosses over a span of a second line, which span in turn crosses a span of a third line, the grade of construction of the uppermost line shall be not less than the highest grade that would be required of either one of the lower lines when crossing the other lower line.
   b. Where communication conductors cross over supply conductors and railroad tracks in the same span, the grades of construction shall be in accordance with Grade B construction. It is recommended that the placing of communication conductors above supply conductors generally be avoided unless the supply conductors are trolley-contact conductors and their associated feeders.

D. Conflicts (see Section 2, Structure Conflict)
The grade of construction of the conflicting structure shall be as required by Rule 243A4.

242. Grades of Construction for Conductors

The grades of construction required for conductors are given in Tables 242-1 and 242-2. For the purpose of these tables certain classes of circuits are treated as follows:

A. Constant-Current Circuit Conductors

The grade of construction for conductors of a constant-current supply circuit involved with a communication circuit and not in Type 1 cable shall be based on either its current rating or on the open-circuit voltage rating of the transformer supplying such circuit, as set forth in Tables 242-1 and 242-2. When the constant current supply circuit is in Type 1 cable, the grade of construction shall be based on its nominal full-load voltage.

B. Railway Feeder and Trolley-Contact Circuit Conductors

Railway feeder and trolley-contact circuit conductors shall be considered as supply conductors for the purpose of determining the required grade of construction.

C. Communication Circuit Conductors Located in the Supply Space

Communication circuit conductors located in the supply space shall have their grade of construction determined as follows:

1. Circuits meeting the requirements of Rule 224A3 may have the same grade of construction as ordinary communication circuits.
2. Circuits not meeting the requirements of Rule 224A3 shall have the same grade of construction as the supply circuits above which they are located.

D. Fire-Alarm Circuit Conductors

Fire-alarm circuit conductors shall meet the strength and loading requirements of communication circuit conductors.

E. Neutral Conductors of Supply Circuits

Supply-circuit neutral conductors, which are effectively grounded throughout their length and are not located above supply conductors of more than 750 V to ground, shall have the same grade of construction as supply conductors of not more than 750 V to ground, except that they need not meet any insulation requirements. Other neutral conductors shall have the same grade of construction as the phase conductors of the supply circuits with which they are associated.

F. Surge-Protection Wires

Surge-protection wires shall be of the same grade of construction as the supply conductors with which they are associated.

Table 242-1
Grades of Construction for Supply Conductors Alone, at Crossing, or on the Same Structures With Other Conductors

(The voltages listed in this table are phase-to-ground values for: effectively grounded ac circuits, two-wire grounded circuits, or center-grounded dc circuits; otherwise phase-to-phase values shall be used. The grade of construction for supply conductors, as indicated across the top of the table, shall also meet the requirements for any lines at lower levels except when otherwise noted.)

<table>
<tr><th rowspan="5">Conductors, tracks, and rights-of-way at lower levels</th><th colspan="13">Supply conductors at higher levels [1]</th></tr>
<tr><th colspan="10">Constant-potential supply conductors</th><th colspan="2" rowspan="3">Constant-current supply conductors</th><th rowspan="3">Communication conductors located in the supply space</th></tr>
<tr><th colspan="2">0 to 750 V</th><th colspan="4">751 V to 8.7 kV</th><th colspan="4">Exceeding 8.7 kV</th></tr>
<tr><th>Urban</th><th>Rural</th><th colspan="2">Urban</th><th colspan="2">Rural</th><th colspan="2">Urban</th><th colspan="2">Rural</th></tr>
<tr><th>Open or Cable</th><th>Open or Cable</th><th>Open</th><th>Cable</th><th>Open</th><th>Cable</th><th>Open</th><th>Cable</th><th>Open</th><th>Cable</th><th>Open</th><th>Cable</th><th>Open or Cable</th></tr>
<tr><td>Exclusive private rights-of-way</td><td>N</td><td>N</td><td>N [2]</td><td>N</td><td>N</td><td>N</td><td>N [2]</td><td>N [2]</td><td>N</td><td>N</td><td colspan="2" rowspan="2">B, C, or N; see Rule 242A</td><td rowspan="2">C or N; see Rule 242C</td></tr>
<tr><td>Common or public rights-of-way</td><td>N</td><td>N</td><td>C</td><td>N</td><td>N</td><td>N</td><td>C [3]</td><td>C</td><td>N</td><td>N</td></tr>
<tr><td>Railroad tracks and limited-access highways [11]</td><td>B</td><td>B</td><td>B</td><td>B</td><td>B</td><td>B</td><td>B</td><td>B</td><td>B</td><td>B</td><td>B</td><td>B</td><td>B</td></tr>
<tr><td>Constant-potential supply conductors 0 to 750 V, open or cable</td><td>N</td><td>N</td><td>C</td><td>N</td><td>N</td><td>N</td><td>C [3]</td><td>C</td><td>C [4]</td><td>N</td><td colspan="2" rowspan="5">B, C, or N; see Rule 242A</td><td rowspan="5">B, C, or N; see Rule 242C</td></tr>
<tr><td>750 V to 8.7 kV Open</td><td>C [5]</td><td>N</td><td>C</td><td>C</td><td>N</td><td>N</td><td>C [3]</td><td>C</td><td>N</td><td>N</td></tr>
<tr><td>Cable</td><td>N</td><td>N</td><td>C</td><td>N</td><td>N</td><td>N</td><td>C [3]</td><td>C</td><td>N</td><td>N</td></tr>
<tr><td>Exceeding 8.7 kV Open</td><td>B [5]</td><td>C [5]</td><td>B</td><td>B</td><td>N</td><td>N</td><td>C [3]</td><td>C</td><td>N</td><td>N</td></tr>
<tr><td>Cable</td><td>C [5]</td><td>N</td><td>C</td><td>N</td><td>N</td><td>N</td><td>C [3]</td><td>C</td><td>N</td><td>N</td></tr>
<tr><td>Constant-current supply conductors: open or cable</td><td colspan="10">B, C, or N; see Rule 242A</td><td colspan="2">B, C, or N: see Rule 242A</td><td>B, C, or N; see Rules 242A and 242C</td></tr>
<tr><td>Communication conductors: open or cable, located in the supply space [10]</td><td colspan="10">B, C, or N; see Rule 242C</td><td colspan="2">B, C, or N; see Rules 242A and 242C</td><td>B, C, or N; see Rule 242C</td></tr>
<tr><td>Communication conductor: urban or rural, open or cable [6]</td><td>N</td><td>N</td><td>B [7, 8]</td><td>C</td><td>B [7, 8]</td><td>C</td><td>B [8]</td><td>C</td><td>B [8]</td><td>C</td><td>B [8, 9]</td><td>C or N; see Rule 242A</td><td>B, C, or N; see Rule 242C</td></tr>
</table>

[1] The words "open" and "cable" appearing in the headings have the following meanings as applied to supply conductors: Cable means the Type 1 cables described in Rule 241A; Open means Type 2 cables described in Rule 241A and open wire.

[2] Lines that can fall outside the exclusive private rights-of-way shall comply with the grades specified for lines not on exclusive private rights-of-way.

[3] Supply conductors shall meet the requirements of Grade B construction if the supply circuits will not be promptly de-energized, both initially and following subsequent breaker operations, in the event of a contact with lower supply conductors or other grounded objects.

[4] Grade N construction may be used if crossing over supply services only.

[5] If the wires are service drops, they may have Grade N sizes and tensions as set forth in Table 263-2.

[6] Grade N construction may be used where the communication conductors consist only of not more than one insulated twisted-pair or parallel-lay conductor, or where service drops only are involved.

[7] Grade C construction may be used if the voltage does not exceed 5.0 kV phase to phase or 2.9 kV phase to ground.

[8] The supply conductors need only meet the requirements of Grade C construction if both of the following conditions are fulfilled:

(a) The supply voltage will be promptly removed from the communications plant by de-energization or other means, both initially and following subsequent circuit-breaker operations in the event of a contact with the communications plant.

(b) The voltage and current impressed on the communications plant in the event of a contact with the supply conductors are not in excess of the safe operating limit of the communications-protective devices.

[9] Grade C construction may be used if the current cannot exceed 7.5 A or the open-circuit voltage of the transformer supplying the circuit does not exceed 2.9 kV.

[10] Communication circuits located below supply conductors shall not affect the grade of construction of the supply circuits.

[11] There is no intent to require Grade B over ordinary streets and highways.

Table 242-2
Grades of Construction for Communication Conductors Alone or in Upper Position of Crossing or on Joint Poles

(The voltages listed in this table are phase-to-ground values for: effectively grounded ac circuits, two-wire grounded circuits, or center-grounded dc circuits; otherwise phase-to-phase values shall be used. The grade of construction for supply conductors, as indicated across the top of the table, shall also meet the requirements for any lines at lower levels except when otherwise noted.) (Placing of communication conductors at higher levels at crossings or on jointly used poles should generally be avoided, unless the supply conductors are trolley-contact conductors and their associated feeders.)

Conductors, tracks, and rights-of-way at lower levels	Communication conductors (communication conductors, rural or urban, open or cable, including those run in the supply space.)
Exclusive private right-of-way	N
Common or public rights-of-way	N
Railroad tracks and limited-access highways[5]	B
Constant-potential supply conductors [1] 0 to 750 V Open or cable	N
750 V to 2.9 kV Open or cable	C
Exceeding 2.9 kV Open Cable	 B C
Constant-current supply conductors [1] 0 to 7.5 A Open [2]	 C
Exceeding 7.5 A Open [2]	B [3]
Communication conductors, open or cable, urban or rural including those run in the supply space	B, C, or N [4]

[1]The words "open" and "cable" appearing in the headlines have the following meaning as applied to supply conductors: Cable means Type 1 cables as described in Rule 241A1; Open means open-wire and also Type 2 cables, as described in Rule 241A2.

[2]Where constant-current circuits are in Type 1 cable, the grade of construction shall be based on the nominal full-load voltage.

[3]Grade C construction may be used if the open-circuit voltage of the transformer supplying the circuit does not exceed 2.9 kV.

[4]See Rule 242C.

[5]There is no intent to require Grade B over ordinary streets and highways.

243. Grades of Construction for Line Supports

A. Structures

The grade of construction shall be that required for the highest grade of conductors supported except as modified by the following:

1. The grade of construction of jointly used structures, or structures used only by communication lines, need not be increased merely because the communication wires carried on such structures cross over trolley-contact conductors of 0 to 750 V to ground.
2. Structures carrying supply service drops of 0 to 750 V to ground shall have a grade of construction not less than that required for supply line conductors of the same voltage.
3. Where the communication lines cross over supply conductors and a railroad in the same span and Grade B is required by Rule 241C3b for the communication conductors, due to the presence of railroad tracks, the grade of the structures shall be B.
4. The grade of construction required for a conflicting structure (first circuit) shall be determined from the requirements of Rule 242 for crossings. The conflicting structure's conductors (first circuit) shall be assumed to cross the other circuit's conductors (second circuit) for the purposes of determining the grade of construction required for the conflicting structure.

   NOTE: The resulting structure grade requirement could result in a higher grade of construction for the structure than for the conductors carried thereon.

B. Crossarms and Support Arms

The grade of construction shall be that required for the highest grade of conductors carried by the arm concerned except as modified by the following:

1. The grade of construction of arms carrying only communication conductors need not be increased merely because the conductors cross over trolley-contact conductors of 0 to 750 V to ground.
2. Arms carrying supply service drops of 0 to 750 V to ground shall have a grade of construction not less than that required for supply line conductors of the same voltage.
3. Where communication lines cross over supply conductors and a railroad in the same span and Grade B is required by Rule 241C3b for the communication conductors due to the presence of railroad tracks, the grade of the arm shall be B.

C. Pins, Armless Construction Brackets, Insulators, and Conductor Fastenings

The grade of construction for pins, armless construction brackets, insulators, and conductor fastenings shall be that required for the conductor concerned except as modified by the following:

1. The grade of construction need not be increased merely because the supported conductors cross over trolley-contact conductors of 0 to 750 V to ground.
2. Supply service drops of 0 to 750 V to ground require only the same grade of construction as supply-line conductors of the same voltage.
3. When Grade B construction is required by Rule 241C3b for the communication conductors due to the presence of railroad tracks, Grade B construction shall be used when supporting communication lines that cross over supply conductors and a railroad in the same span.
4. When communication conductors are required to meet Grade B or C, only the requirements for mechanical strength for these grades are required.
5. Insulators for use on open conductor supply lines shall meet the requirements of Section 27 for all grades of construction.

Section 25.
Loadings for Grades B and C

250. General Loading Requirements and Maps

A. General

1. It is necessary to assume the wind and ice loads that may occur on a line. Two weather loadings are specified in Rules 250B and 250C. Where both rules apply, the required loading shall be the one that has the greater effect.
2. Where construction or maintenance loads exceed those imposed by Rule 250A1, which may occur more frequently in light loading areas, the assumed loadings shall be increased accordingly.
3. It is recognized that loadings actually experienced in certain areas in each of the loading districts may be greater, or in some cases, may be less than those specified in these rules. In the absence of a detailed loading analysis, no reduction in the loadings specified therein shall be made without the approval of the administrative authority.
4. The structural capacity provided by meeting the loading and strength requirements of Sections 25 and 26 provides sufficient capability to resist earthquake ground motions.

B. Combined Ice and Wind Loading

Three general degrees of loading due to weather conditions are recognized and are designated as heavy, medium, and light loading. Figure 250-1 shows the districts where these loadings apply.

NOTE: The localities are classified in the different loading districts according to the relative simultaneous prevalence of the wind velocity and thickness of ice that accumulates on wires. Light loading is for places where little, if any, ice accumulates on wires.

Table 250-1 shows the radial thickness of ice and the wind pressures to be used in calculating loads. Ice is assumed to weigh 913 kg/m^3 (57 lb/ft^3)

C. Extreme Wind Loading

If no portion of a structure or its supported facilities exceeds 18 m (60 ft) above ground or water level, the provisions of this rule are not required, except as specified in Rule 261A1c or Rule 261A2f. Where a structure or its supported facilities exceeds 18 m (60 ft) above ground or water level the structure and its supported facilities shall be designed to withstand the extreme wind load associated with the Basic Wind Speed, as specified by Figure 250-2. The wind pressures calculated shall be applied to the entire structure and supported facilities without ice. The following formula shall be used to calculate wind load.

$$\text{Load in newtons} = 0.613 \cdot (V_{m/s})^2 \cdot k_z \cdot G_{RF} \cdot I \cdot C_d \cdot A(m^2)$$

$$\text{Load in pounds} = 0.00256 \cdot (V_{mi/h})^2 \cdot k_z \cdot G_{RF} \cdot I \cdot C_d \cdot A(ft^2)$$

where:

0.613 0.00256	Ambient Air Density Value, the Constants 0.613, metric, and 0.00256, English, reflects the mass density of air for the standard atmosphere, i.e., temperature of 15 °C (59 °F) and sea level pressure of 760 mm (29.92 in) of mercury,
k_z	Velocity Pressure Exposure Coefficient, as defined in Rule 250C1, Table 250-2,
V	Basic Wind Speed, 3 s gust wind speed in m/s at 10 m (mi/h at 33 ft) above ground, Figure 250-2,
G_{RF}	Gust Response Factor, as defined in Rule 250C2,
I	Importance Factor, 1.0 for utility structures and their supported facilities,
C_d	Shape Factor, as defined in Rule 252B,
A	Projected wind area, m^2 (ft^2).

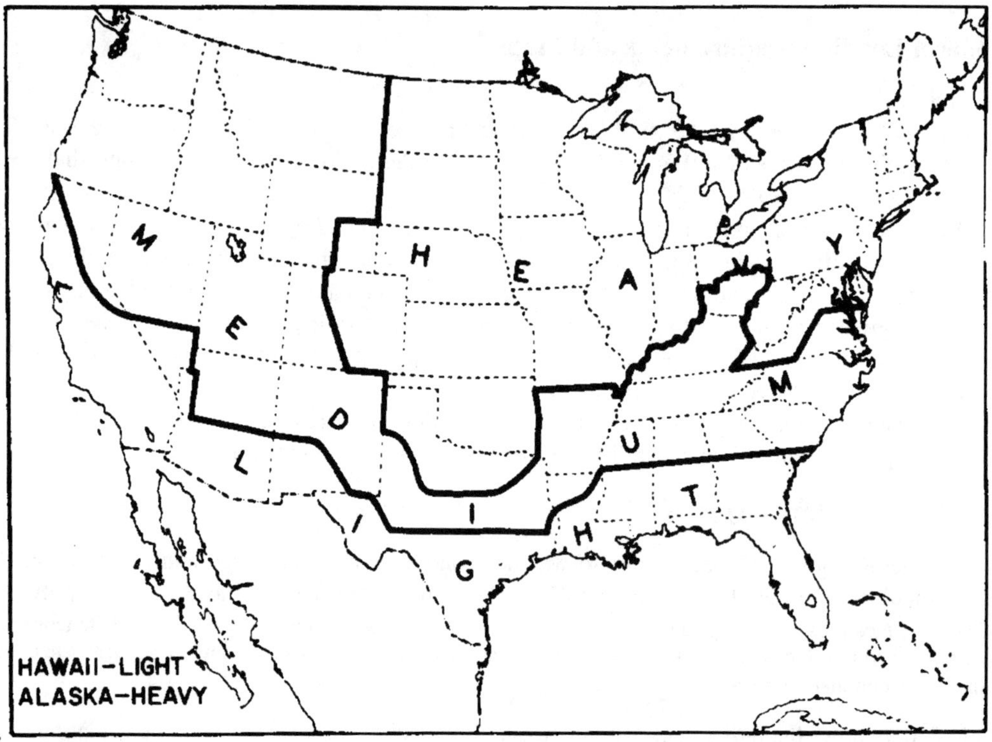

Fig 250-1
General Loading Map of United States with Respect to Loading of Overhead Lines

Table 250-1
Ice, Wind, and Temperature

	Loading districts (For use with Rule 250B)			Extreme wind loading (For use with Rule 250C)
	Heavy	**Medium**	**Light**	
Radial thickness of ice (mm)	12.5	6.5	0	0
(in)	0.50	0.25	0	0
Horizontal wind pressure (Pa)	190	190	430	See Fig 250-2
(lb/ft^2)	4	4	9	See Fig 250-2
Temperature (°C)	–20	–10	–1	+15
(°F)	0	+15	+30	+60

The wind pressure parameters (k_z, V, and G_{RF}) are based on open terrain with scattered obstructions (Exposure Category C as defined in ASCE 7-98). Exposure Category C is the basis of the NESC extreme wind criteria. Topographical features such as ridges, hills, and escarpments may increase the wind loads on site-specific structures. A Topographic Factor, K_{zt}, from ASCE 7-98 may be used to account for these special cases.

1. Velocity Pressure Exposure Coefficient, k_z

   The k_z for the structure is based on the total structure height, h, above the groundline. The k_z for the wire is based on the height, h, of the wire at the structure. The numerical values of k_z, for values of structure and conductor height are tabulated in Table 250-2. In special terrain conditions (i.e., mountainous terrain and canyon) where the height of the conductor above ground at mid-span may be substantially higher than at the attachment point, engineering judgment may be used in determining an appropriate value for the wire k_z. The value of k_z may also be determined using the formulas in Table 250-2. When h is > 75 m (250 ft) the formulas shall be used to determine a k_z value.

2. Gust Response Factor, G_{RF}

   The structure and wire gust response factors are tabulated in Table 250-3. The structure Gust Response Factor, G_{RF}, is determined using the total structure height, h. The wire Gust Response Factor is determined using the height of the wire at the structure, h, and the design wind span, L. The structure and wire gust response factors may also be determined using the formulas in Table 250-3. For values of h > 75 m (250 ft) and L > 600 m (2000 ft) the G_{RF} shall be determined using the formulas in Table 250-3. In special terrain conditions (i.e., mountainous terrain and canyon) where the height of the conductor above ground at midspan may be substantially higher than at the attachment point, engineering judgment may be used in determining an appropriate value for the wire G_{RF}. Wire attachment points that are 18 m (60 ft) or less above ground or water level must be considered if the total structure height is greater than 18 m (60 ft) above ground or water.

Table 250-2
Velocity Pressure Exposure Coefficient k_z, Structure and Wire

Height, h (m)	Height, h (ft)	k_z (Structure)	k_z (Wire)
≤ 10	≤ 33	0.92	1.00
> 10 to 15	> 35 to 50	1.00	1.10
> 15 to 25	> 50 to 80	1.10	1.20
> 25 to 35	> 80 to 115	1.20	1.30
> 35 to 50	> 115 to 165	1.30	1.40
> 50 to 75	> 165 to 250	1.40	1.50
> 75	> 250	Use Formulas	Use Formulas
Formulas (Metric):			
Structure	$k_z = 2.01 \cdot (0.67 \cdot h/275)^{(2/9.5)}$		for 18 m ≤ h ≤ 275 m
Wire	$k_z = 2.01 \cdot (h/275)^{(2/9.5)}$		for 10 m < h ≤ 275 m
Formulas (English):			
Structure	$k_z = 2.01 \cdot (0.67 \cdot h/900)^{(2/9.5)}$		for 60 ft ≤ h ≤ 900 ft
Wire	$k_z = 2.01 \cdot (h/900)^{(2/9.5)}$		for 33 ft < h ≤ 900 ft
h = Structure or Wire height as defined in Rule 250C1			
Minimum k_z = 0.85			
Formulas are for Exposure Category C, ASCE 7-98			

m

Table 250-3
Structure and Wire Gust Response Factors, G_{RF}

Height	Structure	Wire G_{RF}, Span Length, L (m)						
h (m)	G_{RF}	≤75	75<L≤150	150<L≤225	225<L≤300	300<L≤450	450<L≤600	L>600
≤ 10	1.02	0.93	0.86	0.79	0.75	0.73	0.69	(1)
> 10 to 15	0.97	0.88	0.82	0.76	0.72	0.70	0.67	(1)
> 15 to 25	0.93	0.86	0.80	0.75	0.71	0.69	0.66	(1)
> 25 to 35	0.89	0.83	0.78	0.73	0.70	0.68	0.65	(1)
> 35 to 50	0.86	0.82	0.77	0.72	0.69	0.67	0.64	(1)
> 50 to 75	0.83	0.80	0.75	0.71	0.68	0.66	0.63	(1)
> 75	(1)	(1)	(1)	(1)	(1)	(1)	(1)	(1)

Formulas:

Structure $G_{RF} = [1 + (2.7 \cdot E_s \cdot B_s^{0.5})]/k_v^2$

Wire $G_{RF} = [1 + (2.7 \cdot E_w \cdot B_w^{0.5})]/k_v^2$

$E_s = 0.346 \cdot [10/(0.67 \cdot h)]^{1/7}$

$E_w = 0.346 \cdot (10/h)^{1/7}$

$B_s = 1/(1 + 0.375 \cdot h/67)$

$B_w = 1/(1 + 0.8 \cdot L/67)$

Where:

E_w = Wire Exposure Factor

E_s = Structure Exposure Factor

B_w = Dimensionless response term corresponding to the quasi-static background wind loads on the wire

B_s = Dimensionless response term corresponding to the quasi-static background wind loads on the structure

$k_v = 1.43$

h = Structure or Wire height, as defined in Rule 250C2, in meters

L = Design Wind Span, in meters

Formulas are for Exposure Category C, ASCE 7-98

(1) For heights greater than 75 m and/or spans greater than 600 m, the formulas shall be used

ft

Table 250-3
Structure and Wire Gust Response Factors, G_{RF}

Height h (ft)	Structure G_{RF}	Wire G_{RF}, Span Length, L (ft)						
		≤ 250	$250 < L \le 500$	$500 < L \le 750$	$750 < L \le 1000$	$1000 < L \le 1500$	$1500 < L \le 2000$	$L > 2000$[1]
≤ 33	1.02	0.93	0.86	0.79	0.75	0.73	0.69	(1)
> 33 to 50	0.97	0.88	0.82	0.76	0.72	0.70	0.67	(1)
> 50 to 80	0.93	0.86	0.80	0.75	0.71	0.69	0.66	(1)
> 80 to 115	0.89	0.83	0.78	0.73	0.70	0.68	0.65	(1)
> 115 to 165	0.86	0.82	0.77	0.72	0.69	0.67	0.64	(1)
> 165 to 250	0.83	0.80	0.75	0.71	0.68	0.66	0.63	(1)
> 250	(1)	(1)	(1)	(1)	(1)	(1)	(1)	(1)

Formulas:

Structure $G_{RF} = [1 + (2.7 \cdot E_s \cdot B_s^{0.5})]/k_v^2$

Wire $G_{RF} = [1 + (2.7 \cdot E_w \cdot B_w^{0.5})]/k_v^2$

$E_s = 0.346 \cdot [33/(0.67 \cdot h)]^{1/7}$

$E_w = 0.346 \cdot (33/h)^{1/7}$

$B_s = 1/(1 + 0.375 \cdot h/220)$

$B_w = 1/(1 + 0.8 \cdot L/220)$

Where:

E_w = Wire Exposure Factor

E_s = Structure Exposure Factor

B_w = Dimensionless response term corresponding to the quasi-static background wind loads on the wire

B_s = Dimensionless response term corresponding to the quasi-static background wind loads on the structure

$k_v = 1.43$

h = Structure or Wire height, as defined in Rule 250C2, in feet

L = Design Wind Span, in feet

Formulas are for Exposure Category C, ASCE 7-98

(1) For heights greater than 250 ft and/or spans greater than 2000 ft, the formulas shall be used

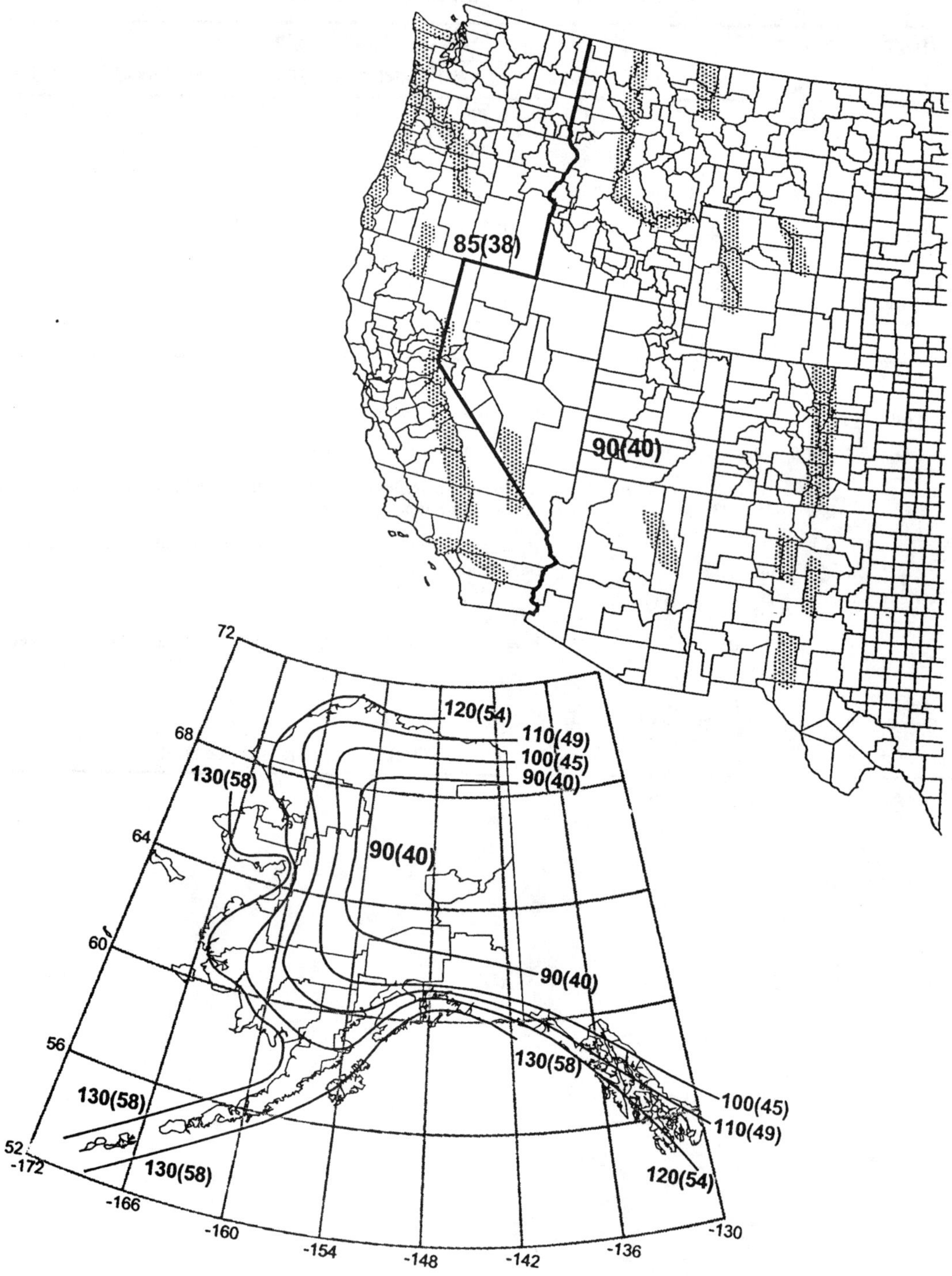

Fig 250-2(a)
Basic Wind Speeds*

*Fig 250-2(a) reprinted with the permission of ASCE Publications, 1801 Alexander Bell Dr., Reston, VA 20191 from ASCE 7-98, Minimum Design Loads for Buildings and Other Structures. Copyright © 1998.

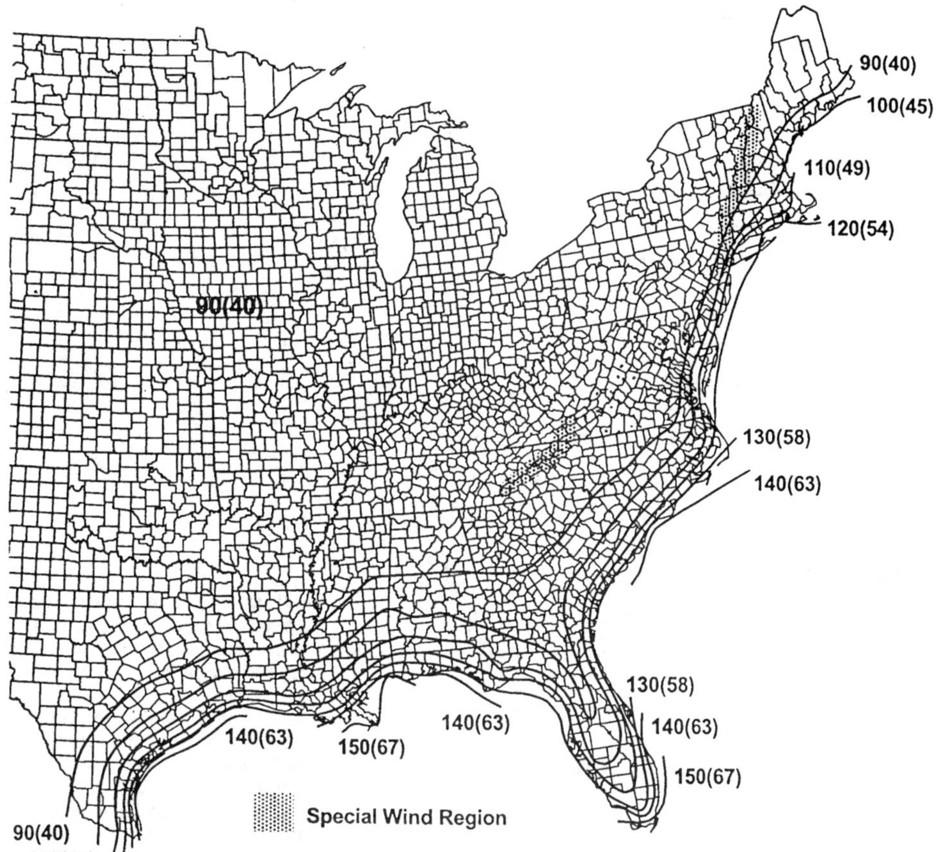

Location	V mph	(m/s)
Hawaii	105	(47)
Puerto Rico	145	(65)
Guam	170	(76)
Virgin Islands	145	(65)
American Samoa	125	(56)

Notes:

1. Values are nominal design 3-second gust wind speeds in miles per hour (m/s) at 33 ft (10 m) above ground for Exposure C category.
2. Linear interpolation between wind contours is permitted.
3. Islands and coastal areas outside the last contour shall use the last wind speed contour of the coastal area.
4. Mountainous terrain, gorges, ocean promontories, and special wind regions shall be examined for unusual wind conditions.

Fig 250-2(b)
Basic Wind Speeds*

*Fig 250-2(b) reprinted with the permission of ASCE Publications, 1801 Alexander Bell Dr., Reston, VA 20191 from ASCE 7-98, Minimum Design Loads for Buildings and Other Structures. Copyright © 1998.

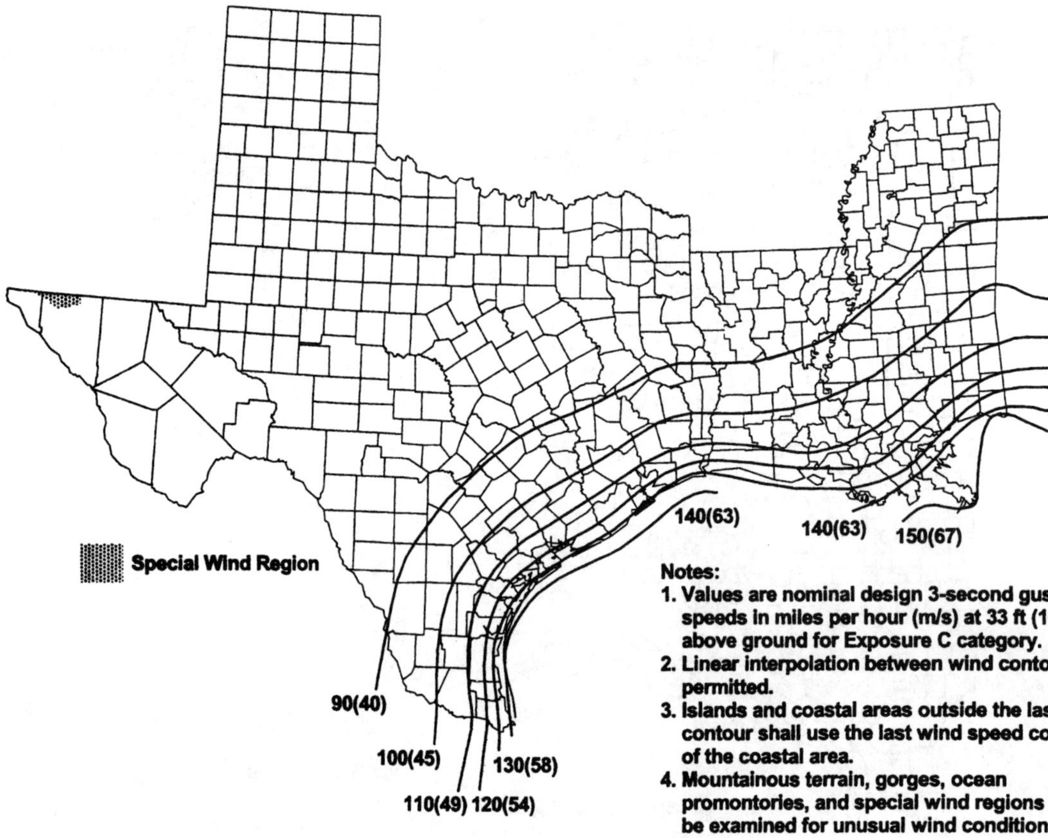

Fig 250-2(c)
Western Gulf of Mexico Hurricane Coastline*

*Fig 250-2(c) reprinted with the permission of ASCE Publications, 1801 Alexander Bell Dr., Reston, VA 20191 from ASCE 7-98, Minimum Design Loads for Buildings and Other Structures. Copyright © 1998.

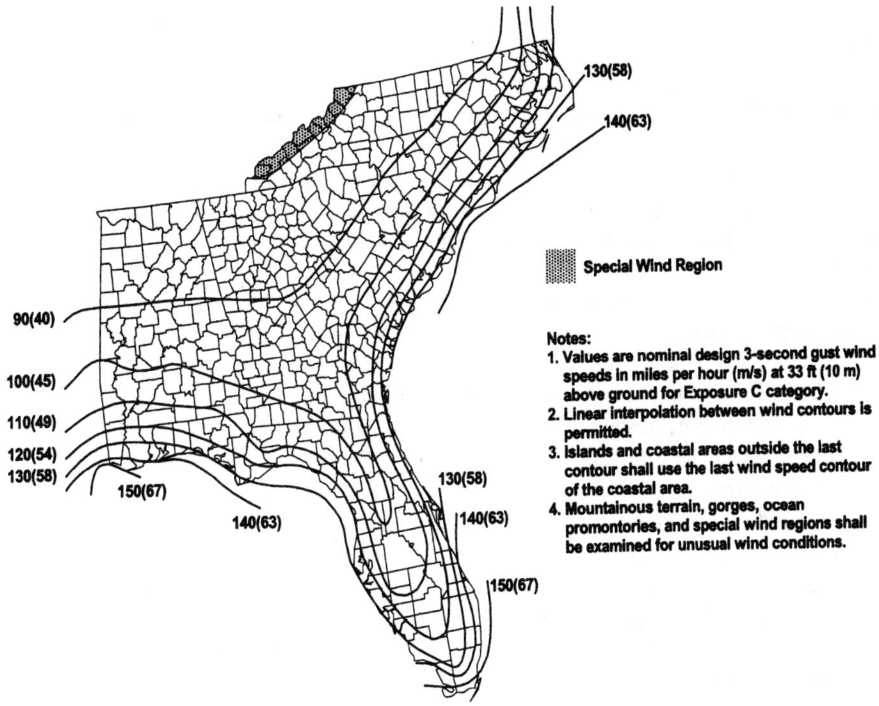

Fig 250-2(d)
Eastern Gulf of Mexico and Southeastern US Hurricane Coastline*

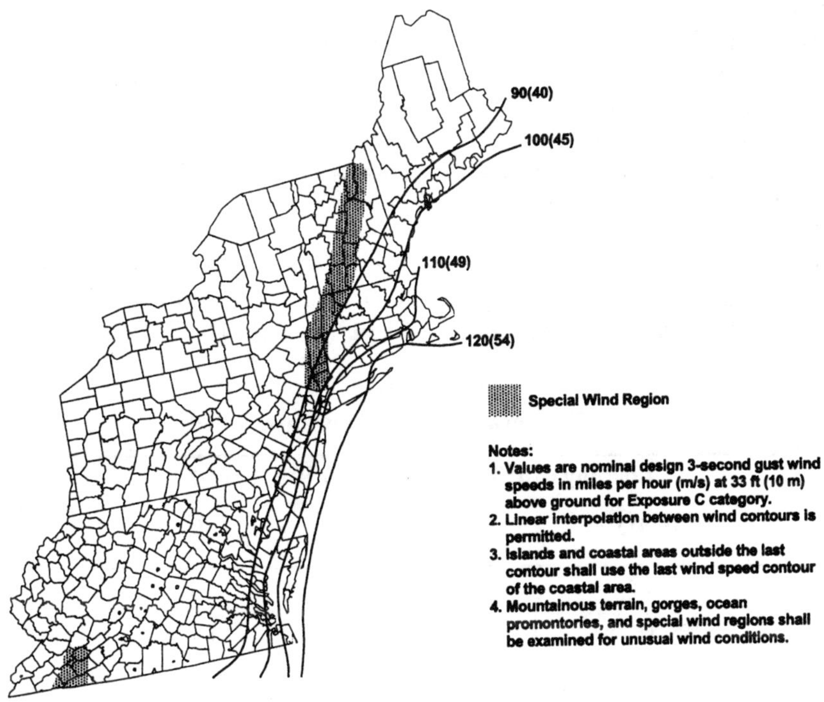

Fig 250-2(e)
Mid and Northern Atlantic Hurricane Coastline*

*Fig 250-2(d) and Fig 250-2(e) reprinted with the permission of ASCE Publications, 1801 Alexander Bell Dr., Reston, VA 20191 from ASCE 7-98, Minimum Design Loads for Buildings and Other Structures. Copyright © 1998.

251. Conductor Loading

A. General

Ice and wind loads are specified in Rule 250.

1. Where a cable is attached to a messenger, the specified loads shall be applied to both cable and messenger.
2. In determining wind loads on a conductor or cable without ice covering, the assumed projected area shall be that of a smooth cylinder whose outside diameter is the same as that of the conductor or cable. The shape factor for cylindrical surfaces is assumed to be 1.0.

   NOTE: Experience has shown that as the size of multiconductor cable decreases, the actual projected area decreases, but the roughness factor increases and offsets the reduction in projected area.
3. An appropriate mathematical model shall be used to determine the wind and weight loads on ice-coated conductors and cables. In the absence of a model developed in accordance with Rule 251A4, the following mathematical model shall be used:
   a. On a conductor, lashed cable, or multiple-conductor cable, the coating of ice shall be considered to be a hollow cylinder touching the outer strands of the conductor or the outer circumference of the lashed cable or multiple-conductor cable.
   b. On bundled conductors, the coating of ice shall be considered as individual hollow cylinders around each subconductor.
4. It is recognized that the effects of conductor stranding or of non-circular cross section may result in wind and ice loadings more or less than those calculated according to assumptions stated in Rule 251A2 and Rule 251A3. No reduction in these loadings is permitted unless testing or a qualified engineering study justifies a reduction.

B. Load Components

The load components shall be determined as follows:

1. Vertical Load Component

   The vertical load on a wire, conductor, or messenger shall be its own weight plus the weight of conductors, spacers, or equipment that it supports, ice covered where required by Rule 250.
2. Horizontal Load Component

   The horizontal load shall be the horizontal wind pressure determined under Rule 250 applied at right angles to the direction of the line using the projected area of the conductor or messenger and conductors, spacers, or equipment that it supports, ice covered where required by Rule 250.
3. Total Load

   The total load on each wire, conductor, or messenger shall be the resultant of components 1 and 2 above, calculated at the applicable temperature in Table 251-1, plus the corresponding constant in Table 251-1. In all cases the conductor or messenger tension shall be computed from this total load.

Table 251-1
Temperatures and Constants

	Loading districts (for use with Rule 250B)			Extreme wind loading (for use with Rule 250C)
	Heavy	Medium	Light	
Temperature (°C)	–20	–10	–1	+15
(°F)	0	+15	+30	+60
Constant to be added to the resultant (all conductors) in N/m	4.4	2.9	0.73	0.0
in lb/ft	0.30	0.20	0.05	0.0

252. Loads on Line Supports

A. Assumed Vertical Loads

The vertical loads on poles, towers, foundations, crossarms, pins, insulators, and conductor fastenings shall be their own weight plus the weight that they support, including all wires and cables, in accordance with Rules 251A and 251B1, together with the effect of any difference in elevation of supports. Loads due to radial ice shall be computed on wires, cables, and messengers, but need not be computed on supports.

B. Assumed Transverse Loads

The total transverse loads on poles, towers, foundations, crossarms, pins, insulators, and conductor fastenings shall include the following:

1. Transverse Loads from Conductors and Messengers

   The transverse loads from conductors and messengers shall be the horizontal load determined by Rule 251.

   EXCEPTION: In medium- and heavy-loading districts, where supporting structures carry ten or more conductors on the same crossarm, not including cables supported by messengers, and where the horizontal pin spacing does not exceed 380 mm (15 in), the transverse wind load may be calculated on two-thirds of the total number of such conductors if at least ten conductors are used in the calculations.

2. Wind Loads on Structures

   The transverse load on structures and equipment shall be computed by applying, at right angles to the direction of the line, the appropriate horizontal wind pressure determined under Rule 250. This load shall be calculated using the projected surfaces of the structures and equipment supported thereon, without ice covering. The following shape factors shall be used.

   a. Cylindrical Structures and Components

      Wind loads on straight or tapered cylindrical structures or structures composed of numerous narrow relatively flat panels that combine to form a total cross section that is circular or elliptical in shape shall be computed using a shape factor of 1.0.

   b. Flat Surfaced (Not Lattice) Structures and Components

      Wind loads on flat surfaced structures, having solid or enclosed flat sides and an overall cross section that is square or rectangular, shall be computed using a shape factor of 1.6.

   c. Lattice Structures

      Wind loads on square or rectangular lattice structures or components shall be computed using a shape factor of 3.2 on the sum of the projected areas of the members of the front face if structural members are flat surfaced or 2.0 if structural surfaces are cylindrical. The total, however, need not exceed the load that would occur on a solid structure of the same outside dimension.

      EXCEPTION: The shape factors listed under Rules 252B2a, 252B2b, and 252B2c may be reduced if wind tunnel tests or aerodynamic analysis justifies a reduction.

3. At Angles

   Where a change in direction of wires occurs, the loads on the structure, including guys, shall be the vector sum of the transverse wind load and the wire tension load. In calculating these loads, a wind direction shall be assumed that will give the maximum resultant load. Proper reduction may be made to the loads to account for the reduced wind pressure on the wires resulting from the angularity of the application of the wind on the wire.

4. Span Lengths

   The calculated transverse load shall be based on the average of the two spans adjacent to the structure concerned.

C. Assumed Longitudinal Loading

1. Change in Grade of Construction

   The longitudinal loads on supporting structures, including poles, towers, and guys at the ends of sections required to be of Grade B construction, when located in lines of lower than Grade B construction, shall be taken as an unbalanced pull in the direction of the higher grade section equal to the larger of the following values:

a. Conductors with Rated Breaking Strength of 13.3 kN (3000 lb) or less

The pull of two-thirds, but not less than two, of the conductors having a rated breaking strength of 13.3 kN (3000 lb) or less. The conductors selected shall produce the maximum stress in the support.

b. Conductors with Rated Breaking Strength of more than 13.3 kN (3000 lb)

The pull resulting from one conductor when there are eight or less conductors (including overhead ground wires) having rated breaking strength of more than 13.3 kN (3000 lb), and the pull of two conductors when there are more than eight conductors. The conductors selected shall produce the maximum stress in the support.

2. Jointly Used Poles at Crossings Over Railroads, Communication Lines, or Limited Access Highways

Where a joint line crosses a railroad, a communication line, or a limited access highway, and Grade B is required for the crossing span, the tension in the communication conductors of the joint line shall be considered as limited to one-half their rated breaking strength, provided they are smaller than Stl WG No. 8 if of steel, or AWG No. 6 if of copper.

3. Deadends

The longitudinal load on a supporting structure at a deadend shall be an unbalanced pull equal to the tensions of all conductors and messengers (including overhead ground wires); except that with spans in each direction from the dead-end structure, the unbalanced pull shall be the difference in tensions.

4. Unequal Spans and Unequal Vertical Loads

The structure should be capable of supporting the unbalanced longitudinal load created by the difference in tensions in the wires in adjacent spans caused by unequal vertical loads or unequal spans.

5. Stringing Loads

Consideration should be given to longitudinal loads that may occur on the structure during wire stringing operations.

6. Longitudinal Capability

It is recommended that structures having a longitudinal strength capability be provided at reasonable intervals along the line.

7. Communication Conductors on Unguyed Supports at Railroad and Limited Access Highway Crossings

The longitudinal load shall be assumed equal to an unbalanced pull in the direction of the crossing of all open-wire conductors supported, the pull of each conductor being taken as 50% of its rated breaking strength in the heavy loading district, 33-1/3% in the medium loading district, and 22-1/4% in the light-loading district.

D. Simultaneous Application of Loads

Where a combination of vertical, transverse, or longitudinal loads may occur simultaneously, the structure shall be designed to withstand the simultaneous application of these loads.

NOTE: Under the extreme wind conditions of Rule 250C, an oblique wind may require greater structural strength than that computed by Rules 252B and 252C.

253. Overload Factors for Structures, Crossarms, Support Hardware, Guys, Foundations, and Anchors

Loads due to the combined ice and wind loads in Rule 250B and the extreme wind loading condition in Rule 250C shall be multiplied by the overload factors in Table 253-1 or the alternate overload factors in Table 253-2. Table 253-1 shall be used with Table 261-1A. Table 253-2 shall be used with Table 261-1B.

For wood and reinforced (not prestressed) concrete, two methods for determining the capacity are included herein. Either method meets the basic requirements for safety.

Table 253-1
Overload Factors for Structures,[1] Crossarms, Support Hardware, Guys, Foundations, and Anchors to Be Used with the Strength Factors of Table 261-1A

	Overload Factors	
	Grade B	**Grade C**
Rule 250B Loads		
Vertical Loads[3]	1.50	1.90[6]
Transverse Loads		
Wind	2.50	2.20[4]
Wire Tension	1.65[2]	1.30[5]
Longitudinal Loads		
At Crossings		
In general	1.10	no requirement
At deadends	1.65[2]	1.30[5]
Elsewhere		
In general	1.00	no requirement
At deadends	1.65[2]	1.30[5]
Rule 250C Loads	1.00	1.00

1 Includes pole.

2 For guys and anchors associated with structures supporting communication conductors and cables only, this factor may be reduced to 1.33.

3 Where vertical loads significantly reduce the stress in a structure member a vertical overload factor of 1.0 should be used for the design of such member. Such member shall be designed for the worst case loading.

4 This factor may be reduced to 1.75 when the span being supported is not at a crossing.

5 For metal or prestressed concrete portions of structures and crossarms, guys, foundations, and anchors, use a value of 1.10.

6 For metal or prestressed concrete portions of structures, crossarms, guys, foundations, and anchors, use a value of 1.50.

Table 253-2
Alternate Overload Factors for Wood and Reinforced (Not Prestressed) Concrete Structures[1, 5] to Be Used with the Strength Factors of Table 261-1B

	Overload Factors			
	Grade B		Grade C	
	When Installed	At Replacement[2, 3]	When Installed	At Replacement[2, 3]
Rule 250B Loads				
Vertical loads[4]	2.20	1.50	2.20	1.50
Transverse loads				
Wind (at crossings)	4.00	2.67	2.67	1.33
Wind (elsewhere)	4.00	2.67	2.00	1.33
Wire tension	2.00	1.33	1.33	1.00
Longitudinal loads				
In general	1.33	1.00	No requirement	No requirement
At deadends	2.00[6]	1.33[7]	1.33	1.00
Rule 250C Loads	1.33	1.00	1.33	1.00

[1] Includes poles.

[2] Where a wood structure is built for temporary service, the overload factors at replacement may be used provided the designated fiber stress is not exceeded during the life of the structure. Where a reinforced concrete (not prestressed) structure is built for temporary service, the overload factors at replacement may be used.

[3] When structure strength deteriorates to the level of the loads multiplied by the overload factors required at replacement, the structure shall be replaced or rehabilitated. If a structure is replaced, it shall meet the "when installed" overload factors at replacement. Rehabilitated portions of structures shall have overload factors at the time of rehabilitation greater than of those required "at replacement."

[4] Where vertical loads significantly reduce the stress in a structural member, a vertical overload factor of 1.0 should be used for the design of such member. Such members shall be designed for the worst-case loading.

[5] Metal portions of a structure may be designed using the overload factors in Table 253-1.

[6] For unguyed wood poles supporting communication conductors and cables only, this factor may be reduced to 1.33.

[7] For unguyed wood poles supporting communication conductors and cables only, this factor may be reduced to 1.0.

Section 26.
Strength Requirements

260. General (see also Section 20)

A. Preliminary Assumptions

1. It is recognized that deformation, deflections, or displacement of parts of the structure may change the effects of the loads assumed. In the calculation of stresses, allowance may be made for such deformation, deflection, or displacement of supporting structures including poles, towers, guys, crossarms, pins, conductor fastenings, and insulators when the effects can be evaluated. Such deformation, deflection, or displacement should be calculated using Rule 250 loads prior to application of the overload factors in Rule 253. For crossings or conflicts, the calculations shall be subject to mutual agreement.
2. It is recognized that new materials may become available. While these materials are in the process of development, they must be tested and evaluated. Trial installations are permitted where qualified supervision is provided.

B. Application of strength factors

1. Structures shall be designed to withstand the appropriate loads multiplied by the overload factors in Section 25 without exceeding their strength multiplied by the strength factors in Section 26.
2. Unless otherwise specified, a strength factor of 0.80 shall be used for the extreme wind loading conditions specified in Rule 250C for all supported facilities.

   NOTE: The latest edition of the following documents are among those available for determining structure design capacity with the specified NESC loads, overload factors, and strength factors:

   ANSI/ASCE-10, Design of Latticed Steel Transmission Structures

   ASCE-91, Design of Guyed Electrical Transmission Structure

   ASCE-PCI, Guide for the Design of Prestressed Concrete Poles

   ASCE-72, Design of Steel Transmission Pole Structures

   PCI, Design Handbook-Precast and Prestressed Concrete

   ACI-318, Building Code Requirements for Structural Concrete

   IEEE Std 751-1990, Trial-Use Design Guide for Wood Transmission Structures

   AISI, Specification for the Design of Cold-Formed Steel Structural Members

   The Aluminum Association, Aluminum Design Manual

261. Grades B and C Construction

A. Supporting Structures

The strength requirements for supporting structures may be met by the structures alone or with the aid of guys or braces or both.

1. Metal, Prestressed-, and Reinforced-Concrete Structures
   a. These structures shall be designed to withstand the loads in Rule 252 multiplied by the appropriate overload factors in Table 253-1 or Table 253-2 without exceeding the permitted load.
   b. The permitted load shall be the strength multiplied by the strength factors in Tables 261-1A or 261-1B (where guys are used, see Rule 261C).
   c. All structures including those below 18 m (60 ft) shall be designed to withstand, without conductors, the extreme wind load in Rule 250C applied in any direction on the structure.
   d. Spliced and Reinforced Structures

      Reinforcements or permanent splices to a supporting structure are permitted provided they develop the required strength of the structure.

2. Wood Structures

Wood structures shall be of material and dimensions to meet the following requirements:

a. Wood structures shall be designed to withstand the loads in Rule 252 multiplied by the appropriate overload factors in Table 253-1 or 253-2, without exceeding the permitted stress level.

NOTE: When determining a fiber stress for column loads, buckling needs to be considered.

EXCEPTION 1: When installed, naturally grown wood poles acting as single-based structures or unbraced multiple-pole structures, shall meet the requirements of Rule 261A2a without exceeding the permitted stress level at the ground line for unguyed poles or at the points of attachment for guyed poles.

EXCEPTION 2: At a Grade B crossing, in a straight section of line, wood structures complying with the transverse strength requirements of Rule 261A2a, without the use of transverse guys, shall be considered as having the required longitudinal strength, providing the longitudinal strength is comparable to the transverse strength of the structure. This *EXCEPTION* does not modify the requirements of this rule for deadends.

EXCEPTION 3: At a Grade B crossing of a supply line over a highway or a communication line where there is an angle in the supply line, wood structures shall be considered as having the required longitudinal strength if all of the following conditions are met:

(a) The angle is not over 20 degrees.

(b) The angle structure is guyed in the plane of the resultant of the conductor tensions. The tension in this guy under the loading in Rule 252 multiplied by an overload factor of 2.0 shall not exceed the rated breaking strength multiplied by the strength factor in Table 261-1A.

(c) The angle structure has sufficient strength to withstand, without guys, the transverse loading of Rule 252 multiplied by the appropriate overload factors in Table 253-1 or 253-2, which would exist if there were no angle at that structure without exceeding the permitted stress level.

b. Permitted Stress Level

(1) Natural Wood Pole

The permitted stress level of natural wood poles of various species meeting the requirements of ANSI O5.1-1992 shall be determined by multiplying the designated fiber stress set forth in that standard by the appropriate strength factors in Tables 261-1A or 261-1B.

(2) Sawn or Laminated Wood Structural Members, Crossarms, and Braces

The permitted stress level of sawn or laminated wood structural members, crossarms, and braces shall be determined by multiplying the appropriate ultimate fiber stress of the material by the appropriate strength factors in Tables 261-1A or 261-1B.

c. Strength of Guyed Poles

Guyed poles shall be designed as columns, resisting the vertical component of the tension in the guy plus any other vertical loads.

d. Spliced and Reinforced Poles

Reinforcements or permanent splices at any section along the pole are permitted provided they develop the required strength of the pole.

e. Average Strength of Three Poles

A pole (single-base structure) not individually meeting the transverse strength requirements will be permitted when reinforced by a stronger pole on each side, if all of the following are met:

(1) The average strength of the three poles meets the transverse strength requirements,

(2) The weak pole shall have not less than 75% of its required strength,

(3) The sag and tension of the wires, conductors, and cables in the adjacent spans shall provide adequate additional support for the weak pole, and

(4) The average of the spans does not exceed 45 m (150 ft).

EXCEPTION 1: The span may exceed 45 m (150 ft), but shall not be greater than 91 m (300 ft), if overhead guys are run between the three poles and the line section is head-guyed and back-guyed.

An extra pole inserted in a normal span for the purpose of supporting a service drop may be ignored in this strength determination.

EXCEPTION 2: This rule does not apply to crossings over railroads, communication lines, or limited access highways.

f. All structures including those below 18 m (60 ft) shall be designed to withstand, without conductors, the extreme wind load in Rule 250C applied in any direction on the structure.

3. Transverse Strength Requirements for Structures Where Side Guying Is Required, But Can Be Installed Only at a Distance

Grade B: If the transverse strength requirements of this section cannot be met except by the use of side guys or special structures, and where it is physically impractical to employ side guys, the transverse strength requirements may be met by side-guying the line at each side of, and as near as practical to, the crossing, or other transversely weak structure, and with a distance between such side-guyed structures of not over 250 m (800 ft), provided that:

a. The side-guyed structures for each such section of 250 m (800 ft) or less shall be designed to withstand the calculated transverse load due to wind on the supports and ice-covered conductors, on the entire section between side-guyed structures.

b. The line between such side-guyed structures shall be substantially in a straight line and the average span between the side-guyed structures shall not exceed 45 m (150 ft).

c. The entire section between the structures with the required transverse strength shall comply with the highest grade of construction concerned in the given section, except as to the transverse strength of the intermediate poles or towers.

Grade C: The above provisions do not apply to Grade C.

4. Longitudinal Strength Requirements for Sections of Higher Grade in Lines of a Lower Grade Construction

a. Methods of Providing Longitudinal Strength

Grade B: The longitudinal strength requirements for sections of line of higher grade in lines of a lower grade (for assumed longitudinal loading, see Rule 252) may be met by placing a structure of the required longitudinal strength at each end of the higher grade section.

Where this is impractical, the structures of the required longitudinal strength may be located away from the section of higher grade, within 150 m (500 ft) on each side and with not more than 250 m (800 ft) between the structures of the required longitudinal strength. This is permitted provided the following conditions are met:

(1) The structures and the line between them meet the requirements for transverse strength and stringing of conductors of the highest grade occurring in the section, and

(2) The line between the structures of the required longitudinal strength is approximately straight or suitably guyed.

The longitudinal strength requirement of the structures may be met by using guys.

Grade C: The above provisions do not apply to Grade C.

b. Flexible Supports

Grade B: When supports of the section of higher grade are capable of considerable deflection in the direction of the line, it may be necessary to increase the clearances required in Section 23 or to provide line guys or special reinforcements to reduce the deflection.

Grade C: The above provision does not apply to Grade C.

B. Strength of Foundations, Settings, and Guy Anchors

Foundations, settings, and guy anchors shall be designed or be determined by experience to withstand the loads in Rule 252 multiplied by the overload factors in Table 253-1 without exceeding the permitted load. The permitted load shall be equal to the strength multiplied by the strength factors in Table 261-1A.

NOTE: Excessive movement of foundations, settings, and guy anchors or errors in settings may reduce clearances or structure capacity.

C. Strength of Guys and Guy Insulators

The strength requirements for guys and guy insulators are covered under Rules 264 and 279A1c, respectively.

1. Metal and Prestressed-Concrete Structures

Guys shall be considered as an integral part of the structure.

2. Wood and Reinforced-Concrete Structures

When guys are used to meet the strength requirements, they shall be considered as taking the entire load in the direction in which they act, the structure acting as a strut only, except for those structures considered to possess sufficient rigidity so that the guy can be considered an integral part of the structure.

NOTE: Excessive movement of guys may reduce clearances or structure capacity.

D. Crossarms and Braces

1. Concrete and Metal Crossarms and Braces

   Crossarms and braces shall be designed to withstand the loads in Rule 252 multiplied by the overload factors in Table 253-1 without exceeding the permitted load. The permitted load shall be equal to the strength multiplied by the strength factors in Table 261-1A.

2. Wood Crossarms and Braces

   a. Strength

      (1) Crossarms and braces shall be designed to withstand the loads in Rule 252 multiplied by the overload factors in Table 253-1 or 253-2 without exceeding their permitted stress level.

      (2) The permitted stress level of solid sawn or laminated wood crossarms and braces shall be determined by multiplying their ultimate fiber stress by the strength factors in Table 261-1A or 261-1B.

   b. Material and Size

      Wood crossarms and braces of select Southern pine or Douglas fir shall have a cross section of not less than those in Table 261-2. Crossarms of other species may be used provided they have equal strength.

3. Crossarms and Braces of Other Materials

   Crossarms and braces should meet the strength requirements of Rule 261D2.

4. Additional Requirements

   a. Longitudinal Strength

      (1) General

         (a) Crossarms shall be designed to withstand a load of 3.1 kN (700 lb) applied at the outer conductor attachment point without exceeding the permitted stress level for wood crossarms or the permitted load for crossarms of other materials, as applicable.

         (b) At each end of a transversely weak section, as described in Rule 261A3, the longitudinal load shall be applied in the direction of the weak section.

      (2) Methods of Meeting Rule 261D2a(1)

         Grade B: Where conductor tensions are limited to a maximum of 9.0 kN (2000 lb) per conductor, double wood crossarms having cross sections in Table 261-2 and properly assembled will comply with the longitudinal strength requirements in Rule 261D2a(1).

         Grade C: This requirement is not applicable.

      (3) Location

         At crossings, crossarms should be mounted on the face of a pole away from the crossing, unless special bracing or double crossarms are used.

   b. Bracing

      Crossarms shall be supported by bracing, if necessary, to support expected loads, including line personnel working on them.

   c. Double Crossarms or Brackets

      Grade B: Where pin-type construction is used, double crossarms, each crossarm having the strength required by Rule 261D2a, or a support assembly of equivalent strength shall be used at each crossing structure, at ends of joint use or conflict sections, at deadends, and at corners where the angle of departure from a straight line exceeds 20 degrees. Under similar conditions, where a bracket supports a conductor operated at more than 750 V to ground and there is no crossarm below, double brackets shall be used.

      EXCEPTION: The above does not apply where communication cables or conductors cross below supply conductors and either are attached to the same pole, or where supply conductors are continuous and of uniform tension in the crossing span and each adjacent span. This exception does not apply to railroad crossings and limited access highways except by mutual agreement.

      Grade C: The above requirement is not applicable.

E. Insulators

The strength requirements for insulators are covered under Rules 277 and 279.

F. Strength of Pin-Type or Similar Construction and Conductor Fastenings

1. Longitudinal Strength

a. General

Pin-type or similar construction and ties or other conductor fastenings shall be designed to withstand the applicable longitudinal loads in Rule 252, multiplied by the overload factors in Table 253-1, or 3.1 kN (700 lb) applied at the pin, whichever is greater.

b. Method of Meeting Rules 261F1a

Grade B: Where conductor tensions are limited to 9.0 kN (2000 lb) and such conductors are supported on pin insulators, double wood pins and ties or their equivalent will be considered to meet the requirements of Rule 261F1a.

Grade C: No requirement.

c. At Deadends and at Ends of Higher Grade Construction in Line of Lower Grade

Grade B: Pins and ties or other conductor fastenings connected to the structure at a deadend or at each end of the higher grade section shall be designed to withstand an unbalanced pull due to the conductor load in Rule 251 multiplied by the overload factors in Rule 253-1.

Grade C: This requirement is not applicable except for deadends.

d. At Ends of Transverse Sections Described in Rule 261A3

Grade B: Pins and ties or other conductor fastenings connected to the structure at ends of the transverse section as described in Rule 261A3 shall be designed to withstand the unbalanced pull in the direction of that transverse section under the load in Rule 252 multiplied by the overload factors in Rule 253-1.

Grade C: No requirement.

2. Double Pins and Conductor Fastenings

Grade B: Double pins and conductor fastenings shall be used where double crossarms or brackets are required by Rule 261D4c.

EXCEPTION: The above does not apply where communication cables or conductors cross below supply conductors and either are attached to the same pole, or where supply conductors are continuous and of uniform tension in a crossing span and each adjacent span. This exception does not apply in the case of railroad crossings and limited access highway crossings except by mutual agreement.

Grade C: No requirement.

3. Single Supports Used in Lieu of Double Wood Pins

A single conductor support and its conductor fastening, when used in lieu of double wood pins, shall develop strength equivalent to double wood pins and their conductor fastenings as specified in Rule 261F1a.

G. Armless Construction

1. General

Open conductor armless construction is a type of open conductor supply line construction in which conductors are individually supported at the structure without the use of crossarms.

2. Insulating Material

Strength of insulating material shall meet the requirements of Section 27.

3. Other Components

Strengths of other components shall meet the requirements of Rules 260 and 261.

H. Open Supply Conductors and Overhead Shield Wires

1. Sags and Tensions

a. The supply conductor and overhead shield wire tensions shall be not more than 60 percent of their rated breaking strength for the load of Rule 250B in Rule 251 multiplied by an overload factor of 1.0.

b. The tension at 15 °C (60 °F), without external load, shall not exceed the following percentages of their rated breaking strength:

Initial unloaded tension	35%
Final unloaded tension	25%

EXCEPTION: In the case of conductors with a generally triangular cross section, such as cables composed of three wires, the final unloaded tension at 15 °C (60 °F) shall not exceed 30% of the rated breaking strength of the conductor.

NOTE: The above limitations are based on the use of recognized methods for avoiding fatigue failures by minimizing chafing and stress concentration. If such practices are not followed, lower tensions should be employed.

2. Splices, Taps, Dead-End Fittings, and Associated Attachment Hardware

a. Splices should be avoided in crossings and adjacent spans. If it is impractical to avoid such splices, they shall have sufficient strength to withstand the maximum tension resulting from the loads of Rule 250B in Rule 251 multiplied by an overload factor of 1.65. If Rule 250C is applicable, splices shall not be stressed beyond 80% of their rated breaking strength under the loads of Rule 250C in Rule 251 multiplied by an overload factor of 1.0.

b. Taps should be avoided in crossing spans but, if required, shall be of a type that will not impair the strength of the conductors to which they are attached.

c. Dead-end fittings, including the associated attachment hardware, shall have sufficient strength to withstand the maximum tension resulting from the loads of Rule 250B in Rule 251 multiplied by an overload factor of 1.65. If Rule 250C is applicable, deadend fittings shall not be stressed beyond 80% of their rated breaking strength under the loads of Rule 250C in Rule 251 multiplied by an overload factor of 1.0.

3. Trolley-Contact Conductors

In order to provide for wear, no trolley-contact conductor shall be installed of less size than AWG No. 0, if of copper, or AWG No. 4, if of silicon bronze.

I. Supply Cable Messengers

Messengers shall be stranded and shall not be stressed beyond 60% of their rated breaking strength under the loads of Rule 250B in Rule 251 multiplied by an overload factor of 1.0. If Rule 250C is applicable, messengers shall not be stressed beyond 80% of their rated breaking strength under the loads of Rule 250C in Rule 251 multiplied by an overload factor of 1.0.

NOTE: There are no strength requirements for cables supported by messengers.

J. Open-Wire Communication Conductors

Open-wire communication conductors in Grade B or C construction shall have the tensions and sags in Rule 261H2 for supply conductors of the same grade.

EXCEPTION: Where supply conductors are trolley-contact conductors of 0 to 750 V to ground, WG No. 12 Stl may be used for communication conductors for spans of 0 to 30 m (0 to 100 ft), and Stl WG No. 9 may be used for spans of 38 to 45 m (125 to 150 ft).

K. Communication Cables

1. Communication Cables

There are no strength requirements for communication cables supported by messengers. See Rule 261K2 for the strength requirements for messengers supporting communication cables.

2. Messenger

The messenger shall not be stressed beyond 60% of its rated breaking strength under the loads of Rule 250B in Rule 251 multiplied by an overload factor of 1.0. If Rule 250C is applicable, messengers shall not be stressed beyond 80% of their rated breaking strength under the loads of Rule 250C in Rule 251 multiplied by an overload factor of 1.0.

L. Paired Communication Conductors

1. Paired Conductors Supported on Messenger

a. Use of Messenger
A messenger may be used for supporting paired conductors in any location, but is required for paired conductors crossing over trolley-contact conductors of more than 7.5 kV to ground.

b. Sag of Messenger
Messenger used for supporting paired conductors required to meet Grade B construction because of crossing over trolley-contact conductors shall meet the sag requirements for Grade B.

c. Size and Sag of Conductors
There are no requirements for paired conductors when supported on messenger.

2. Paired Conductors Not Supported on Messenger

a. Above Supply Lines
Grade B: Sizes and sags shall be not less than those in Rule 261H1 for supply conductors of similar grade.
Grade C: Sizes and sags shall be not less than the following:
Spans 0 to 30 m (0 to 100 ft)—No sag requirements.
Each conductor shall have a rated breaking strength of not less than 0.75 kN (170 lb).
Spans 30 to 45 m (100 to 150 ft)—Sizes and sags shall be not less than required for Grade B communication conductors.
Spans exceeding 45 m (150 ft)—Sizes and sags shall be not less than required for Grade C supply conductors. (See Rule 261H2).

b. Above Trolley-Contact Conductors
Grade B: Sizes and sags shall be not less than the following:
Spans 0 to 30 m (0 to 100 ft)—No size requirements. Sags shall be not less than for AWG No. 8 hard-drawn copper. (See Rule 261H2.)
Spans exceeding 30 m (100 ft)—Each conductor shall have a rated breaking strength of not less than 0.75 kN (170 lb). Sags shall be not less than for AWG No. 8 hard-drawn copper. (See Rule 261H2.)
Grade C: Sizes and sags shall be as follows:
Spans 0 to 30 m (0 to 100 ft)—No requirements.
Spans exceeding 30 m (100 ft)—No sag requirements.
Each conductor shall have a rated breaking strength of not less than 0.75 kN (170 lb).

M. Support and Attachment Hardware
The strength required for all support and attachment hardware not covered by Rule 261F or Rule 261H2 shall be not less than the load times the appropriate overload factor given in Section 25. For appropriate strength factors, see Rule 260B.

Table 261-1A

Strength Factors for Structures,[1] Crossarms, Support Hardware, Guys, Foundations, and Anchors for Use with Overload Factors of Table 253-1

[It is recognized that structures will experience some level of deterioration after installation, depending upon materials, maintenance, and service conditions. The table values specify strengths required at installation. Footnotes specify deterioration allowed, if any. When new or changed facilities add loads to existing structures (a) the strength of the structure when new shall have been great enough to support the additional loads and (b) the strength of the deteriorated structure shall exceed the strength required at replacement. If either (a) or (b) cannot be met, the structure must be replaced, augmented, or rehabilitated.]

	Grade B	Grade C
Strength factors for use with loads of Rule 250B		
Metal and Prestressed-Concrete Structures[6]	1.0	1.0
Wood and Reinforced-Concrete Structures[2, 4]	0.65	0.85
Support Hardware	1.0	1.0
Guy Wire[5, 6]	0.9	0.9
Guy Anchor and Foundation[6]	1.0	1.0
Strength factors for use with loads of Rule 250C		
Metal and Prestressed-Concrete Structures[6]	1.0	1.0
Wood and Reinforced-Concrete Structures[3, 4]	0.75	0.75
Support Hardware	1.0	1.0
Guy Wire[5, 6]	0.9	0.9
Guy Anchor and Foundation[6]	1.0	1.0

[1] Includes poles.

[2] Wood and reinforced concrete structures shall be replaced or rehabilitated when deterioration reduces the structure strength to 2/3 of that required when installed. If a structure is replaced, it shall meet the strength required by Table 261-1A. Rehabilitated portions of structures shall have strength greater than 2/3 of that required when installed.

[3] Wood and reinforced concrete structures shall be replaced or rehabilitated when deterioration reduces the structure strength to 3/4 of that required when installed. If a structure is replaced, it shall meet the strength required by Table 261-1A. Rehabilitated portions of structures shall have strength greater than 3/4 of that required when installed.

[4] Where a wood or reinforced concrete structure is built for temporary service, the structure strength may be reduced to values as low as those permitted by footnotes (2) and (3) provided the structure strength does not decrease below the minimum required during the planned life of the structure.

[5] For guy insulator requirements, see Rule 279.

[6] Deterioration during service shall not reduce strength capability below the required strength.

Table 261-1B

Strength Factors for Structures[1, 2] and Crossarms for Use with Overload Factors of Table 253-2

[It is recognized that structures will experience some levels of deterioration after installation, depending upon materials, maintenance, and service conditions. The table values specify strengths required at installation. Footnotes specify deterioration allowed for wood and reinforced concrete structures. When new or changed facilities add loads to existing structures (a) the strength of the structure when new shall have been great enough to support the additional loads, and (b) the strength of the deteriorated structure shall exceed the strength required at replacement. If either (a) or (b) cannot be met, the structure must be replaced, augmented, or rehabilitated.]

	Grade B	Grade C
Strength factors for use with loads of Rule 250B and Rule 250C		
Wood and Reinforced-Concrete Structures	1.0	1.0

[1] Includes poles.

[2] Where a wood or reinforced-concrete structure is built for temporary service, the structure strength may be reduced to values as low as those permitted by the *at replacement* overload factors in Table 253-2, footnotes (2) and (3) provided the structure strength does not decrease below the minimum required during the planned life of the structure.

Table 261-2
Dimensions of Crossarm Cross Section of Select Southern Pine and Douglas Fir

Crossarm Length		Grades of Construction	
		Grade B	Grade C
1.20 m or less	mm:	75 × 100	70 × 95
4 ft or less	in:	3 × 4	2-3/4 × 3-3/4
2.45 m	mm:	82 × 108	75 × 100
8 ft	in:	3-1/4 × 4-1/4	3 × 4
3.0 m	mm:	82 × 108	75 × 100
10 ft	in:	3-1/4 × 4-1/4	3 × 4

262. Number 262 not used in this edition.

263. Grade N Construction

The strength of Grade N construction need not be equal to or greater than Grade C.

A. Poles
Poles used for lines for which neither Grade B nor C is required shall be of initial size or guyed or braced to withstand expected loads, including line personnel working on them.

B. Guys
The general requirements for guys are covered in Rules 264 and 279A.

C. Crossarm Strength
Crossarms shall be securely supported by bracing, if necessary, to withstand expected loads, including line personnel working on them.
NOTE: Double crossarms are generally used at crossings, unbalanced corners, and dead ends, in order to permit conductor fastenings at two insulators to limit the opportunity for slipping, although single crossarms might provide sufficient strength. To secure extra strength, double crossarms are frequently used, and crossarm guys are sometimes used.

D. Supply-Line Conductors
1. Size
Supply-line conductors shall be not smaller than the sizes listed in Table 263-1.
RECOMMENDATION: It is recommended that these sizes for copper and steel not be used in spans longer than 45 m (150 ft) for the heavy-loading district, and 53 m (175 ft) for the medium- and light-loading districts.

E. Service Drops
1. Size of Open-Wire Service Drops
a. Not over 750 V.
Service drops shall be as required by (1) or (2):
(1) Spans not exceeding 45 m (150 ft)
Sizes shall be not smaller than those in Table 263-2.
(2) Spans exceeding 45 m (150 ft)
Sizes shall be not smaller than 8 AWG.

b. Exceeding 750 V

Sizes of service drops of more than 750 V shall be not less than required for supply-line conductors of the same voltage.

2. Tension of Open-Wire Service Drops

The tension of the service drop conductors shall not exceed the strength of the conductor attachment or its support under the expected loads.

3. Cabled Service Drops

Service conductors may be grouped together in a cable, provided the following requirements are met:

a. Size

The size of each conductor shall be not less than required for drops of separate conductors (Rule 263E1).

b. Tension of Cabled Service Drops

The tension of the service drop conductors shall not exceed the strength of the conductor attachment or its support under the expected loads.

F. Trolley-Contact Conductors

In order to provide for wear, trolley-contact conductors shall be not smaller than size AWG No. 0, if of copper, or AWG No. 4, if of silicon bronze.

G. Communication Conductors

There are no specific requirements for Grade N communication line conductors or service drops.

H. Street and Area Lighting Equipment

The lowering rope or chain for luminaires arranged to be lowered for examination or maintenance shall be of a material and strength designed to withstand climatic conditions and to sustain the luminaire safely.

I. Insulators

The strength requirements for insulators are covered under Rules 277 and 279.

Table 263-1
Sizes for Grade N Supply Line Conductors
(AWG for Copper and Aluminum, Stl WG for Steel)

	Urban	**Rural**
Soft copper	6	8
Medium- or hard-drawn copper	8	8
Steel	9	9
	Spans 45 m (150 ft) or less	Spans exceeding 45 m (150 ft)
Stranded aluminum:		
EC	4	2
ACSR	6	4
ALLOY	4	4
ACAR	4	2

Table 263-2
Sizes of Service Drops of 750 V or Less
(Voltages of trolley-contact conductors are voltage to ground.
AWG used for aluminum and copper wires; Stl WG used for steel wire.)

Situation	Copper Wire			
	Soft-Drawn	Medium- or Hard-Drawn	Steel Wire	EC aluminum wire[2]
Alone	10	12	12	4
Concerned with communication conductor	10	12	12	4
Over supply conductors of				
0 to 750 V	10	12	12	4
750 V to 8.7 kV[1]	8	10	12	4
Exceeding 8.7 kV[1]	6	8	9	4
Over trolley-contact conductors				
0 to 750 V ac or dc	8	10	12	4
Exceeding 750 V ac or dc	6	8	9	4

[1] Installation of service drops of not more than 750 V above supply lines of more than 750 V should be avoided where practical.
[2] ACSR or high-strength aluminum alloy conductor size shall be not less than No. 6.

264. Guying and Bracing

A. Where Used
When the loads are greater than can be supported by the structure alone, additional strength shall be provided by the use of guys, braces, or other suitable construction. Such measures shall also be used where necessary to limit the increase of sags in adjacent spans and provide sufficient strength for those supports on which the loads are sufficiently unbalanced, for example, at corners, angles, dead ends, large differences in span lengths, and changes of grade of construction.

B. Strength
Guys shall be designed to withstand the loads in Rule 252 multiplied by the overload factors in Table 253-1 without exceeding the permitted load. The permitted load shall be equal to the strength multiplied by the strength factors in Table 261-1A. For guy wires conforming to ASTM Standards, the minimum breaking strength value therein defined shall be the rated breaking strength required in this code.

C. Point of Attachment
The guy or brace should be attached to the structure as near as is practical to the center of the conductor load to be sustained. However, on lines exceeding 8.7 kV, the location of the guy or brace may be adjusted to minimize the reduction of the insulation offered by nonmetallic support arms and supporting structures.

D. Guy Fastenings
Guys having a rated breaking strength of 9.0 kN (2000 lb) or more and that are subject to small radius bends should be stranded and should be protected by suitable guy thimbles or their equivalent. Any

guy having a design loading of 44.5 kN (10 000 lb) or more wrapped around cedar or similar softwood poles should be protected by the use of suitable guy shims.

Where there is a tendency for the guy to slip off the shim, guy hooks or other suitable means of limiting the likelihood of this action should be used. Shims are not necessary in the case of supplementary guys, such as storm guys.

E. Guy Markers and Protection

1. The ground end of anchor guys exposed to pedestrian traffic shall be provided with a substantial and conspicuous marker.

   NOTE: Visibility of markers can be improved by the use of color or color patterns that provide contrast with the surroundings.
2. Where an anchor is located in an established parking area, the guy shall either be protected from vehicle contact or marked.
3. Nothing in this rule is intended to require protection or marking of structural components located outside of the traveled ways of roadways or established parking areas. Experience has shown that it is not practical to protect structures from contact by out of control vehicles operating outside of established traveled ways.

F. Electrolysis

Where anchors and rods are subject to electrolysis, suitable measures should be taken to minimize corrosion from this source.

G. Anchor Rods

1. Anchor rods should be installed so as to be in line with the pull of the attached guy when under load.

   EXCEPTION: This is not required for anchor rods installed in rock or concrete.
2. The anchor and rod assembly shall have an ultimate strength not less than that required of the guy(s) by Rule 264B.

Section 27.
Line Insulation

270. Application of Rule

These requirements apply only to open-conductor supply lines.

NOTE 1: See Rule 243C6.

NOTE 2: See Rule 242E for insulation requirements for neutral conductors.

271. Material and Marking

Insulators for operation of supply circuits shall be made of wet-process porcelain or other material that will provide equivalent or better electrical and mechanical performance. Insulators for use at or above 2.3 kV between conductors shall be marked by the maker with its name or trademark and an identification mark or markings that will permit determination of the electrical and mechanical properties. The marking shall be applied so as not to reduce the electrical or mechanical strength of the insulator.

NOTE: The identifying marking can be either a catalog number, trade number, or other means so that properties of the unit can be determined either through catalogs or other literature.

272. Ratio of Flashover to Puncture Voltage

Insulators shall be designed so that the ratio of their rated low-frequency dry-flashover voltage to low-frequency puncture voltage is in conformance with applicable American National Standards. When a standard does not exist, this ratio shall not exceed 75%.

The applicable American National Standards are as follows:

ANSI C29.1-1988
ANSI C29.2-1992
ANSI C29.3-1986
ANSI C29.4-1989
ANSI C29.5-1984
ANSI C29.6-1996
ANSI C29.7-1996

EXCEPTION: Insulators specifically designed for use in areas of high atmospheric contamination may have a rated low-frequency dry-flashover voltage not more than 80% of their low-frequency puncture voltage.

273. Insulation Level

The rated dry flashover voltage of the insulator or insulators, when tested in accordance with ANSI C29.1-1988, shall be not less than that shown in Table 273-1, unless based on a qualified engineering study. Higher insulation levels than those shown in Table 273-1, or other effective means, shall be used where severe lightning, high atmospheric contamination, or other unfavorable conditions exist. Insulation levels for system voltages in excess of those shown shall be based on a qualified engineering study.

274. Factory Tests

Each insulator or insulating part thereof for use on circuits operating at or above 2.3 kV between conductors shall be tested by the manufacturer in accordance with applicable American National Standards, or, where such standards do not exist, other good engineering practices to ensure their performance.

The applicable American National Standards are listed in Rule 272.

275. Special Insulator Applications

A. Insulators for Constant-Current Circuits

Insulators for use on constant-current circuits shall be selected on the basis of the rated full-load voltage of the supply transformer.

Table 273-1
Insulation Level Requirements

Nominal voltage (between phases) (kV)	Rated dry flashover voltage of insulators [1] (kV)	Nominal voltage (between phases) (kV)	Rated dry flashover voltage of insulators [1] (kV)
0.75	5	115	315
2.4	20	138	390
6.9	39	161	445
13.2	55	230	640
23.0	75	345	830
34.5	100	500	965
46	125	765	1145
69	175		

[1] Interpolate for intermediate values.

B. Insulators for Single-Phase Circuits Directly Connected to Three-Phase Circuits

Insulators used on single-phase circuits directly connected to three-phase circuits (without intervening isolating transformers) shall have an insulation level not less than that required for the three-phase circuit.

276. Number 276 not used in this edition.

277. Mechanical Strength of Insulators

Insulators shall withstand all applicable loads specified in Rules 250, 251, and 252 except those of Rule 250C without exceeding the following percentages of their rated ultimate strength for the specified application:

Cantilever	40%
Compression	50%
Tension	50%

Proper allowance should be made for the loads in Rule 250C.

NOTE 1: The rated ultimate strength of suspension type insulators (used in tangent, angle, or dead-end applications) is considered to be:

a. For porcelain or toughed glass insulators, the manufacturer's "combined mechanical and electrical strength" as determined per ANSI C29.1-1988.

b. For composite insulators, the manufacturer's "specified cantilever load" for cantilever or "specified mechanical load" for tension, per IEEE Std 1024-1988.

NOTE 2: The rated ultimate strength of porcelain post insulators is considered to be the manufacturer's strength rating per ANSI C29.7-1996 and ANSI C29.9-1983 [B7].

278. Aerial Cable Systems

A. Electrical Requirements

1. Covered or insulated conductors not meeting the requirements of Rule 230C1, 230C2, or 230C3 shall be considered as bare conductors for all insulation requirements.
2. The insulators or insulating supports shall meet the requirements of Rule 273.
3. The systems shall be so designed and installed as to minimize long-term deterioration from electrical stress.

B. Mechanical Requirements

1. Insulators other than spacers used to support aerial cable systems shall meet the requirements of Rule 277.
2. Insulating spacers used in spacer cable systems shall withstand the loads specified in Section 25 (except those of Rule 250C) without exceeding 50% of their rated ultimate strength.

279. Guy and Span Insulators

A. Insulators

1. Properties of Guy Insulators

a. Material

Insulators shall be made of wet-process porcelain, wood, fiberglass-reinforced plastic, or other material of suitable mechanical and electrical properties.

b. Electrical Strength

The guy insulator shall have a rated dry flashover voltage at least double, and a rated wet flashover voltage at least as high as, the nominal line voltage between conductors of the guyed circuit. A guy insulator may consist of one or more units.

c. Mechanical Strength

The rated ultimate strength of the guy insulator shall be at least equal to the required strength of the guy in which it is installed.

2. Use of Guy Insulators

a. Ungrounded guys shall be insulated if attached to a supporting structure carrying any supply conductors of more than 300 V or if vulnerable to accidental energization by such conductors due to a slack conductor or guy.

NOTE: Guys grounded in accordance with Rule 215C2 need not be insulated.

EXCEPTION: A guy insulator is not required if the guy is attached to a supporting structure on private right-of-way if all the supply circuits exceeding 300 V meet the requirements of Rule 220B2.

b. Insulators shall be installed as follows:

(1) All insulators shall be located at a position that maintains the bottom of the insulator not less than 2.45 m (8 ft) above the ground if the guy is broken below the insulator.

(2) Insulators shall be so placed that, in case any guy contacts, or is contacted by, an energized conductor or part, the voltage will not be transferred to other facilities on the structure(s).

EXCEPTION: Guy insulators are not required if both of the following conditions are fulfilled.

(a) The supply voltage will be promptly removed by de-energization or by other means, both initially and following subsequent circuit breaker operations in the event of a contact.

(b) The voltage and current impressed on the other facilities in the event of a contact are not in excess of the safe operating limit of the protective devices of such facility.

(3) Insulators shall be so placed that in case any guy sags down upon another, the insulators will not become ineffective.

3. Corrosion Protection

An insulator in the guy strand used exclusively for the elimination of corrosion of metal in ground rods, anchors, anchor rods, or pipe in an effectively grounded system shall not be classified as a guy insulator and shall not reduce the mechanical strength of the guy.

B. Span-Wire Insulators

1. Properties of Span-Wire Insulators

a. Material

Insulators shall be made of wet-process porcelain, wood, fiberglass, or other material of suitable mechanical and electrical properties.

b. Insulation Level

The insulation level of span-wire insulators shall meet the requirements of Rule 274.

A hanger insulator, where used to provide single insulation as permitted by Rule 279B2, shall meet the requirements of Rule 274.

c. Mechanical Strength

The rated ultimate strength of the span-wire insulator shall be at least equal to the required strength of the span wire in which it is located.

2. Use of Span-Wire Insulators

a. All span wires, including bracket span wires, shall have a suitable insulator (in addition to an insulated hanger if used) inserted between each point of support of the span wire and the luminaire or trolley-contact conductor supported.

EXCEPTION 1: Single insulators, as provided by an insulated hanger, may be permitted when the span wire or bracket is supported on wood poles supporting only trolley, railway feeder, or communication conductors used in the operation of the railway concerned.

EXCEPTION 2: Insulators are not required if the span wire is effectively grounded.

EXCEPTION 3: This rule does not apply to insulated feeder taps used as span wires.

b. In case insulated hangers are not used, the insulator shall be located so that in the event of a broken wire the energized part of the span wire cannot be reached from the ground.

Section number 28 not used in this edition.

Section number 29 not used in this edition.

Part 3.
Safety Rules for the Installation and Maintenance of Underground Electric Supply and Communication Lines

Section 30.
Purpose, Scope, and Application of Rules

300. Purpose

The purpose of Part 3 of this code is the practical safeguarding of persons during the installation, operation, or maintenance of underground or buried supply and communication cables and associated equipment.

301. Scope

Part 3 of this code covers supply and communication cables and equipment in underground or buried systems. The rules cover the associated structural arrangements and the extension of such systems into buildings. It also covers the cables and equipment employed primarily for the utilization of electric power when such cables and equipment are used by the utility in the exercise of its function as a utility. They do not cover installations in electric supply stations.

302. Application of Rules

The general requirements for application of these rules are contained in Rule 013.

Section 31.
General Requirements
Applying to Underground Lines

310. Referenced Sections

The Introduction (Section 1), Definitions (Section 2), List of Referenced Documents (Section 3), and Grounding Methods (Section 9) of this code shall apply to the requirements of Part 3.

311. Installation and Maintenance

A. Persons responsible for underground facilities shall be able to indicate the location of their facilities.

B. Reasonable advance notice should be given to owners or operators of other proximate facilities that may be adversely affected by new construction or changes in existing facilities.

312. Accessibility

All parts that must be examined or adjusted during operation shall be arranged so as to be readily accessible to authorized persons by the provision of adequate working spaces, working facilities, and clearances.

313. Inspection and Tests of Lines and Equipment

A. When In Service

1. Initial Compliance With Safety Rules
   Lines and equipment shall comply with these safety rules upon being placed in service.
2. Inspection
   Accessible lines and equipment shall be inspected by the responsible party at such intervals as experience has shown to be necessary.
3. Tests
   When considered necessary, lines and equipment shall be subjected to practical tests to determine required maintenance.
4. Record of Defects
   Any defects affecting compliance with this code revealed by inspection, if not promptly corrected, shall be recorded; such record shall be maintained until the defects are corrected.
5. Remedying Defects
   Lines and equipment with recorded defects that would endanger life or property shall be promptly repaired, disconnected, or isolated.

B. When Out of Service

1. Lines Infrequently Used
   Lines and equipment infrequently used shall be inspected or tested as necessary before being placed into service.
2. Lines Temporarily Out of Service
   Lines and equipment temporarily out of service shall be maintained in a safe condition.
3. Lines Permanently Abandoned
   Lines and equipment permanently abandoned shall be removed or maintained in a safe condition.

314. Grounding of Circuits and Equipment

A. Methods

The methods to be used for grounding of circuits and equipment are given in Section 9.

B. Conductive Parts to Be Grounded

Cable sheaths and shields (except conductor shields), equipment frames and cases (including pad-mounted devices), and conductive lighting poles shall be effectively grounded. Conductive-material ducts and riser guards that enclose electric supply lines or are exposed to contact with open supply conductors of greater than 300 V shall be effectively grounded.

EXCEPTION: This rule does not apply to parts that are 2.45 m (8 ft) or more above readily accessible surfaces or are otherwise isolated or guarded.

C. Circuits
   1. Neutrals
      Primary neutrals, secondary and service neutrals, and common neutrals shall be effectively grounded as specified in Rule 314A.
      EXCEPTION: Circuits designed for ground-fault detection and impedance current-limiting devices.
   2. Other Conductors
      Conductors, other than neutral conductors, that are intentionally grounded, shall be effectively grounded as specified in Rule 314A.
   3. Surge Arresters
      Surge arresters shall be effectively grounded as specified in Rule 314A.
   4. Use of Earth as Part of Circuit
      a. Supply circuits shall not be designed to use the earth normally as the sole conductor for any part of the circuit.
      b. Monopolar operation of a bipolar HVDC system is permissible for emergencies and limited periods for maintenance.

315. Communications Protective Requirements

A. Where Required
   Where communications apparatus is handled by other than qualified persons, it shall be protected by one or more of the means listed in Rule 315B if such apparatus is permanently connected to lines subject to any of the following:
   1. Lightning
   2. Contact with supply conductors with voltages exceeding 300 V
   3. Transient rise in ground potential exceeding 300 V
   4. Steady-state induced voltage of a level that may cause personal injury
      NOTE: When communication cables will be in the vicinity of supply stations where large ground currents may flow, the effect of these currents on communication circuits should be evaluated.

B. Means of Protection
   Where communications apparatus is required to be protected under Rule 315A, protective means adequate to withstand the voltage expected to be impressed shall be provided by insulation, protected where necessary by surge arresters. Severe conditions may require the use of additional devices such as auxiliary arresters, drainage coils, neutralizing transformers, or isolating devices.

316. Induced Voltage

Rules covering supply-line influence and communication-line susceptiveness have not been detailed in this code. Cooperative procedures are recommended to minimize steady-state voltages induced from proximate facilities. Therefore, reasonable advance notice should be given to owners or operators of other known proximate facilities that may be adversely affected by new construction or changes in existing facilities.

Section 32.
Underground Conduit Systems

NOTE 1: While it is often the practice to use *duct* and *conduit* interchangeably, *duct,* as used herein, is a single enclosed raceway for conductors or cable; *conduit* is a structure containing one or more ducts; and *conduit system* is the combination of conduit, conduits, manholes, handholes, and/or vaults joined to form an integrated whole.

NOTE 2: For cables installed in a single duct not part of a conduit system, the rules in Section 35 apply.

320. Location

A. Routing

1. General

   a. Conduit systems should be subject to the least disturbance practical. Conduit systems extending parallel to other subsurface structures should not be located directly over or under other subsurface structures. If this is not practical, the rule on separation, as stated in Rule 320B, should be followed.

   b. Conduit alignment should be such that there are no protrusions that would be harmful to the cable.

   c. Where bends are required, the bending radius shall be sufficiently large to limit the likelihood of damage to cable being installed in the conduit.

2. Natural Hazards

   Routes through unstable soils such as mud, shifting soil, etc., or through highly corrosive soils, should be avoided. If construction is required in these soils, the conduit should be constructed in such a manner as to minimize movement or corrosion or both.

3. Highways and Streets

   Where conduit must be installed longitudinally under the roadway, it should be installed in the shoulder. If this is not practical, the conduit should be installed within the limits of one lane of traffic.

4. Bridges and Tunnels

   The conduit system shall be located so as to limit the likelihood of damage by traffic. It should be located to provide safe access for inspection or maintenance of both the structure and the conduit system.

5. Crossing Railroad Tracks

   a. The top of the conduit system should be located not less than 900 mm (36 in) below the top of the rails of a street railway or 1.27 m (50 in) below the top of the rails of a railroad. Where unusual conditions exist or where proposed construction would interfere with existing installations, a greater depth than specified above may be required.

      EXCEPTION: Where this is impractical, or for other reasons, this separation may be reduced by agreement between the parties concerned. In no case, however, shall the top of the conduit or any conduit protection extend higher than the bottom of the ballast section that is subject to working or cleaning.

   b. At crossings under railroads, manholes, handholes, and vaults should not, where practical, be located in the roadbed.

6. Submarine Crossing

   Submarine crossings should be routed, installed, or both so they will be protected from erosion by tidal action or currents. They should not be located where ships normally anchor.

B. Separation From Other Underground Installations

1. General

   The separation between a conduit system and other underground structures paralleling it should be as large as necessary to permit maintenance of the system without damage to the paralleling structures. A conduit that crosses over another subsurface structure shall have a separation sufficient to limit the likelihood of damage to either structure. These separations should be determined by the parties involved.

EXCEPTION: When conduit crosses a manhole, vault, or subway tunnel roof, it may be supported directly on the roof with the concurrence of all parties involved.

2. Separations Between Supply and Communication Conduit Systems

   Conduit systems to be occupied by communication conductors shall be separated from conduit systems to be used for supply systems by

   a. 75 mm (3 in) of concrete
   b. 100 mm (4 in) of masonry
   c. 300 mm (12 in) of well-tamped earth

   EXCEPTION: Lesser separations may be used where the parties concur.

3. Sewers, Sanitary and Storm

   a. If conditions require a conduit to be installed parallel to and directly over a sanitary or storm sewer, it may be done provided both parties are in agreement as to the method.
   b. Where a conduit run crosses a sewer, it shall be designed to have suitable support on each side of the sewer to limit the likelihood of transferring any direct load onto the sewer.

4. Water Lines

   Conduit should be installed as far as is practical from a water main in order to protect it from being undermined if the main breaks. Conduit that crosses over a water main shall be designed to have suitable support on each side as required to limit the likelihood of transferring any direct loads onto the main.

5. Gas and Other Fuel Lines

   Conduit should have sufficient separation from gas and other fuel lines to permit the use of pipe maintenance equipment. Conduit shall not enter the same manhole, handhole, or vault with gas or other fuel lines.

6. Steam Lines

   Conduit should be installed so as to limit the likelihood of detrimental heat transfer between the steam and conduit systems.

321. Excavation and Backfill

A. Trench

The bottom of the trench should be undisturbed, tamped, or relatively smooth earth. Where the excavation is in rock, the conduit should be laid on a protective layer of clean tamped backfill.

B. Quality of Backfill

All backfill should be free of materials that may damage the conduit system.

RECOMMENDATION: Backfill within 150 mm (6 in) of the conduit should be free of solid material greater than 100 mm (4 in) in maximum dimension or with sharp edges likely to damage it. The balance of backfill should be free of solid material greater than 200 mm (8 in) in maximum dimension. Backfill material should be adequately compacted.

322. Ducts and Joints

A. General

1. Duct material shall be corrosion-resistant and suitable for the intended environment.
2. Duct materials, the construction of the conduit, or both shall be designed so that a cable fault in one duct would not damage the conduit to such an extent that it would cause damage to cables in adjacent ducts.
3. The conduit system shall be designed to withstand external forces to which it may be subjected by the surface loadings set forth in Rule 323A, except that impact loading may be reduced one third for each 300 mm (12 in) of cover so no impact loading need be considered when cover is 900 mm (3 ft) or more.
4. The internal surface of the duct shall be free of sharp edges or burrs, which could damage supply cable.

B. Installation

1. Restraint

   Conduit, including terminations and bends, should be suitably restrained by backfill, concrete envelope, anchors, or other means to maintain its design position under stress of installation procedures, cable pulling operations, and other conditions such as settling and hydraulic or frost uplift.

2. Joints

   Ducts shall be joined in a manner so as to limit solid matter from entering the conduit line. Joints shall form a sufficiently continuous smooth interior surface between joining duct sections so that supply cable will not be damaged when pulled past the joint.

3. Externally Coated Pipe

   When conditions are such that externally coated pipe is required, the coating shall be corrosion resistant and should be inspected, tested, or both, to see that the coating is continuous and intact prior to backfill. Precautions shall be taken to prevent damage to the coating when backfilling.

4. Building Walls

   Conduit installed through a building wall shall have internal and external seals intended to limit the likelihood of the entrance of gas into the building. The use of seals may be supplemented by gas-venting devices in order to minimize building up of positive gas pressures in the conduit.

5. Bridges

   a. Conduit installed in bridges shall include the capability to allow for expansion and contraction of the bridge.
   b. Conduits passing through a bridge abutment should be installed so as to avoid or resist any shear due to soil settlement.
   c. Conduit of conductive material installed on bridges shall be effectively grounded.

6. In Vicinity of Manholes

   Conduit should be installed on compacted soil or otherwise supported when entering a manhole to limit the likelihood of detrimental shear stress on the conduit at the point of manhole entrance.

323. Manholes, Handholes, and Vaults

A. Strength

Manholes, handholes, and vaults shall be designed to sustain all expected loads that may be imposed upon the structure. The horizontal design loads, vertical design loads, or both shall consist of dead load, live load, equipment load, impact, load due to water table, frost, and any other load expected to be imposed upon the structure, to occur adjacent to the structure, or both. The structure shall sustain the combination of vertical and lateral loading that produces the maximum shear and bending moments in the structure.

1. In roadway areas, the live load shall consist of the weight of a moving tractor-semitrailer truck illustrated in Fig 323-1. The vehicle wheel load shall be considered applied to an area as indicated in Fig 323-2. In the case of multilane pavements, the structure shall sustain the combination of loadings that results in vertical and lateral structure loadings that produce the maximum shear and bending moments in the structure.

   NOTE: Loads imposed by equipment used in road construction may exceed loads to which the completed road may be subjected.

2. In designing structures not subject to vehicular loading, the design live load shall be not less than 14.5 kPa (300 lb/ft^2).
3. Live loads shall be increased by 30% for impact.
4. When hydraulic, frost, or other uplift will be encountered, the structure shall either be of sufficient weight or so restrained as to withstand this force. The weight of equipment installed in the structure is not to be considered as part of the structure weight.
5. Where pulling iron facilities are furnished, they should be installed to withstand twice the expected load to be applied to the pulling iron.

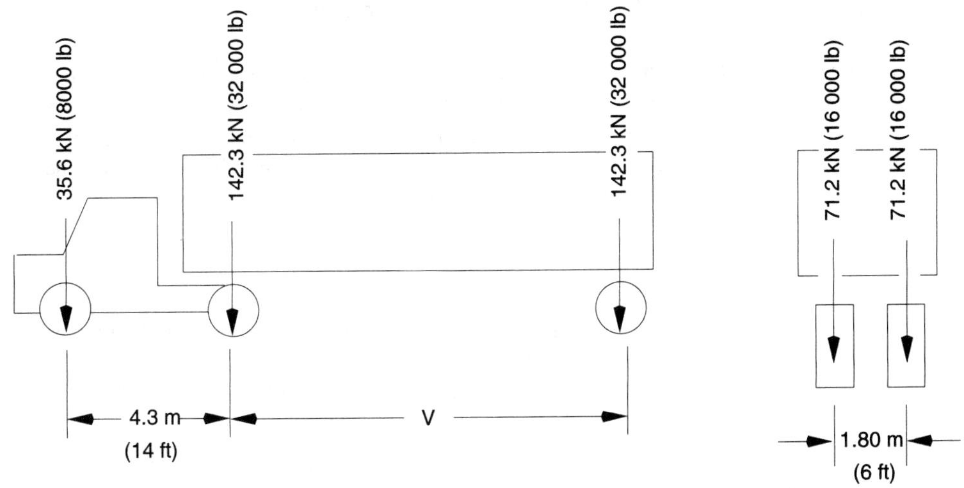

V = Variable spacing, 4.3 m to 9.0 m (14 ft to 30 ft), inclusive. Spacing to be used is that which results in vertical and lateral structure loading that produces the maximum shear and bending moments in the structure.

Fig 323-1
Roadway Vehicle Load

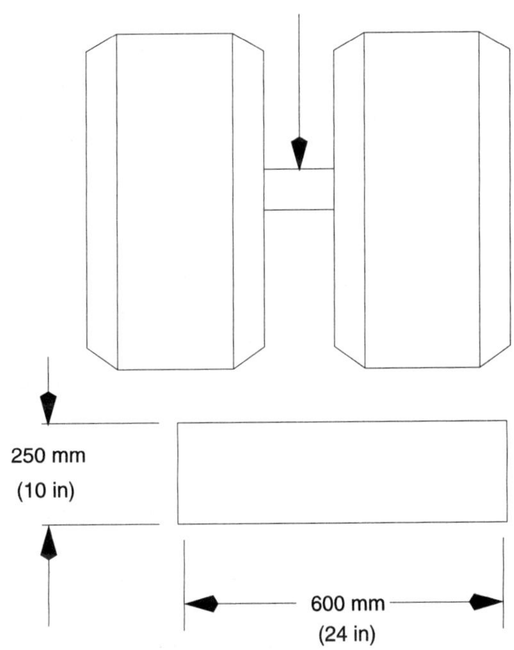

Fig 323-2
Wheel Load Area

B. Dimensions

Manholes shall meet the following requirements: A clear working space sufficient for performing the necessary work shall be maintained. The horizontal dimensions of the clear working space shall be not less than 900 mm (3 ft). The vertical dimensions shall be not less than 1.83 m (6 ft) except in manholes where the opening is within 300 mm (1 ft), horizontally, of the adjacent interior side wall of the manhole.

EXCEPTION 1: Where one boundary of the working space is an unoccupied wall and the opposite boundary consists of cables only, the horizontal working space between these boundaries may be reduced to 750 mm (30 in).

EXCEPTION 2: In manholes containing only communication cables, equipment, or both, one horizontal dimension of the working space may be reduced to not less than 600 mm (2 ft), provided the other horizontal dimension is increased so that the sum of the two dimensions is at least 1.83 m (6 ft).

C. Manhole Access

1. Round access openings in a manhole containing supply cables shall be not less than 650 mm (26 in) in diameter. Round access openings in any manhole containing communication cables only, or manholes containing supply cables and having a fixed ladder that does not obstruct the opening, shall be not less than 600 mm (24 in) in diameter. Rectangular access openings should have dimensions not less than 650 mm × 560 mm (26 in × 22 in).
2. Openings shall be free of protrusions that will injure personnel or prevent quick egress.
3. Manhole openings shall be located so that safe access can be provided. When in the highway, they should be located outside of the paved roadway when practical. They should be located outside the area of street intersections and crosswalks whenever practical to reduce the traffic hazards to the workers at these locations.
4. Personnel access openings should be located so that they are not directly over the cable or equipment. When these openings interfere with curbs, etc., they can be located over the cable if one of the following is provided:
   a. A conspicuous safety sign

      NOTE: ANSI Z535.1-1998, ANSI Z535.2-1998, ANSI Z535.3-1998, ANSI Z535.4-1998, and ANSI Z535.5-1998 contain information regarding safety signs.
   b. A protective barrier over the cable
   c. A fixed ladder
5. Any manhole greater than 1.25 m (4 ft) in depth shall be designed so it can be entered by means of a ladder or other suitable climbing device. Equipment, cable, and hangers are not suitable climbing devices.

D. Covers

1. Manholes and handholes, when not being worked in, shall be securely closed by covers of sufficient weight or proper design so they cannot be easily removed without tools.
2. Covers should be suitably designed or restrained so that they cannot fall into manholes or protrude into manholes sufficiently far to contact cable or equipment.
3. Strength of covers and their supporting structure shall be at least sufficient to sustain the applicable loads of Rule 323A.

E. Vault and Utility Tunnel Access

1. Access openings shall be located so that safe access can be provided.
2. Personnel access openings in vaults should be located so that they are not directly over or do not directly open into equipment or cable. In vaults, other types of openings (not personnel access) may be located over equipment to facilitate work on, replacement, or installation of equipment.
3. Where accessible to the public, access doors to utility tunnels and vaults shall be locked unless qualified persons are in attendance to restrict entry by unqualified persons. When vaults and utility tunnels contain exposed live parts, a prominent safety sign shall be visibly posted before entering the vault.

   NOTE: ANSI Z535.1-1998, ANSI Z535.2-1998, ANSI Z535.3-1998, ANSI Z535.4-1998, and ANSI Z535.5-1998 contain information regarding safety signs.

4. Such doors shall be designed so that a person on the inside may exit when the door is locked from the outside.

   EXCEPTION: This rule does not apply where the only means of locking is by padlock and the latching system is so arranged that the padlock can be closed on the latching system to prevent locking from the outside.

F. Ladder Requirements

Fixed ladders shall be corrosion-resistant. Portable ladders shall be used in accordance with Rule 420J.

RECOMMENDATION: Ladders should conform to ANSI A14.1-1994 [B1], ANSI A14.2-1990 [B2], ANSI A14.3-1992 [B3], or ANSI A14.5-1992 [B4].

G. Drainage

Where drainage is into sewers, suitable traps or other means should be provided to limit the likelihood of sewer gas entering into manholes, vaults, or tunnels.

H. Ventilation

Adequate ventilation to open air shall be provided for manholes, vaults, and tunnels, having an opening into enclosed areas used by the public. Where such enclosures house transformers, switches, regulators, etc., the ventilating system shall be cleaned at necessary intervals.

EXCEPTION: This does not apply to enclosed areas under water or in other locations where it is impractical to comply.

I. Mechanical Protection

Supply cables and equipment should be installed or guarded in such a manner as to avoid damage by objects falling or being pushed through the grating.

J. Identification

Manhole and handhole covers should have an identifying mark that will indicate ownership or type of utility.

Section 33.
Supply Cable

330. General

RECOMMENDATION: Cable should be capable of withstanding tests applied in accordance with an applicable standard issued by a recognized organization such as the American National Standards Institute (ANSI), the Association of Edison Illuminating Companies (AEIC), the Insulated Cable Engineers Association (ICEA), the National Electrical Manufacturers Association (NEMA), or the American Society for Testing and Materials (ASTM).

A. The design and construction of conductors, insulation, sheath, jacket, and shielding shall include consideration of mechanical, thermal, environmental, and electrical stresses that are expected during installation and operation.

B. Cable shall be designed and manufactured to retain specified dimensions and structural integrity during manufacture, reeling, storage, handling, and installation.

C. Cable shall be designed and constructed in such a manner that each component is protected from harmful effects of other components.

D. The conductor, insulation, and shielding shall be designed to withstand the effects of the expected magnitude and duration of fault current, except in the immediate vicinity of the fault.

331. Sheaths and Jackets

Sheaths, jackets, or both shall be provided when necessary to protect the insulation or shielding from moisture or other adverse environmental conditions.

332. Shielding

A. General
1. Conductor shielding should, and insulation shielding shall, be provided as specified by an applicable document issued by a nationally recognized cable standardization organization.
   NOTE: Typical cable standardization organizations include: the AEIC, the ICEA, and the NEMA.
   EXCEPTION: Shielding is not required for short jumpers that do not contact a grounded surface within enclosures or vaults, provided the jumpers are guarded or isolated.
2. Insulation shielding may be sectionalized provided that each section is effectively grounded.

B. Material
1. The shielding system may consist of semiconducting materials, nonmagnetic metal, or both. The shielding adjacent to the insulation shall be designed to remain in intimate contact with the insulation under all operating conditions.
2. Shielding material shall either be designed to resist excessive corrosion under the expected operating conditions or shall be protected.

333. Cable Accessories and Joints

A. Cable accessories and joints shall be designed to withstand the mechanical, thermal, environmental, and electrical stresses expected during operation.

B. Cable accessories and joints shall be designed and constructed in such a manner that each component of the cable and joint is protected from harmful effects of the other components.

C. Cable accessories and joints shall be designed and constructed to maintain the structural integrity of the cables to which they are applied and to withstand the magnitude and duration of the fault current expected during operation, except in the immediate vicinity of the fault.

D. For insulating joints, see Rule 332A2.

Section 34. Cable in Underground Structures

340. General

A. Section 33 shall apply to supply cable in underground structures.

B. On systems operating above 2 kV to ground, the design of the conductors or cables installed in nonmetallic conduit should consider the need for an effectively grounded shield, a sheath, or both.

341. Installation

A. General

1. Bending of the supply cable during handling, installation, and operation shall be controlled to avoid damage.
2. Pulling tensions and sidewall pressures on the supply cable should be limited to avoid damage.
   NOTE: Manufacturers' recommendations may be used as a guide.
3. Ducts should be cleaned of foreign material that could damage the supply cable during pulling operations.
4. Cable lubricants shall not be detrimental to cable or conduit systems.
5. On slopes or vertical runs, consideration should be given to restraining cables to limit the likelihood of downhill movement.
6. Supply cables shall not be installed in the same duct with communication cables unless all of the cables are operated and maintained by the same utility.
7. Communication cables may be installed together in the same duct provided all utilities involved are in agreement.

B. Cable in Manholes and Vaults

1. Supports
   a. Cable supports shall be designed to withstand both live and static loading and should be compatible with the environment.
   b. Supports shall be provided to maintain specified clearance between cables.
   c. Horizontal runs of supply cables shall be supported at least 75 mm (3 in) above the floor, or shall be suitably protected.
      EXCEPTION: This rule does not apply to grounding or bonding conductors.
   d. The installation should allow cable movement without destructive concentration of stresses. The cable should remain on supports during operation.
      NOTE: Special protection may be necessary at the duct entrance.
2. Clearance
   a. Adequate working space shall be provided in accordance with Rule 323B.
   b. Between supply and communications facilities (cable, equipment, or both):
      (1) Where cable, equipment, or both are to be installed in a joint-use manhole or vault, it shall be done only with the concurrence of all parties concerned.
      (2) Supply and communication cables should be racked from separate walls. Crossings should be avoided.
      (3) Where supply and communication cables must be racked from the same wall, the supply cables should be racked below the communication cables.
      (4) Supply and communications facilities shall be installed to permit access to either without moving the other.
      (5) Clearances shall be not less than those specified in Table 341-1.

Table 341-1
Clearance Between Supply and Communications Facilities in Joint-Use Manholes and Vaults

Phase-to-phase supply voltage	Surface to surface	
	(mm)	(in)
0 to 15 000	150	6
15 001 to 50 000	230	9
50 001 to 120 000	300	12
120 001 and above	600	24

EXCEPTION 1: These clearances do not apply to grounding conductors.
EXCEPTION 2: These clearances may be reduced by mutual agreement between the parties concerned when suitable barriers or guards are installed.

3. Identification
   a. General
      (1) Cables shall be permanently identified by tags or otherwise at each manhole or other access opening of the conduit system.
      EXCEPTION: This requirement does not apply where the position of a cable, in conjunction with diagrams or maps supplied to workers, gives sufficient identification.
      (2) All identification shall be of a corrosion-resistant material suitable for the environment.
      (3) All identification shall be of such quality and located so as to be readable with auxiliary lighting.
   b. Joint-Use Manholes and Vaults
   Cables in a manhole or vault that are operated and maintained by different utilities shall be permanently identified by markings or tags denoting the utility name and type of cable use.

342. Grounding and Bonding

A. Insulation shielding of cable and joints shall be effectively grounded.

B. Cable sheaths or shields that are connected to ground at a manhole shall be bonded or connected to a common ground.

C. Bonding and grounding leads shall be of a corrosion-resistant material suitable for the environment or suitably protected.

343. Fireproofing

Although fireproofing is not a requirement, it may be provided in accordance with each utility's normal service reliability practice to provide protection from external fire.

344. Communication Cables Containing Special Supply Circuits

A. Special circuits operating at voltages in excess of 90 V ac or 150 V dc and used for supplying power solely to communications equipment may be included in communication cables under the following conditions:
   1. Such cables shall have a conductive sheath or shield that shall be effectively grounded and each such circuit shall be carried on conductors that are individually enclosed with an effectively grounded shield.
   2. All circuits in such cables shall be owned or operated by one party and shall be maintained only by qualified personnel.

3. Supply circuits included in such cables shall be terminated at points accessible only to qualified employees.
4. Communication circuits brought out of such cables, if they do not terminate in a repeater station or terminal office, shall be protected or arranged so that in event of a failure within the cable, the voltage on the communication circuit will not exceed 400 V to ground.
5. Terminal apparatus for the power supply shall be so arranged that live parts are inaccessible when such supply circuits are energized.
6. Such cables shall be identified, and the identification shall meet the pertinent requirements of Rule 341B3.

   EXCEPTION: The requirements of Rule 344A do not apply to communication circuits where the transmitted power does not exceed 150 W.

Section 35.
Direct-Buried Cable

350. General

A. Section 33 shall apply to direct-buried supply cable.

B. Cables operating above 600 V to ground shall have a continuous metallic shield, sheath, or concentric neutral that is effectively grounded.

EXCEPTION: At a splice or joint, the current path of the metallic shield, sheath, or neutral shall be made continuous but need not be concentric.

C. Cables meeting Rule 350B of the same supply circuit may be buried with no deliberate separation.

D. Cables of the same circuit operating below 600 V to ground and without an effectively grounded shield or sheath shall be placed in close proximity (no intentional separation) to each other.

E. Communication cables containing special circuits supplying power solely to communications equipment shall comply with the requirements of Rules 344A1 through 344A5.

F. Bonding should be provided between all above ground metallic power and communications apparatus (pedestals, terminals, apparatus cases, transformer cases, etc.) that are separated by a distance of 1.8 m (6 ft) or less.

G. All direct-buried jacketed supply cable meeting Rule 350B and all direct-buried communication cables shall be legibly marked as follows:

The appropriate identification symbol shown in Fig 350-1 shall be indented or embossed in the outermost cable jacket at a spacing of not more than 1 m (40 in). The symbol may be separate or sequentially combined with other data, or symbols, or both, printed on the jacket. If the symbol is sequentially combined, it shall be separated as indicated in Fig 350-1.

This rule became effective for cable installed on or after 1 January 1996.

EXCEPTION 1: Cables with jackets that cannot be effectively marked in accordance with Rule 350G need not be marked.

EXCEPTION 2: Unmarked cable from stock existing prior to 1 January 1996 may be used to repair unmarked direct-buried jacketed supply cables and communication cables.

H. The rules in this section also apply to direct-buried supply and communication cables installed in duct that is not part of a conduit system.

351. Location and Routing

A. General

1. Cables should be located so as to be subject to the least disturbance practical. When cables are to be installed parallel to and directly over or under other subsurface structures, the rules on separation in Rule 353 or 354, as applicable, shall be followed.
2. Cables should be installed in as straight and direct a line as practical. Where bends are required, the bending radius shall be sufficiently large to limit the likelihood of damage to the cable being installed.
3. Cable systems should be routed so as to allow safe access for construction, inspection, and maintenance.
4. The location of structures in the path of the projected cable route shall, as far as practical, be determined prior to trenching, plowing, or boring operation.

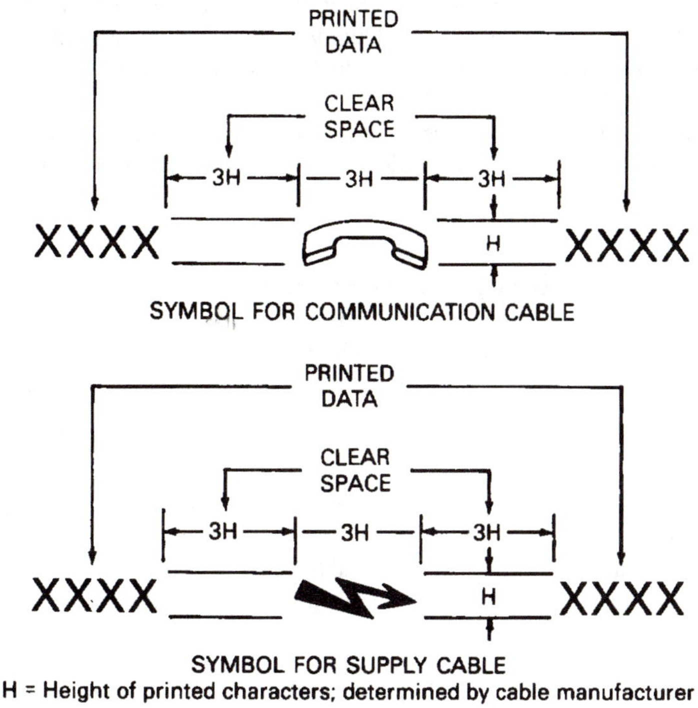

Fig 350-1
Symbols for Identification of Buried Cables

B. Natural Hazards

Routes through unstable soil such as mud, shifting soils, corrosive soils, or other natural hazards should be avoided. If burying is required through areas with natural hazards, the cables shall be constructed and installed in such a manner as to protect them from damage. Such protective measures should be compatible with other installations in the area.

C. Other Conditions

1. Swimming Pools

   Supply cable should not be installed within 1.5 m (5 ft) of a swimming pool or its auxiliary equipment. If 1.5 m (5 ft) is not attainable, supplemental mechanical protection shall be provided.

2. Buildings and Other Structures

   Cable should not be installed directly under building or storage tank foundations. Where a cable must be installed under such a structure, the structure shall be suitably supported to limit the likelihood of transfer of a detrimental load onto the cable.

3. Railroad Tracks

   a. The installation of cable longitudinally under the ballast section for railroad tracks should be avoided. Where cable must be installed longitudinally under the ballast section of a railroad, it should be located at a depth of not less than 1.27 m (50 in) below the top of the rail.

      EXCEPTION: Where this is impractical, or for other reasons, this clearance may be reduced by agreement between the parties concerned.

      NOTE: Where unusual conditions exist or where proposed construction would interfere with existing installations, a greater depth than specified above would be required.

   b. Where a cable crosses under railroad tracks, the same clearances indicated in Rule 320A5 shall apply.

4. Highways and Streets
   The installation of cable longitudinally under traveled surfaces of highways and streets should be avoided. When cable must be installed longitudinally under the roadway, it should be installed in the shoulder or, if this is not practical, within the limits of one lane of traffic to the extent practical.
5. Submarine Crossings
   Submarine crossings should be routed, installed, or both, so they will be protected from erosion by tidal action or currents. They should not be located where ships normally anchor.

352. Installation

A. Trenching

The bottom of the trench receiving direct-buried cable should be relatively smooth, undisturbed earth; well-tamped earth; or sand. When excavation is in rock or rocky soils, the cable should be laid on a protective layer of well-tamped backfill. Backfill within 100 mm (4 in) of the cable should be free of materials that may damage the cable. Backfill should be adequately compacted. Machine compaction should not be used within 150 mm (6 in) of the cable.

B. Plowing

1. Plowing in of cable in soil containing rock or other solid material should be done in such a manner that the solid material will not damage the cable, either during the plowing operation or afterward.
2. The design of cable-plowing equipment and the plowing-in operation should be such that the cable will not be damaged by bending, side-wall pressure, or excessive cable tension.

C. Boring

Where a cable system is to be installed by boring and the soil and surface loading conditions are such that solid material in the region may damage the cable, the cable shall be adequately protected.

D. Depth of Burial

1. The distance between the top of a cable and the surface under which it is installed (depth of burial) shall be sufficient to protect the cable from damage imposed by expected surface usage.
2. Burial depths as indicated in Table 352-1 are considered adequate for supply cables or conductors, except as noted in a, b, or c following:
   a. In areas where frost conditions could damage cables, greater burial depths than indicated above may be desirable.
   b. Lesser depths than indicated above may be used where supplemental protection is provided. The supplemental protection should be sufficient to protect the cable from damage imposed by expected surface usage.
   c. Where the surface under which a cable is to be installed is not to final grade, the cable should be placed so as to meet or exceed the requirements indicated above, both at the time of installation and subsequent thereto.

Table 352-1
Supply Cable or Conductor Burial Depth
(See Rule 352D)

Voltage (Phase-to-phase)	Depth of burial	
	(mm)	**(in)**
0 to 600	600	24
601 to 50 000	750	30
50 001 and above	1070	42

EXCEPTION: Where conflicts with other underground facilities exist, street and area lighting cables operating at not more than 150 V to ground may be buried at a depth not less than 450 mm (18 in).

353. Deliberate Separations—Equal to or Greater than 300 mm (12 in) From Underground Structures or Other

A. General

1. These rules apply to a radial separation of supply and communication cables or conductors from each other and from other underground structures such as sewers, water lines, gas and other fuel lines, building foundations, steam lines, etc., when separation is equal to or greater than 300 mm (12 in).

   NOTE: For radial separation less than 300 mm (12 in) see Rule 354.

2. The radial separation should be adequate to permit access to and maintenance of either facility to limit damage to the other.

B. Crossings

1. Where a cable crosses under another underground structure, the structure shall be suitably supported to limit the likelihood of transfering of a detrimental load onto the cable system.
2. Where a cable crosses over another underground structure, the cable shall be suitably supported to limit the likelihood of transfering a detrimental load onto the structure.
3. Adequate support may be provided by installing the facilities with sufficient vertical separation.

C. Parallel Facilities

Where a cable system is to be installed directly over and parallel to another underground structure (or another underground structure installed directly over and parallel to a cable), it may be done providing all parties are in agreement as to the method. Adequate vertical separation shall be maintained to permit access to and maintenance of either facility without damage to the other.

D. Thermal Protection

Cable should be installed with sufficient separation from other underground structures, such as steam or cryogenic lines, to avoid thermal damage to the cable. Where it is not practical to provide adequate clearance, a suitable thermal barrier shall be placed between the two facilities.

354. Random Separation—Separation Less Than 300 mm (12 in) from Underground Structures or Other Cables

A. General

1. These rules apply to a radial separation of supply and communication cables or conductors from each other and from other underground structures when the radial separation between them will be less than 300 mm (12 in).
2. Radial separation of supply and communications cables or conductors from steam lines, gas, and other fuel lines shall be not less than 300 mm (12 in) and shall meet Rule 353.
3. Supply circuits operating above 300 V to ground or 600 V between conductors shall be so constructed, operated, and maintained that when faulted, they shall be promptly de-energized initially or following subsequent protective device operation (phase-to-ground faults for grounded circuits, phase-to-phase faults for ungrounded circuits).
4. Communication cables and conductors, and supply cables and conductors buried in random separation may be treated as one system when considering separation from other underground structures or facilities.

B. Supply Cables or Conductors

The cables or conductors of a supply circuit and those of another supply circuit may be buried together at the same depth with no deliberate separation between facilities, provided all parties involved are in agreement.

C. Communication Cables or Conductors

The cables or conductors of a communication circuit and those of another communication circuit may be buried together and at the same depth with no deliberate separation between facilities, provided all parties involved are in agreement.

D. Supply and Communication Cables or Conductors

Supply cables or conductors and communication cables or conductors may be buried together at the same depth, with no deliberate separation between facilities, provided all parties involved are in agreement and the applicable rules in 354D1 are met and either Rule 354D2 or 354D3 is met.

EXCEPTION: Entirely dielectric fiber-optic communication cables may be buried together at the same depth with no deliberate separation from supply cables or conductors provided all parties involved are in agreement and Rules 354D1a, b, c, and d are met.

1. General
   a. Grounded supply systems shall not be operated in excess of 22 000 V to ground.
   b. Ungrounded supply systems shall not be operated in excess of 5300 V phase to phase.
   c. Cables of an ungrounded supply system operating above 300 V shall be of effectively grounded concentric shield construction. Such cables shall be maintained in close proximity to each other.
   d. Ungrounded supply circuits operating above 300 V between conductors and in random separation with communication conductors shall be equipped with a ground-fault indication system.
   e. Communication cables and communication service wire having metallic conductors or metallic components shall have a continuous metallic shield under the outer jacket.

      EXCEPTION: This requirement does not apply to Rule 354C.
   f. Communications-protective devices shall be adequate for the voltage and currents expected to be impressed on them in the event of contact with the supply conductors.
   g. Adequate bonding shall be provided between the effectively grounded supply conductor or conductors and the communication cable shield or sheath at intervals that should not exceed 300 m (1000 ft).
   h. In the vicinity of supply stations where large ground currents may flow, the effect of these currents on communication circuits should be evaluated before communication cables are placed in random separation with supply cables.
2. Grounded Bare or Semiconducting Jacketed Neutral Supply Cables
   a. A supply facility operating above 300 V to ground shall include a bare or semiconducting jacketed grounded conductor in continuous contact with the earth. This conductor, adequate for the expected magnitude and duration of the fault current that may be imposed, shall be one of the following:
      (1) A sheath, an insulation shield, or both
      (2) Multiple concentric conductors closely spaced circumferentially
      (3) A separate conductor in contact with the earth and in close proximity to the cable, where such cable or cables also have a grounded sheath or shield not necessarily in contact with the earth. The sheath, shield, or both, as well as the separate conductor, shall be adequate for the expected magnitude and duration of the fault currents that may be imposed.

         NOTE: This is applicable when a cable in nonmetallic duct is considered as a direct-buried cable installation and random separation is desired.

         EXCEPTION: Where buried cable passes through a short section of conduit such as under a roadway, the contact with earth of the grounded conductor can be omitted, provided the grounded conductor is continuous through the conduit.
   b. The bare conductor or conductors in contact with the earth shall be of suitable corrosion-resistant material. The conductor covered by a semiconducting jacket shall be compatible with the jacketing compound.

      NOTE: Experience has shown that in many geographic areas, bare concentric copper neutral conductors experience severe corrosion.
   c. The radial resistivity of the semiconducting jacket shall be not more than 100 Ω•m and shall remain essentially stable in service. The radial resistivity of the jacket material is that value calculated from measurements on a unit length of cable, of the resistance between the concentric neutral and a surrounding conducting medium. Radial resistivity is equal to the resistance of a unit length times the surface area of the jacket divided by the average thickness of the jacket over the neutral conductors. All dimensions are to be expressed in meters.

3. Insulating Jacketed Grounded Neutral Supply Cables
   Each phase conductor of a multi-grounded supply system operating above 300 V to ground and having an overall insulating jacket shall have an effectively grounded copper concentric conductor meeting all of the following requirements:
   a. A conductance not less than one half that of the phase conductor.
   b. A conductance adequate for the expected magnitude and duration of fault current that may be imposed.
   c. A conductance grounded in accordance with Rule 314 except that the grounding interval required by Rule 96C shall be not less than eight in each 1.6 km (1 mile) of the random buried section, not including grounds at individual services.
4. Insulating Jacketed Grounded Neutral Supply Cables in Nonmetallic Duct
   Insulating jacketed grounded neutral supply cables meeting the rules of 354D3, when installed in nonmetallic duct, may be random-laid with communication cables.

E. Supply and Communication Cables or Conductors and Non-Metallic Water and Sewer Lines
   1. Supply cables and conductors and non-metallic water and sewer lines may be buried together with no deliberate separation between facilities and at the same depth, provided all parties involved are in agreement.
   2. Communication cables and conductors and non-metallic water and sewer lines may be buried together with no deliberate separation between facilities and at the same depth, provided all parties involved are in agreement.
   3. Supply cables or conductors, communication cables or conductors, non-metallic water and sewer lines may be buried together with no deliberate separation between facilities and at the same depth, provided the applicable rules in Rule 354D are met and all parties involved are in agreement.

Section 36. Risers

360. General

A. Mechanical protection for supply conductors or cables shall be provided as required by Rule 239D of this code. This protection should extend at least 300 mm (1 ft) below ground level.

B. Supply conductors or cable should rise vertically from the cable trench with only such deviation as necessary to permit a reasonable cable-bending radius.

C. Exposed conductive pipes or guards containing supply conductors or cables shall be grounded in accordance with Rule 314.

361. Installation

A. The installation should be designed so that water does not stand in riser pipes above the frost line.

B. Conductors or cables shall be supported in a manner designed to limit the likelihood of damage to conductors, cables, or terminals.

C. Where conductors or cables enter the riser pipe or elbow, they shall be installed in such a manner that shall minimize the possibility of damage due to relative movement of the cable and pipe.

362. Pole Risers—Additional Requirements

A. Risers should be located on the pole in the safest available position with respect to climbing space and exposure to traffic damage.

B. The number, size, and location of riser ducts or guards shall be limited to allow adequate access for climbing.

363. Pad-Mounted Installations

A. Supply conductors or cables rising from the trench to transformers, switchgear, or other equipment mounted on pads shall be so placed and arranged that they will not bear on the edges of holes through the pad nor the edges of bends or other duct work below the pad.

B. Cable entering pad-mounted equipment shall be maintained substantially at adequate depth for the voltage class until it becomes protected by being directly under the pad, unless other suitable mechanical protection is provided.

Section 37. Supply Cable Terminations

370. General

A. Cable terminations shall be designed and constructed to meet the requirements of Rule 333.

B. Riser terminations not located within a vault, pad-mounted equipment, or similar enclosure shall be installed in a manner designed to ensure that the clearance specified in Parts 1 and 2 of this code are maintained.

C. A cable termination shall be designed to limit the likelihood of moisture penetration into the cable where such penetration is detrimental to the cable.

D. Where clearances between parts at different potentials are reduced below those adequate for the voltage and BIL (basic impulse insulation level), suitable insulating barriers or fully insulated terminals shall be provided to meet the required equivalent clearances.

371. Support at Terminations

A. Cable terminations shall be installed in a manner designed to maintain their installed position.

B. Where necessary, cable shall be supported or secured in a manner designed to limit the likelihood of the transfer of damaging mechanical stresses to the termination, equipment, or structure.

372. Identification

Suitable circuit identification shall be provided for all terminations.

EXCEPTION: This requirement does not apply where the position of the termination, in conjunction with diagrams or maps supplied to workers, gives sufficient identification.

373. Clearances in Enclosures or Vaults

A. Adequate electrical clearances of supply terminations shall be maintained, both between conductors and between conductors and ground, consistent with the type of terminator used.

B. Where exposed live parts are in an enclosure, clearances or insulating barriers adequate for the voltages and the design BIL shall be provided.

C. Where a termination is in a vault, uninsulated live parts are permissible provided they are guarded or isolated.

374. Grounding

A. All exposed conducting surfaces of the termination device, other than live parts and equipment to which it is attached, shall be effectively grounded, bonded, or both.

B. Conductive structures supporting cable terminations shall be effectively grounded.

EXCEPTION: Grounding, bonding, or both is not required where the above parts are isolated or guarded.

Section 38. Equipment

380. General

A. Equipment includes:
   1. Buses, transformers, switches, etc., installed for the operation of the electric supply system
   2. Repeaters, loading coils, etc., installed for the operation of the communications system
   3. Auxiliary equipment, such as sump pumps, convenience outlets, etc., installed incidental to the presence of the supply or communications systems

B. Where equipment is to be installed in a joint-use manhole, it shall be done with the concurrence of all parties concerned.

C. Supporting structures, including racks, hangers, or pads and their foundations, shall be designed to sustain all loads and stresses expected to be imposed by the supported equipment including those stresses caused by its operation.

D. Pad-mounted equipment, pedestals, and other above ground enclosures, should be located not less than 1.2 m (4 ft) from fire hydrants.
EXCEPTION: Where conditions do not permit a clearance of 1.2 m (4 ft), a clearance of not less than 900 mm (3 ft) is allowed.

381. Design

A. The expected thermal, chemical, mechanical, and environmental conditions at the location shall be considered in the design of all equipment and mountings.

B. All equipment, including auxiliary devices, shall be designed to withstand the effects of normal, emergency, and fault conditions expected during operation.

C. Switches shall be provided with clear indication of contact position, and the handles or activating devices clearly marked to indicate operating directions.
RECOMMENDATION: The handles or control mechanism of all switches throughout the system should operate in a like direction to open and in a uniformly different direction to close in order to minimize errors.

D. Remotely controlled or automatic devices shall have local provisions to render remote or automatic controls inoperable if such operation may result in a hazard to the worker.

E. Enclosures containing fuses and interrupter contacts shall be designed to withstand the effects of normal, emergency, and fault conditions expected during operation.

F. When tools are to be used to connect or disconnect energized devices, space or barriers shall be designed to provide adequate clearance from ground or between phases.

G. Pad-Mounted and Other Above-Ground Equipment
   1. Pad-mounted and other above-ground equipment shall have an enclosure that is either locked or otherwise secured against unauthorized entry.
   2. Access to exposed live parts in excess of 600 V shall require two separate conscious acts. The first shall be the opening of a door or barrier that is locked or otherwise secured against unauthorized entry as required by Rule 381G1. The second act shall be either the opening of a door or the removal of a barrier.
   RECOMMENDATION: A prominent and appropriate safety sign should be visible when the first door or barrier is opened or removed.
   NOTE: ANSI Z535.1-1998, ANSI Z535.2-1998, ANSI Z535.3-1998, ANSI Z535.4-1998, and ANSI Z535.5-1998 contain information regarding safety signs.

382. Location in Underground Structures

A. Equipment shall not obstruct personnel access openings in manholes or vaults, nor shall it impede egress by persons working in the structures containing the equipment.

B. Equipment shall not be installed closer than 200 mm (8 in) to the back of fixed ladders and shall not interfere with the proper use of such ladders.

C. Equipment should be arranged in a manhole or vault to permit installation, operation, and maintenance of all items in such structures.

D. Switching devices that have provision for manual or electrical operation shall be operable from a safe position. This may be accomplished by use of portable auxiliary devices, temporarily attached.

E. Equipment should not interfere with drainage of the structure.

F. Equipment shall not interfere with the ability to ventilate any structure or enclosure.

383. Installation

A. Provisions for lifting, rolling to final position, and mounting shall be adequate for the weight of the device.

B. Live parts shall be guarded or isolated to limit the likelihood of contact by persons in a normal position adjacent to the equipment.

C. Operating levers, inspection facilities, and test facilities shall be visible and readily accessible when equipment is in final location without moving permanent connections.

D. Live parts shall be isolated or protected from exposure to conducting liquids or other material expected to be present in the structure containing the equipment.

E. Operating controls of supply equipment, readily accessible to unauthorized personnel, shall be secured by bolts, locks, or seals.

384. Grounding

A. Cases and enclosures made of conductive material shall be effectively grounded or guarded.

B. Guards constructed of conductive material shall be effectively grounded.

385. Identification

Where transformers, regulators, or other similar equipment operate in multiple, tags, diagrams, or other suitable means shall be used to indicate that fact.

Section 39.
Installation in Tunnels

390. General

A. The installation of supply and communications facilities in tunnels shall meet the applicable requirements contained elsewhere in Part 3 of this code as supplemented or modified by this section.

B. Where the space occupied by supply or communications facilities in a tunnel is accessible to other than qualified persons, or where supply conductors do not meet the requirements of Part 3 of this code for cable systems, the installation shall be in accordance with the applicable requirements of Part 2 of this code.

C. All parties concerned must be in agreement with the design of the structure and designs proposed for installations within it.

391. Environment

A. When the tunnel is accessible to the public or when workers must enter the structure to install, operate, or maintain the facilities in it, the design shall provide a controlled safe environment including, where necessary, barriers, detectors, alarms, ventilation, pumps, and adequate safety devices for all facilities. Controlled safe environment shall include the following:

1. Design to avoid poisonous or suffocation atmosphere
2. Design to protect persons from pressurized lines, fire, explosion, and high temperatures
3. Design to avoid unsafe conditions due to induced voltages
4. Design to limit the likelihood of hazards due to flooding
5. Design to ensure egress; two directions for egress shall be provided for all points in tunnels
6. Working space, in accordance with Rule 323B, the boundary of which shall be not less than 600 mm (2 ft) from a vehicular operating space or from exposed moving parts of machinery
7. Safeguards designed to protect workers from hazards due to the operation of vehicles or other machinery in tunnels
8. Unobstructed walkways for workers in tunnels

B. A condition of occupancy in multiple-use tunnels by supply and communications facilities shall be that the design and installation of all facilities is coordinated to provide a safe environment for the operation of supply facilities, communications facilities, or both. Safe environment for facilities shall include the following:

1. Means to protect equipment from harmful effects of humidity or temperature
2. Means to protect equipment from harmful effects of liquids or gases
3. Coordinated design and operation of corrosion-control systems

Part 4.
Rules for the Operation of Electric Supply and Communications Lines and Equipment

Section 40.
Purpose and Scope

400. Purpose

The purpose of Part 4 of this code is to provide practical work rules as one of the means of safeguarding employees and the public from injury. It is not the intent of these rules to require unreasonable steps to comply; however, all reasonable steps shall be taken.

401. Scope

Part 4 of this code covers work rules to be followed in the installation, operation, and maintenance of electric supply and communications systems.

402. Referenced Sections

The Introduction (Section 1), Definitions (Section 2), References (Section 3), and Grounding Methods (Section 9) of this code shall apply to the requirements of Part 4.

The standards listed in Section 3 shall be used with Part 4 where applicable.

NOTE: After ANSI C2-1973 [B6] was originally approved, 30 June 1972, the US Occupational Safety and Health Administration (OSHA) issued 29 CFR 1926, Subpart V [B42], applying to employee safety in *construction*. The differences between this document and the editions of the NESC through the 1993 Edition were noted in footnotes to the text of Part 4.

In 1989, OSHA published in the *Federal Register* [01/31/89], vol. 54, no. 19, pp. 4974 to 5024 inclusive, Docket S—015 [B43], a Notice of Proposed Rulemaking to issue new regulations as 29 CFR 1910.137 and 29 CFR 1910.269 to address the work practices to be used during the operation and maintenance of generation, transmission, and distribution facilities. Coordination between these OSHA proposals and the 1997 Edition of the NESC has been effected such that the technical content of each document is harmonized at this time to the extent possible with the closure of the rulemaking hearing docket. In some cases, it was thus not possible for the OSHA final rule published as 29 CFR 1910.137 and 29 CFR 1910.269 (as published in the *Federal Register,* [01/31/94] vol. 59, no. 20, pp. 4321 to 4477 [B44]) to address all the technical information considered in this revision of NESC Part 4.

Section 41. Supply and Communications Systems— Rules for Employers

410. General Requirements

A. General

1. The employer shall inform each employee working on or about communications equipment or electric supply equipment and the associated lines, of the safety rules governing the employee's conduct while so engaged.

   When deemed necessary, the employer shall provide a copy of such rules.
2. The employer shall provide training to all employees who work in the vicinity of exposed energized facilities. The training shall include applicable work rules required by this Part and other mandatory referenced standards or rules. The employer shall ensure that each employee has demonstrated proficiency in required tasks. The employer shall provide retraining for any employee who, as a result of routine observance of work practices, is not following work rules.
3. Employers shall utilize positive procedures to secure compliance with these rules. Cases may arise, however, where the strict enforcement of some particular rule could seriously impede the safe progress of the work at hand; in such cases the employee in charge of the work to be done should make such temporary modification of the rules as will accomplish the work without increasing the hazard.
4. If a difference of opinion arises with respect to the application of these rules, the decision of the employer or the employer's authorized agent shall be final. This decision shall not result in any employee performing work in a manner that is unduly hazardous to the employee or to the employee's fellow workers.

B. Emergency Procedures and First Aid Rules

1. Employees shall be informed of procedures to be followed in case of emergencies and rules for first aid, including approved methods of resuscitation. Copies of such procedures and rules should be kept in conspicuous locations in vehicles and places where the number of employees and the nature of the work warrants.
2. Employees working on communications or electric supply equipment or lines shall be regularly instructed in methods of first aid and emergency procedures, if their duties warrant such training.

C. Responsibility

1. A designated person shall be in charge of the operation of the equipment and lines and shall be responsible for their safe operation.
2. If more than one person is engaged in work on or about the same equipment or line, one person shall be designated as in charge of the work to be performed. Where there are separate work locations, one person may be designated at each location.

411. Protective Methods and Devices

A. Methods

1. Access to rotating or energized equipment shall be restricted to authorized personnel.
2. Diagrams, showing plainly the arrangement and location of the electric supply equipment and lines, shall be maintained on file and shall be readily available to authorized personnel for that portion of the system for which they are responsible.
3. Employees shall be instructed as to the character of the equipment or lines and methods to be used before any work is undertaken thereon.
4. Employees should be instructed to take additional precautions to ensure their safety when conditions create unusual hazards.

B. Devices and Equipment

An adequate supply of protective devices and equipment, sufficient to enable employees to meet the requirements of the work to be undertaken, and first aid equipment and materials shall be available in readily accessible and, where practical, conspicuous places.

Protective devices and equipment shall conform to the applicable standards listed in Section 3.

NOTE: The following is a list of some common protective devices and equipment, the number and kinds of which will depend upon the requirements of each case:

1. Insulating wearing apparel such as rubber gloves, rubber sleeves, and headgear
2. Insulating shields, covers, mats, and platforms
3. Insulating tools for handling or testing energized equipment or lines
4. Protective goggles
5. *Person at work* tags, portable danger signs, traffic cones, and flashers
6. Line worker's body belts, lanyards, and positioning straps
7. Fire-extinguishing equipment designed for safe use on energized parts or plainly marked that they must not be so used
8. Protective grounding materials and devices
9. Portable lighting equipment
10. First aid equipment and materials

C. Inspection and Testing of Protective Devices

1. Protective devices and equipment shall be inspected or tested to ensure that they are in safe working condition.
2. Insulating gloves, sleeves, and blankets shall be inspected before use. Insulating gloves and sleeves shall be tested as frequently as their use requires.
3. Line-worker's body belts, lanyards, and positioning straps and other personal equipment, whether furnished by employer or employee, shall be inspected to ensure that they are in safe working condition.

D. Signs and Tags for Employee Safety

Safety signs and tags required by Part 4 shall comply with the provisions of ANSI Z535.1-1998 through ANSI Z535.5-1998, inclusive.

E. Identification and Location

Means shall be provided so that identification of supply and communication lines can be determined before work is undertaken. Persons responsible for underground facilities shall be able to indicate the location of their facilities.

F. Fall Protection

1. Employers shall develop, implement, and maintain an effective fall protection program applicable to climbing or otherwise accessing and working from elevated work locations, which shall include all of the following:
   a. Training, retraining, and documentation
   b. Guidance on equipment selection, inspection, care, and maintenance
   c. Considerations concerning structural design and integrity, with particular reference to anchorages and their availability
   d. Rescue plans and related training
   e. Hazard recognition
2. The employer shall not permit employees the use of 100% leather positioning straps.

Section 42.
General Rules for Employees

420. Personal General Precautions

A. Rules and Emergency Methods

1. Employees shall carefully read and study the safety rules, and may be called upon at any time to show their knowledge of the rules.
2. Employees shall familiarize themselves with approved methods of first aid, rescue techniques, and fire extinguishment.

B. Qualifications of Employees

1. Employees whose duties require working on or in the vicinity of energized equipment or lines shall perform only those tasks for which they are trained, equipped, authorized, and so directed. Inexperienced employees shall: (a) work under the direction of an experienced and qualified person at the site, and (b) perform only directed tasks.
2. If an employee is in doubt as to the safe performance of any assigned work, the employee shall request instructions from the employee's supervisor or person in charge.
3. Employees who do not normally work on or in the vicinity of electric supply lines and equipment but whose work brings them into these areas for certain tasks shall proceed with this work only when authorized by a qualified person.

C. Safeguarding Oneself and Others

1. Employees shall heed safety signs and signals and warn others who are in danger or in the vicinity of energized equipment or lines.
2. Employees shall report promptly to the proper authority any of the following:
   a. Line or equipment defects such as abnormally sagging wires, broken insulators, broken poles, or lamp supports
   b. Accidentally energized objects such as conduits, light fixtures, or guys
   c. Other defects that may cause a dangerous condition
3. Employees whose duties do not require them to approach or handle electric equipment and lines shall keep away from such equipment or lines and should avoid working in areas where objects and materials may be dropped by persons working overhead.
4. Employees who work on or in the vicinity of energized lines shall consider all of the effects of their actions, taking into account their own safety as well as the safety of other employees on the job site, or on some other part of the affected electric system, the property of others, and the public in general.
5. No employee shall approach or bring any conductive object, without a suitable insulating handle, closer to any exposed energized part than allowed by Rule 431 (communication) or Rule 441 (supply), as applicable.
6. Employees should exercise care when extending metal ropes, tapes, or wires parallel to and in the proximity of energized high-voltage lines because of induced voltages. When it is necessary to measure clearances from energized objects, only devices approved for the purpose shall be used.

D. Energized or Unknown Conditions

Employees shall consider electric supply equipment and lines to be energized, unless they are positively known to be de-energized. Before starting work, employees shall perform preliminary inspections or tests to determine existing conditions. Operating voltages of equipment and lines should be known before working on or in the vicinity of energized parts.

E. Ungrounded Metal Parts

Employees shall consider all ungrounded metal parts of equipment or devices such as transformer cases and circuit breaker housings, to be energized at the highest voltage to which they are exposed, unless these parts are known by test to be free from such voltage.

F. Arcing Conditions

Employees should keep all parts of their bodies as far away as practical from switches, brushes, commutators, circuit breakers, or other parts at which arcing may occur during operation or handling.

G. Liquid-Cell Batteries

1. Employees shall ascertain that battery areas are adequately ventilated before performing work.
2. Employees should avoid smoking, using open flames, or using tools that may produce sparks in the vicinity of liquid-cell batteries.
3. Employees shall use eye and skin protection when handling an electrolyte.
4. Employees shall not handle energized parts of batteries unless necessary precautions are taken to avoid short circuits and electrical shocks.

H. Tools and Protective Equipment

Employees shall use the personal protective equipment, the protective devices, and the special tools provided for their work. Before starting work, these devices and tools shall be carefully inspected to make sure that they are in good condition.

I. Clothing

1. Employees shall wear clothing suitable for the assigned task and the work environment. See Rule 410A2.
2. When working in the vicinity of energized lines or equipment, employees should avoid wearing exposed metal articles.

J. Ladders and Supports

1. Employees shall not support themselves, or any material or equipment, on any portion of a tree, pole structure, scaffold, ladder, walkway, or other elevated structure or aerial device, etc., without it first being determined, to the extent practical, that such support is adequately strong, in good condition, and properly secured in place.
2. Portable wood ladders intended for general use shall not be painted except with a clear nonconductive coating, nor shall they be longitudinally reinforced with metal.
3. Portable metal ladders intended for general use shall not be used when working on or in the vicinity of energized parts.
4. If portable ladders are made partially or entirely conductive for specialized work, necessary precautions shall be taken to ensure that their use will be restricted to the work for which they are intended.

K. Fall Protection

1. At elevated locations above 3 m (10 ft), climbers shall be attached to equipment or structures by a fall protection system while at the worksite, at a rest site, in aerial devices, helicopters, cable carts, and a boatswain's chair.
2. Qualified climbers may be permitted to be unattached to equipment or structures while climbing, transferring, or transitioning across obstacles on structures. Unqualified climbers shall be attached while performing these activities.
3. Fall protection equipment shall be inspected before use by the employee to ensure that the equipment is in safe working condition.
4. Fall arrest equipment shall be attached to a suitable anchorage.
5. The employee shall determine that all components of the fall protection system are properly engaged and that the employee is secure in the line-worker's body belt, harness, or any other fall protection system.

   NOTE: Climbers need to be aware of accidental disengagement of fall protection components. Accidental disengagement is the sudden, unexpected release of a positioning strap snaphook from the D-ring of the line-worker's body belt without the user directly manipulating the latch of the snaphook. In general, there are two primary reasons for this occurrence.

a. Foreign objects may open the latch of the snaphook during normal use. It is possible for the snaphook to come in contact with such things as hand lines, guy wires, or other apparatus. These items may place pressure on the latch, causing the snaphook to separate from the D-ring without the user's knowledge. This could cause an accident. The worker must take care to keep the snaphooks away from any potential causes of release. Locking snaphooks reduce the possibility of this occurrence.
b. Roll-out is the sudden separation of the snaphook/D-ring combination when the snaphook is twisted in the D-ring, but the user does not deliberately open the latch. This occurs when a twist is introduced into a positioning strap with a snaphook/D-ring combination that is incompatible. However, compatible hardware, when properly maintained, will not separate in this fashion.

6. Snaphooks shall be dimensionally compatible with the member to which they are connected so as to prevent unintentional disengagement of the connection.
NOTE:
a. The possibility exists for some snaphooks to roll out of D-rings. Attachment of a mismatched or multiple snaphooks, either of the nonlocking or locking type, to a single D-ring needs to be avoided. Multiple locking snaphooks may be attached to a single D-ring if they have been evaluated in the combination to be used. Locking snaphooks reduce the potential for roll-out.
b. Disengagement through contact of the snaphook keeper with the connected member may be prevented by the use of a locking snaphook.
c. Hardware compatibility can be verified. Simply attach the snaphook to the D-ring, then roll the snaphook placing the latch towards the body of the D-ring. This is similar to the action that occurs when the strap is twisted. If the rivet falls beyond the edge of the inside of the D-ring, placing pressure on the latch, the hardware is not compatible, and a roll-out potential exists.
d. Other factors may increase the potential for accidental disengagement even if the hardware is compatible (e.g., foreign objects carried on the D-rings, condition of the snaphook, the shape of the D-ring).
7. Snaphooks shall not be connected to each other.
8. One hundred percent leather positioning straps shall not be used.
9. Wire rope lanyards shall be used in operations where the lanyard is subject to being cut. Wire rope lanyards shall not be used in the vicinity of energized lines or equipment.

L. Fire Extinguishers

In fighting fires or in the vicinity of exposed energized parts of electric supply systems, employees shall use fire extinguishers or materials that are suitable for the purpose. If this is not possible, all adjacent and affected equipment should first be de-energized.

M. Machines or Moving Parts

Employees working on normally moving parts of remotely controlled equipment shall be protected against accidental starting by proper tags installed on the starting devices, or by locking or blocking where practical. Employees shall, before starting any work, satisfy themselves that these protective devices have been installed. When working or in the vicinity of automatically or remotely operated equipment such as circuit breakers that may operate suddenly, employees shall avoid being in a position where they might be injured from such operation.

N. Fuses

When fuses must be installed or removed with one or both terminals energized, employees shall use special tools or gloves insulated for the voltage involved. When installing expulsion-type fuses, employees shall wear personal eye protection and take precautions to stand clear of the exhaust path of the fuse barrel.

O. Cable Reels

Cable reels shall be securely blocked so they cannot roll or rotate accidentally.

P. Street and Area Lighting
   1. The lowering rope or chain, its supports, and fastenings shall be examined periodically.
   2. A suitable device shall be provided by which each lamp on series-lighting circuits of more than 300 V may be safely disconnected from the circuit before the lamp is handled.
      EXCEPTION: This rule does not apply where the lamps are always worked on from suitable insulated platforms or aerial lift devices, or handled with suitable insulated tools, and treated as under full voltage of the circuit concerned.

421. General Operating Routines

A. Duties of a First-Level Supervisor or Person in Charge
   This individual shall:
   1. Adopt such precautions as are within the individual's authority to prevent accidents.
   2. See that the safety rules and operating procedures are observed by the employees under the direction of this individual.
   3. Make all the necessary records and reports, as required.
   4. Prevent unauthorized persons from approaching places where work is being done, as far as practical.
   5. Prohibit the use of tools or devices unsuited to the work at hand or that have not been tested or inspected as required.

B. Area Protection
   1. Areas Accessible to Vehicular and Pedestrian Traffic
      a. Before engaging in work that may endanger the public, safety signs or traffic control devices, or both, shall be placed conspicuously to alert approaching traffic. Where further protection is needed, suitable barrier guards shall be erected. Where the nature of work and traffic requires it, a person shall be stationed to warn traffic while the hazard exists.
      b. When openings or obstructions in the street, sidewalk, walkways, or on private property are being worked on or left unattended during the day, danger signals, such as warning signs and flags, shall be effectively displayed. Under these same conditions at night, warning lights shall be prominently displayed and excavations shall be enclosed with protective barricades.
   2. Areas Accessible to Employees Only
      a. If the work exposes energized or moving parts that are normally protected, safety signs shall be displayed. Suitable barricades shall be erected to restrict other personnel from entering the area.
      b. When working in one section where there is a multiplicity of such sections, such as one panel of a switchboard, one compartment of several, or one portion of a substation, employees shall mark the work area conspicuously and place barriers to prevent accidental contact with energized parts in that section or adjacent sections.
   3. Locations With Crossed or Fallen Wires
      An employee, finding crossed or fallen wires that are creating, or may create, a hazard, shall remain on guard or adopt other adequate means to prevent accidents. The proper authority shall be notified. If the employee is qualified, and can observe the rules for safely handling energized parts by the use of insulating equipment, this employee may correct the condition.

C. Escort
   Persons accompanying nonqualified employees or visitors or in the vicinity of electric equipment or lines shall be qualified to safeguard the people in their care, and see that the safety rules are observed.

422. Overhead Line Operating Procedures

Employees working on or with overhead lines shall observe the following rules in addition to applicable rules contained elsewhere in Sections 43 and 44.

A. Setting, Moving, or Removing Poles In or Near Energized Electric Supply Lines

1. When setting, moving, or removing poles in or in the vicinity of energized lines, precautions shall be taken to avoid direct contact of the pole with the energized conductors. Employees shall wear suitable insulating gloves or use other suitable means where voltages may exceed rating of gloves in handling poles where conductors energized at potentials above 750 V can be contacted. Employees performing such work shall not contact the pole with uninsulated parts of their bodies.
2. Contact with trucks, or other equipment that is not bonded to an effective ground being used to set, move, or remove poles in or in the vicinity of energized lines shall be avoided by employees standing on the ground or in contact with grounded objects unless employees are wearing suitable protective equipment.

B. Checking Structures Before Climbing

1. Before climbing poles, ladders, scaffolds, or other elevated structures, employees shall determine, to the extent practical, that the structures are capable of sustaining the additional or unbalanced stresses to which they will be subjected.
2. Where there are indications that poles and structures may be unsafe for climbing, they shall not be climbed until made safe by guying, bracing, or other means.

C. Installing and Removing Wires or Cables

1. Precautions shall be taken to prevent wires or cables that are being installed or removed from contacting energized wires or equipment. Wires or cables that are not bonded to an effective ground and which are being installed or removed in the vicinity of energized conductors shall be considered as being energized.
2. Sag of wire or cables being installed or removed shall be controlled to prevent danger to pedestrian and vehicular traffic.
3. Before installing or removing wires or cables, the strains to which poles and structures will be subjected shall be considered and necessary action taken to prevent failure of supporting structures.
4. Employees should avoid contact with moving winch lines, especially in the vicinity of sheaves, blocks, and take-up drums.
5. Employees working on or in the vicinity of equipment or lines exposed to voltages higher than those guarded against by the safety appliances provided shall take steps to be assured that the equipment or lines on which the employees are working are free from dangerous leakage or induction or have been effectively grounded.

423. Underground Line Operating Procedures

Employees working on or with underground lines shall observe the following rules in addition to applicable rules contained elsewhere in Sections 43 and 44.

A. Guarding Manhole and Street Openings

When covers of manholes, handholes, or vaults are removed, the opening shall be promptly protected with a barrier, temporary cover, or other suitable guard.

B. Testing for Gas in Manholes and Unventilated Vaults

1. The atmosphere shall be tested for combustible or flammable gas(es) before entry.
2. Where combustible or flammable gas(es) are detected, the work area shall be ventilated and made safe before entry.
3. Unless forced continuous ventilation is provided, a test shall also be made for oxygen deficiency.
4. Provision shall be made for an adequate continuous supply of air.

   NOTE: The term *adequate* includes evaluation of both the quantity and quality of the air.

C. Flames
1. Employees shall not smoke in manholes.
2. Where open flames must be used in manholes or vaults, extra precautions shall be taken to ensure adequate ventilation.
3. Before using open flames in an excavation in areas where combustible gases or liquids may be present, such as in the vicinity of gasoline service stations, the atmosphere of the excavation shall be tested and found safe or cleared of the combustible gases or liquids.
4. When a torch or open flame is used (as in heat shrink splicing) in proximity to a visibly exposed gas or other fuel line(s), adequate air space or a barrier shall be provided to protect the gas or fuel line(s) from the heat source.

D. Excavation
1. Cables and other buried utilities in the immediate vicinity shall be located, to the extent practical, prior to excavating.
2. Hand tools used for excavating in the vicinity of energized supply cables shall be equipped with handles made of nonconductive material.
3. Mechanized equipment should not be used to excavate in close proximity to cables and other buried utilities.
4. If a gas or fuel line is broken or damaged, employees shall:
   a. Leave the excavation open
   b. Extinguish flames that could ignite the escaping gas or fuel
   c. Notify the proper authority
   d. Keep the public away until the condition is under control
5. When a worker is required to perform tasks in trenches or excavations where a cave-in hazard exists or the trench or excavation is in excess of 1.5 m (5 ft) in depth, shoring, sloping, or shielding methods shall be used to provide employee protection.

E. Identification
1. When underground facilities are exposed, they should be identified and shall be protected as necessary to avoid damage.
2. Where multiple cables exist in an excavation, cables other than the one being worked on shall be protected as necessary.
3. Before cutting into a cable or opening a splice, the cable should be identified and verified to be the proper cable.
4. When multiple cables exist in an excavation, the cable to be worked on shall be positively identified.

F. Operation of Power-Driven Equipment

Employees should avoid being in manholes where power-driven rodding equipment is in operation.

Section 43.
Additional Rules for Communications Employees

430. General

Communications employees shall observe the following rules in addition to the rules contained in Section 42.

431. Approach to Energized Conductors or Parts

A. No employee shall approach, or bring any conductive object, within the distances to any exposed energized part as listed in Table 431-1. When repairing storm damage to communication lines that are joint use with electric supply lines at that or another point, employees shall:
   1. Treat all such supply and communication lines as energized to the highest voltage to which they are exposed, or
   2. Assure that the supply lines involved are de-energized and grounded in accordance with Section 44.

B. Altitude Correction
   The distances in Tables 431-1 shall be used at elevations below 900 m (3000 ft). Altitude correction factors as indicated in Table 441-5 shall be applied above that altitude. Altitude correction factors shall be applied only to the electrical component of the minimum approach distance.

432. Joint-Use Structures

When working on jointly used poles or structures, employees shall not approach closer than distances specified in Table 431-1 and shall not position themselves above the level of the lowest electric supply conductor exclusive of vertical runs and street lighting.

EXCEPTION: This rule does not apply where communications facilities are attached above electric supply conductors if a rigid fixed barrier has been installed between the supply and communications facilities.

433. Attendant on Surface at Joint-Use Manhole

While personnel are in a joint-use manhole, an employee shall be available on the surface in the immediate vicinity to render assistance as may be required.

434. Sheath Continuity

Metallic or semiconductive sheath continuity shall be maintained by bonding across the opening, or by equivalent means, when working on buried cable or on cable in manholes.

Table 431-1
Overhead Supply Lines and Equipment Minimum Approach Distances to Exposed Energized Parts
(See Rule 431 in its entirety)

	Distance to Employee	
Voltage range (phase-to-phase, rms)[1]	**Phase-to-ground (m)**	**Phase-to-ground (ft–in)**
0 V to 50 V[2]	not specified	not specified
51 V to 300 V[2]	avoid contact	avoid contact
301 V to 750 V[2]	0.31	1–0
751 V to 15 kV	0.65	2–2
15.1 kV to 36.0 kV	0.91	3–0
36.1 kV to 46.0 kV	1.06	3–6
46.1 kV to 121.0 kV	1.21	4–0
121 kV to 140.0 kV	1.38	4–6

[1] For single-phase lines off three-phase systems, use the phase-to-phase voltage of that system.
[2] For single-phase systems, use the highest voltage available.

Section 44.
Additional Rules for Supply Employees

440. General

Supply employees shall observe the following rules in addition to the rules contained in Section 42.

441. Energized Conductors or Parts

Employees shall not approach, or knowingly permit others to approach, any exposed ungrounded part normally energized except as permitted by this rule.

A. Minimum Approach Distance to Live Parts

1. General

Employees shall not approach or bring any conductive object within the distances to exposed parts that operate at the voltages listed in Table 441-1 or Table 441-4 unless one of the following is met:

a. The line or part is de-energized and grounded per Rule 444D.

b. The employee is insulated from the energized line or part. Electrical protective equipment insulated for the voltage involved, such as tools, gloves, rubber gloves, or rubber gloves with sleeves, shall be considered effective insulation for the employee from the energized part being worked on.

c. The energized line or part is insulated from the employee and from any other line or part at a different potential.

2. Precautions for Approach—Voltages from 51 V to 300 V

Employees shall not contact exposed energized parts operating at 51 V to 300 V, unless the provisions of Rule 441A1 are met.

3. Precautions for Approach—Voltages from 301 V to 72.5 kV

At voltages from 301 V to 72.5 kV, employees shall be protected from phase-to-phase and phase-to-ground differences in electric potential.

a. When exposed grounded lines, conductors, or parts are in the work area, they shall be guarded or insulated.

b. When the Rubber Glove Work Method is employed, rubber insulating gloves, insulated for the maximum use voltage as listed in Table 441-6, shall be worn whenever employees are in the vicinity of energized conductors or parts, supplemented by one of the following two protective methods:

(1) The employee shall wear rubber insulating sleeves, insulated for the maximum use voltage as listed in Table 441-6, in addition to the rubber insulating gloves.

EXCEPTION: When work is performed on electric supply equipment energized at 750 V or less, rubber sleeves are not required if only the live parts being worked on are exposed.

(2) All exposed energized lines or parts, other than those temporarily exposed to perform work and maintained under positive control, located within maximum reach of the employee's work position, shall be covered with insulating protective equipment.

EXCEPTION: When work is being performed on parts energized between 300 V and 750 V within enclosed spaces, (e.g., control panels and relay cabinets), insulating or guarding of all exposed grounded lines, conductors, or parts in the work area is not required provided that employees use insulated tools and/or gloves and that exposed grounded lines, conductors, or parts are covered to the extent feasible.

c. When the Rubber Glove Work Method is employed at voltages above 15 kV phase-to-phase, supplementary insulation (e.g., insulated aerial device or structure-mounted insulating work platform), tested for the voltage involved shall be used to support the worker.

d. Cover-up rated for the voltage involved, when used, shall be applied to the exposed facilities as the employee first approaches the facilities from any direction, be that from the structure or from an aerial device, and shall be removed in the reverse order. This protective cover-up shall extend beyond the reach of the employee's anticipated work position or extended reach distance.

4. Transient Overvoltage Control Above 72.5 kV

   a. For 121 kV to 362 kV single break per pole switching devices with three phase reclosing into trapped charges the maximum per-unit values given in the table (3 per unit for 121 to 362 kV) may increase significantly. Minimum approach distances for these devices shall be determined by a transient overvoltage study.

      NOTE: These overvoltages will not exceed the values of Table 441-1 if reclosing is blocked.

   b. For voltages above 72.5 kV, the minimum approach distance may be reduced if the maximum anticipated transient overvoltage is known for the work site. Engineering analysis is required when transient overvoltage control techniques are employed. When preinsertion resistors are employed, they shall be operational. The minimum approach distances derived from Tables 441-2, 441-3, and 441-4 may be used. When a reduced clearance distance from Tables 441-2, 441-3, and 441-4 is used for a specific per-unit transient overvoltage, the maximum transient overvoltage shall be controlled at the work site by one of the following methods:

      1. The operation of a circuit breaker or other switching device shall be modified, including blocking reclosing.
      2. The overvoltage itself shall be forcibly held to an acceptable level by the installation of temporary transient voltage protective devices such as surge arresters or temporary protective gaps.
      3. The operation of the system shall be changed to restrict potential overvoltages resulting from the effect of activity on the system (e.g., capacitor switching, tap changing, cable deenergization, etc.).

5. Altitude Correction

   The distances in Tables 441-1, 441-2, 441-3, and 441-4 shall be used at elevations below 900 m (3000 ft). Above that altitude, the minimum approach distance shall be increased by:

   a. Multiplying the electrical component of the minimum approach distance by the applicable altitude correction factors of Table 441-5, and
   b. Adding the result to the values for inadvertent movement required by Rule 441A6a(3).

      NOTE: The electrical component of clearance included in Tables 441-1, 441-2, 441-3, and 441-4 is the table value less the value for inadvertent movement for that voltage shown in Rule 441A6a(3).

6. Calculation of Approach Distances

   a. Minimum approach distances shown in Tables 441-2, 441-3, and 441-4 are calculated in feet to two decimal places. The second decimal place is rounded up if the third decimal place is other than zero. The metric values shown in these tables are derived directly from rod gap data originally recorded in metric measurements. For this reason, the english-unit tables do not exactly convert into the values shown in the metric tables. Because the original rod gap data for voltages from 1.1 to 72 kV is measured in metric units, the values in Table 441-1 are derived from metric and converted to feet and inches. The following processes are used:

      1. When converting to feet and inches, the decimal part of a foot is converted to inches and rounded up if the first decimal place is other than zero.
      2. When converting from feet to metric, the feet dimension from Rule 441A5a above is converted to meters. The second decimal place shall be rounded up if the third decimal place is other than zero.

3. Minimum approach distances calculated under this rule for 0.301 kV to 0.750 kV contain the electric component plus 0.31 m (1 ft) for inadvertent movement. Voltages 0.751 kV to 72.5 kV contain the electric component plus 0.61 m (2 ft) for inadvertent movement. Above 72.5 kV, the inadvertent movement distance is 0.31 m (1 ft).

4. Table 4 (Alternating Current) and Table 5 (Direct Current) of IEEE Std 516-1987 are the electrical basis for approach distances for voltages above 72.5 kV. IEEE Std 516-1987 includes the formula used to derive electrical clearance distances. IEEE Std 4-1995 is the basis for the approach distances for voltages below 72.5 kV.

5. The voltage ranges are contained in ANSI C84.1-1995, Table 1.

b. Interpolation between the values contained in Tables 441-2 through 441-4 is not permitted. Minimum approach distances for maximum phase-to-phase voltages other than those contained in Tables 441-2 through 441-4 shall be determined using the process outlined in Rule 441A5a. The approach distances of Table 441-1 shall be used unless the per-unit transient overvoltage is known and controlled.

Table 441-1
AC Live Work Minimum Approach Distance
(See Rule 441 in its entirety.)

Voltage in kilovolts phase to phase[1,2]	Distance to employee			
	Phase to ground		Phase to phase	
	(m)	(ft–in)	(m)	(ft–in)
0 to 0.050[1]	not specified		not specified	
0.051 to 0.300[1]	avoid contact		avoid contact	
0.301 to 0.750[1]	0.31	1–0	0.31	1–0
0.751 to 15	0.65	2–2	0.67	2–3
15.1 to 36.0	0.77	2–7	0.86	2–10
36.1 to 46.0	0.84	2–9	0.96	3–2
46.1 to 72.5	1.00[3]	3–3[3]	1.20	3–11
72.6 to 121	0.95[3]	3–2[3]	1.29	4–3
138 to 145	1.09	3–7	1.50	4–11
161 to 169	1.22	4–0	1.71	5–8
230 to 242	1.59	5–3	2.27	7–6
345 to 362	2.59	8–6	3.80	12–6
500 to 550	3.42	11–3	5.50	18–1
765 to 800	4.53	14–11	7.91	26–0

[1] For single-phase systems, use the highest voltage available.

[2] For single-phase lines off three phase systems, use the phase-to-phase voltage of the system.

[3] The 46.1 to 72.5 kV phase-to-ground 3-3 distance contains a 1-3 electrical component and a 2-0 inadvertent movement component while the 72.6 to 121 kV phase-to-ground 3-2 distance contains a 2-2 electrical component and a 1-0 inadvertent movement component.

m

Table 441-2
AC Live Work Minimum Approach Distance With Transient Overvoltage Factor
(See Rule 441 in its entirety.)

Maximum anticipated per–unit transient overvoltage	Distance to employee in meters, phase-to-ground Air, bare-hand, and clear live-line tool Maximum phase-to-phase voltage in kilovolts						
	121	**145**	**169**	**242**	**362**	**550**	**800**
1.5						1.82	2.95
1.6						1.97	3.23
1.7						2.13	3.54
1.8						2.29	3.86
1.9						2.47	4.19
2.0	0.74	0.83	0.92	1.16	1.59	2.65	4.53
2.1	0.76	0.85	0.95	1.21	1.65	2.83	
2.2	0.78	0.88	0.98	1.25	1.74	3.01	
2.3	0.80	0.91	1.01	1.29	1.84	3.20	
2.4	0.82	0.93	1.04	1.33	1.94	3.42	
2.5	0.84	0.96	1.07	1.38	2.04		
2.6	0.86	0.98	1.10	1.42	2.14		
2.7	0.88	1.01	1.13	1.45	2.25		
2.8	0.91	1.03	1.16	1.50	2.36		
2.9	0.93	1.06	1.19	1.54	2.47		
3.0	0.95	1.09	1.22	1.59	2.59		

ft

Table 441-2
AC Live Work Minimum Approach Distance With Transient Overvoltage Factor
(See Rule 441 in its entirety.)

Maximum anticipated per-unit transient overvoltage	Distance to employee in feet-inches, phase-to-ground Air, bare-hand, and clear live-line tool Maximum phase-to-phase voltage in kilovolts						
	121	**145**	**169**	**242**	**362**	**550**	**800**
1.5						6–0	9–8
1.6						6–6	10–8
1.7						7–0	11–8
1.8						7–7	12–8
1.9						8–1	13–9
2.0	2–5	2–9	3–0	3–10	5–3	8–9	14–11
2.1	2–6	2–10	3–2	4–0	5–5	9–4	
2.2	2–7	2–11	3–3	4–1	5–9	9–11	
2.3	2–8	3–0	3–4	4–3	6–1	10–6	
2.4	2–9	3–1	3–5	4–5	6–4	11–3	
2.5	2–9	3–2	3–6	4–6	6–8		
2.6	2–10	3–3	3–8	4–8	7–1		
2.7	2–11	3–4	3–9	4–10	7–5		
2.8	3–0	3–5	3–10	4–11	7–9		
2.9	3–1	3–6	3–11	5–1	8–2		
3.0	3–2	3–7	4–0	5–3	8–6		

m

Table 441-3
AC Live Work Minimum Approach Distance With Transient Overvoltage Factor
(See Rule 441 in its entirety.)

Maximum anticipated per-unit transient overvoltage	Distance to employee in meters, phase-to-phase						
	Air, bare-hand, and clear live-line tool						
	Maximum phase-to-phase voltage in kilovolts						
	121	**145**	**169**	**242**	**362**	**550**	**800**
1.5						2.24	3.67
1.6						2.65	4.42
1.7						3.08	5.23
1.8						3.53	6.07
1.9						4.01	6.97
2.0	1.08	1.24	1.41	1.85	2.61	4.52	7.91
2.1	1.10	1.27	1.44	1.89	2.68	4.75	
2.2	1.12	1.29	1.47	1.93	2.78	4.98	
2.3	1.14	1.32	1.50	1.97	2.90	5.21	
2.4	1.16	1.35	1.53	2.01	3.02	5.50	
2.5	1.18	1.37	1.56	2.06	3.14		
2.6	1.21	1.40	1.59	2.10	3.27		
2.7	1.23	1.43	1.62	2.13	3.40		
2.8	1.25	1.45	1.65	2.19	3.53		
2.9	1.27	1.48	1.68	2.22	3.67		
3.0	1.29	1.50	1.71	2.27	3.80		

ft

Table 441-3
AC Live Work Minimum Approach Distance With Transient Overvoltage Factor
(See Rule 441 in its entirety.)

Maximum anticipated per-unit transient overvoltage	Distance to employee in feet-inches, phase-to-phase Air, bare-hand, and clear live-line tool Maximum phase-to-phase voltage in kilovolts						
	121	**145**	**169**	**242**	**362**	**550**	**800**
1.5						7–4	12–1
1.6						8–9	14–6
1.7						10–2	17–2
1.8						11–7	19–11
1.9						13–2	22–11
2.0	3–7	4–1	4–8	6–1	8–7	14–10	26–0
2.1	3–7	4–2	4–9	6–3	8–10	15–7	
2.2	3–8	4–3	4–10	6–4	9–2	16–4	
2.3	3–9	4–4	4–11	6–6	9–6	17–2	
2.4	3–10	4–5	5–0	6–7	9–11	18–1	
2.5	3–11	4–6	5–2	6–9	10–4		
2.6	4–0	4–7	5–3	6–11	10–9		
2.7	4–1	4–8	5–4	7–0	11–2		
2.8	4–1	4–9	5–5	7–2	11–7		
2.9	4–2	4–10	5–6	7–4	12–1		
3.0	4–3	4–11	5–8	7–6	12–6		

m

Table 441-4
DC Live Work Minimum Approach Distance With Transient Overvoltage Factor
(See Rule 441 in its entirety.)

Maximum anticipated per-unit transient overvoltage	Distance to employee in meters, conductor to ground Air, bare-hand, and clear live-line tool Maximum conductor-to-ground voltage in kilovolts				
	250	**400**	**500**	**600**	**750**
1.5 or lower	1.12	1.60	2.06	2.62	3.61
1.6	1.17	1.69	2.24	2.86	3.98
1.7	1.23	1.82	2.42	3.12	4.37
1.8	1.28	1.95	2.62	3.39	4.79

ft

Table 441-4
DC Live Work Minimum Approach Distance With Transient Overvoltage Factor
(See Rule 441 in its entirety.)

Maximum anticipated per-unit transient overvoltage	Distance to employee in meters, conductor to ground Air, bare-hand, and clear live-line tool Maximum conductor-to-ground voltage in kilovolts				
	250	**400**	**500**	**600**	**750**
1.5 or lower	3–8	5–3	6–9	8–7	11–10
1.6	3–10	5–7	7–4	9–5	13–1
1.7	4–1	6–0	7–11	10–3	14–4
1.8	4–3	6–5	8–7	11–2	15–9

Table 441-5
Altitude Correction Factor
(See Rule 441 in its entirety.)

Altitude		Correction factor
(m)	(ft)	
900	3000	1.00
1200	4000	1.02
1500	5000	1.05
1800	6000	1.08
2100	7000	1.11
2400	8000	1.14
2700	9000	1.17
3000	10 000	1.20
3600	12 000	1.25
4200	14 000	1.30
4800	16 000	1.35
5400	18 000	1.39
6000	20 000	1.44

Table 441-6
Maximum Use Voltage for Rubber Insulating Equipment

Class of Equipment	Maximum Use Voltage[1]
00	500
0	1000
1	7500
2	17 000
3	26 500
4	36 000

[1] The maximum use voltage is the ac voltage (rms) rating of the protective equipment that designates the maximum nominal design voltage of the energized system that may be safely worked. The nominal design voltage is equal to the phase-to-phase voltage on multiphase circuits.

EXCEPTION 1: If there is no multiphase exposure in a system area (at the worksite) and the voltage exposure is limited to the phase (polarity on dc systems) to ground potential, the phase (polarity on dc systems) to ground potential shall be considered to be the nominal design voltage.

EXCEPTION 2: If electric equipment and devices are insulated, isolated, or both, such that the multiphase exposure on a grounded wye circuit is removed and if supplemental insulation (e.g., insulated aerial device or structure-mounted insulating work platform) is used to insulate the employee from ground, then the nominal design voltage may be considered as the phase-to-ground voltage on that circuit.

B. Additional Approach Requirements

1. The clear insulation distance associated with insulators shall be the shortest straight-line air-gap distance from the nearest energized part to the nearest grounded part.
2. When working on insulators under live work procedures employing rubber gloves or live-line tools (hot sticks), the clear insulation distance shall be not less than the straight-line distance in air required by Tables 441-1, 441-2, 441-3, and 441-4.
3. Work may be performed at the grounded end of an open switch if all of the following conditions are met:
   a. The air-gap distance of the switch shall not be reduced in any manner. This distance shall be not less than the electrical basis for approach distances determined by Rule 441A6a(4) for the maximum anticipated transient overvoltage. The inadvertent movement values of Rule 441A6a(3) are not required in this distance.
   b. The minimum approach distance to the energized part of the switch shall be not less than that required by Rule 441A.
4. Special Rules for Working on Insulator Assemblies Operating Above 72.5 kV
   a. When work is to be performed at the ground end of an insulator assembly, the minimum approach distance to the nearest energized part may equal the straight-line distance measured along the insulators.
   b. For suspension insulator assembly installations (see ANSI C29.2-1992) operating above 72.5 kV, the first insulator at the grounded end may be temporarily shorted out as part of the work procedure.

      EXCEPTION: For voltages at 230 kV and above, up to three insulator units may be temporarily shorted out as part of the work procedure, provided that the minimum approach distance requirements of Rule 441 are met.
   c. When performing live work employing the bare-hand technique on installations operating above 72.5 kV, the first insulator at the energized (hot) end of a suspension insulator assembly (see ANSI C29.2-1992) may be shorted out during the work.

      EXCEPTION: For voltages at 230 kV and above, up to three insulator units may be temporarily shorted out as part of the work procedure, provided that the minimum approach distance requirements of Rule 441 are met.

      (1) The minimum approach distance to the grounded end of the insulator assembly may be equal to the straight-line distance from the nearest energized part to the closest grounded part across the insulators.

      (2) The straight-line insulation distance shall be not less than the values required by Tables 441-1, 441-2, 441-3, and 441-4.

C. Live-Line Tool Clear Insulation Length

1. Clear Live-Line Tool Length.

   The clear live-line tool distance shall be not less than the distance measured longitudinally along the live-line tool from the conductive part at the working end of the tool and any part of the employee. Distances for conducting sections (such as metallic splices and hardware) shall be subtracted from the clear live-line length. The clear live-line tool length shall equal or exceed the values for the minimum approach distance in Tables 441-1, 441-2, 441-3, and 441-4 for the indicated voltage ranges. The minimum clear live-line tool distance shall be the distance measured longitudinally along the live-line tool from the conductive part at the working end of the tool to any part of the employee.
2. Live-Line Conductor Support Tool Length

   Conductor support tools such as link sticks, strain carriers, and insulator cradles may be used provided that the clear insulating distance is at least as long as the insulator string or the maximum distance specified in Rule 441A. When installing this equipment, the employee shall maintain the minimum approach distance required equal to the clear insulating length for the support tools.

   NOTE: Conductive components of tools disturb the field in the gap and decrease the insulation value of the tool more than the linear subtraction of the length(s) of the conductive components.

442. Switching Control Procedures

A. Designated Person

A designated person shall:

1. Keep informed of operating conditions affecting the safe and reliable operation of the system.
2. Maintain a suitable record showing operating changes in such conditions.
3. Issue or deny authorization for switching, as required, for safe and reliable operation.

B. Specific Work

Authorization from the designated person shall be secured before work is begun on or in the vicinity of station equipment, transmission, or interconnected feeder circuits and where circuits are to be de-energized at stations. The designated person shall be notified when such work ceases.

EXCEPTION 1: In an emergency, to protect life or property, or when communication with the designated person is difficult because of storms or other causes, any qualified employee may make repairs on or in the vicinity of the equipment or lines covered by this rule without special authorization if the qualified employee can clear the trouble promptly with available help in compliance with the remaining rules. The designated person shall thereafter be notified as soon as possible of the action taken.

EXCEPTION 2: Suspension of normal rule or rules under disaster conditions: Where catastrophic service disruptions occur (e.g., earthquake, hurricane) and where multiple employer crews may be imported to assist in service restorations, the normal use of Rule 442 procedures may be suspended provided that:

1. Each individual involved in system repairs is informed of the suspension of normal rules.
2. Employees are required to observe all requirements of Rule 443 and Rule 444, including protection designated from step and touch potentials.
3. Equipment used to de-energize or re-energize circuits at designated points of control (e.g., station breakers), is operated in conformance with Rule 442A, and Rule 442D.
4. Tagging requirements under Rule 444C, for this *EXCEPTION*, shall include, and may be limited to, designated points of control.

C. Operations at Stations

Qualified employees shall obtain authorization from the designated person before switching sections of circuits.

In the absence of specific operating schedules, employees shall secure authorization from the designated person before opening and closing supply circuits or portions thereof or starting and stopping equipment affecting system operation at stations.

EXCEPTION 1: Sections of distribution circuits are excepted if the designated person is notified as soon as possible after the action is taken.

EXCEPTION 2: In an emergency, to protect life or property, any qualified employee may open circuits and stop moving equipment without special authorization if, in the judgment of the qualified employee, this action will promote safety, but the designated person shall be notified as soon as possible of such action, with reasons therefore.

D. Re-energizing After Work

Instructions to re-energize equipment or lines that have been de-energized by permission of the designated person shall not be issued by the designated person until all employees who requested the line to be de-energized have reported clear. Employees who have requested equipment or lines de-energized for other employees or crews shall not request that equipment or lines be re-energized until all of the other employees or crews have reported clear. The same procedure shall be followed when more than one location is involved.

E. Tagging Electric Supply Circuits Associated With Work Activities

1. Equipment or circuits that are to be treated as de-energized and grounded per Rule 444D shall have suitable tags attached to all points where such equipment or circuits can be energized.
2. When the automatic reclosing feature of a reclosing device is disabled during the course of work

on energized equipment or circuits, a tag shall be placed at the reclosing device location.

EXCEPTION: If the automatic reclosing feature of a reclosing device is disabled by a Supervisory Control and Data Acquisition System (SCADA), the system shall provide for the following:

a. At the SCADA Operating Point

(1) A signal is received by the SCADA operator confirming that the disabling operation has occurred at the reclosing device location, and

(2) A readily visible tag or electronic display is used to inform any potential SCADA operator that a disabling operation has been initiated, and

(3) The tag or electronic display is removed before action is taken to reenable the automatic reclosing feature.

b. At the Reclosing Device Location

(1) The reclosing feature is disabled in such a manner as to prevent manual override of the normal control by any potential on-site operator, or

(2) A signal, flag, or other display is used in such a manner as to alert any potential on-site operator that the reclosing feature has been disabled.

3. The required tags shall be placed to clearly identify the equipment or circuits on which work is being performed.

F. Restoration of Service After Automatic Trip

1. When circuits or equipment upon which tags have been placed open automatically, the circuits or equipment shall be left open until reclosing has been authorized.
2. When circuits open automatically, local operating rules shall determine in what manner and how many times they may be closed with safety.

G. Repeating Oral Messages

Each employee receiving an oral message concerning the switching of lines and equipment shall immediately repeat it back to the sender and obtain the identity of the sender. Each employee sending such an oral message shall require it to be repeated back by the receiver and secure the latter's identity.

443. Work on Energized Lines and Equipment

A. General Requirements

1. When working on energized lines and equipment, one of the following safeguards shall be applied:
   a. Insulate employee from energized parts
   b. Isolate or insulate the employee from ground and grounded structures, and potentials other than the one being worked on.
2. Employees shall not place dependence for their safety on the covering (nonrated insulation) of wires. All precautions (see Section 44) for working on energized parts shall be observed.
3. All employees working on or in the vicinity of lines or equipment exposed to voltages higher than those guarded against by the safety protective equipment provided shall assure themselves that the equipment or lines on which they are working are free from dangerous leakage or induction, or have been effectively grounded.
4. Cutting Into Insulating Coverings of Energized Conductors
   a. A supply cable to be worked on as de-energized that cannot be positively identified or determined to be de-energized shall be pierced or severed at the work location with a tool designed for the purpose.
   b. Before cutting into an energized supply cable, the operating voltage shall be determined and appropriate precautions taken for handling conductors at that voltage.

c. When the insulating covering on energized wires or cables must be cut into, the employee shall use a tool designed for the purpose. While doing such work, suitable eye protection and insulating gloves with protectors shall be worn. Employees shall exercise extreme care to prevent short-circuiting conductors when cutting into the insulation.

5. Metal measuring tapes, and tapes or ropes containing metal threads or strands, shall not be used closer to exposed energized parts than the distance specified in Rule 441A. Care should be taken when extending metallic ropes or tapes parallel to and in the proximity of high-voltage lines because of the effect of induced voltages.

6. Equipment or material of a noninsulating substance that is not bonded to an effective ground and which extends into an energized area, and which could approach energized equipment closer than the distance specified in Rule 441A, shall be treated as though it is energized at the same voltage as the line or equipment to which it is exposed.

B. Requirement for Assisting Employee

In inclement weather or at night, no employee shall work alone outdoors on or dangerously in the vicinity of energized conductors or parts of more than 750 V between conductors.

EXCEPTION: This shall not preclude a qualified employee, working alone, from cutting trouble in the clear, switching, replacing fuses, or similar work if such work can be performed safely.

C. Opening and Closing Switches

Manual switches and disconnectors shall always be closed by a continuous motion. Care should be exercised in opening switches to avoid serious arcing.

D. Working Position

Employees should avoid working on equipment or lines in any position from which a shock or slip will tend to bring the body toward exposed parts at a potential different than the employee's body. Work should, therefore, generally be done from below, rather than from above.

E. Protecting Employees by Switches and Disconnectors

When equipment or lines are to be disconnected from any source of electric energy for the protection of employees, the switches, circuit breakers, or other devices designated and designed for operation under the load involved at sectionalizing points shall be opened or disconnected first. When re-energizing, the procedure shall be reversed.

F. Making Connections

In connecting de-energized equipment or lines to an energized circuit by means of a conducting wire or device, employees should first attach the wire to the de-energized part. When disconnecting, the source end should be removed first. Loose conductors should be kept away from exposed energized parts.

G. Switchgear

Switchgear shall be de-energized and grounded per Rule 444D prior to performing work involving removal of protective barriers unless other suitable means are provided for employee protection. The personnel safety features in switchgear shall be replaced after work is completed.

H. Current Transformer Secondaries

The secondary of a current transformer shall not be opened while energized. If the entire circuit cannot be properly de-energized before working on an instrument, a relay, or other section of a current transformer secondary circuit, the employee shall bridge the circuit with jumpers so that the current transformer secondary will not be opened.

I. Capacitors

Before employees work on capacitors, the capacitors shall be disconnected from the energizing source, short-circuited, and grounded. Any line to which capacitors are connected shall be short-circuited and grounded before it is considered de-energized. Since capacitor units may be connected in series-parallel, each unit shall be shorted between all insulated terminals and the capacitor tank before handling. Where the tanks of capacitors are on ungrounded racks, the racks shall also be grounded. The internal resistor shall not be depended upon to discharge capacitors.

J. Gas-Insulated Equipment

Employees working on gas-insulated cable systems or circuit breakers shall be instructed concerning the special precautions required for possible presence of arcing by-products of sulfur-hexafluoride (SF_6).

NOTE: By-products resulting from arcing in sulfur-hexafluoride (SF_6) gas-insulated systems are generally toxic and irritant. Gaseous by-products can be removed for maintenance on the compartments by purging with air or dry nitrogen. The solid residue that must be removed is mostly metallic fluoride. This fine powder absorbs moisture and produces fluorides of sulfur and hydrofluoric acid, which are toxic and corrosive.

K. Attendant on Surface

While electric supply personnel are in a manhole, an employee shall be available on the surface in the immediate vicinity to render assistance from the surface. This shall not preclude the employee on the surface from entering the manhole to provide short-term assistance.

EXCEPTION: This shall not preclude a qualified employee, working alone, from entering a manhole where energized cables or equipment are in service, for the purpose of inspection, housekeeping, taking readings, or similar work if such work can be performed safely.

L. Unintentional Grounds on Delta Circuits

Unintentional grounds on delta circuits shall be removed as soon as practical.

444. De-energizing Equipment or Lines to Protect Employees

A. Application of Rule

1. When employees must depend on others to operate switches or otherwise de-energize circuits on which they are to work, or must secure special authorization before they operate such switches themselves, the precautionary measures that follow shall be taken in the order given before work is begun.
2. If the employee under whose direction a section of a circuit is disconnected is in sole charge of the section and of the means of disconnection, those portions of the following measures that pertain to dealing with the designated person may be omitted.
3. Records shall be kept on all contractual utility interactive systems on any electric supply lines. When these lines are de-energized according to Rule 444C, the utility interactive system shall be visibly disconnected from the lines.

B. Employee's Request

The employee in charge of the work shall apply to the designated person to have the particular section of equipment or lines de-energized, identifying it by position, letter, color, number, or other means.

C. Operating Switches, Disconnectors, and Tagging

The designated person shall direct the operation of all switches and disconnectors through which electric energy may be supplied to the particular section of equipment and lines to be de-energized, and shall direct that such switches and disconnectors be rendered inoperable and tagged. If switches that are controlled automatically or remotely or both can be rendered inoperable, they shall be tagged at the switch location. If it is impractical to render such switches and disconnectors inoperable, then these remotely controlled switches shall also be tagged at all points of control. A record shall be made when placing the tag, giving the time of disconnection, the name of the person making the disconnection, the name of the employee who requested the disconnection, and the name or title or both, of the designated person.

D. Employee's Protective Grounds

When all the switches and disconnectors designated have been operated, rendered inoperable where practical, and tagged in accordance with Rule 444C, and the employee has been given permission to work by the designated person, the employee in charge should immediately proceed to make the employee's own protective grounds or verify that adequate grounds have been applied (see Rule 445) on the disconnected lines or equipment. During the testing for potential and/or application of grounds, distances not less than those shown in Tables 441-1 to 441-3, as applicable, shall be maintained.

Grounds shall be placed at each side of the work location and as close as practical to the work location, or a single point ground shall be placed at the work location. If work is to be performed at more than one location on a line section, the line section shall be grounded and short-circuited at one location in the line section and the conductor to be worked on shall be grounded at each work location.

The distance in Tables 441-1, 441-2, or 441-3, as applicable, shall be maintained from ungrounded conductors at the work location. Where the making of a ground is impractical, or the conditions resulting therefrom are more hazardous than working on the lines or equipment without grounding, the ground may be omitted by special permission of the designated person.

E. Proceeding With Work

1. After the equipment or lines have been de-energized and grounded per Rule 444D, the employee in charge, and those under the direction of the employee in charge, may proceed with work on the de-energized parts.

   Equipment may be re-energized for testing purposes only under the supervision of the employee in charge and subject to authorization by the designated person.
2. Each additional employee in charge desiring the same equipment or lines to be de-energized and grounded per Rule 444D for the protection of that person, or the persons under direction, shall follow these procedures to secure similar protection.

F. Reporting Clear—Transferring Responsibility

1. The employee in charge, upon completion of the work and after ensuring that all persons assigned to this employee in charge are in the clear, shall remove protective grounds and shall report to the designated person that all tags protecting that person may be removed.
2. The employee in charge who received the permission to work may, if specifically permitted by the designated person, transfer the permission to work and the responsibility for persons by personally informing the affected persons of the transfer.

G. Removal of Tags

1. The designated person shall then direct the removal of tags and the removal shall be reported back to the designated person by the persons removing them. Upon the removal of any tag, there shall be added to the record containing the name of the designated person or title or both, and the person who requested the tag, the name of the person requesting removal, the time of removal, and the name of the person removing the tag.
2. The name of the person requesting removal shall be the same as the name of the person requesting placement, unless responsibility has been transferred according to Rule 444F.

H. Sequence of Re-energizing

Only after all protective grounds have been removed from the circuit or equipment and after protective tags have been removed in accordance with Rule 444G at a specific location, may the designated person direct the operation of switches and disconnectors at that location.

445. Protective Grounds

A. Installing Grounds

When placing protective grounds on a previously energized part, the following sequence and precautionary measures shall be observed.

EXCEPTION: In certain situations, such as when grounding conductors are supported on some high-voltage towers, it may be appropriate to perform the voltage test before bringing the grounding device into the work area.

1. Current-Carrying Capacity of Grounds

   The grounding device shall be of such size as to carry the induced current and anticipated fault current that could flow at the point of grounding for the time necessary to clear the line.

2. Initial Connections

   Before grounding any previously energized part, the employee shall first securely connect one end of the grounding device to an effective ground. Grounding switches may be employed to connect the equipment or lines being grounded to the actual ground connections.

3. Test for Voltage

   The previously energized parts that are to be grounded shall be tested for voltage except where previously installed grounds are clearly in evidence. The employee shall keep every part of the body at the required distance by using insulating handles of proper length or other suitable devices.

4. Completing Grounds

   a. If the part shows no voltage, the grounding may be completed.
   b. If voltage is present, the source shall be determined to ensure that presence of this voltage does not prohibit completion of the grounding.
   c. After the initial connections are made to ground, the grounding device shall next be brought into contact with the previously energized part using insulating handles or other suitable devices and securely clamped or otherwise secured thereto. Where bundled conductor lines are being grounded, grounding of each subconductor should be made. Only then may the employee come within the distances from the previously energized parts specified in Rule 441A or proceed to work upon the parts as upon a grounded part.

B. Removing Grounds

The employee shall first remove the grounding devices from the de-energized parts using insulating handles or other suitable devices.

Extreme caution shall be exercised that the proper sequence of installing or removing grounds is followed. The connection to the effective ground shall not be removed first. If done, electric shock and injury may result.

446. Live Work

All employees using live work practices shall observe the following rules in addition to applicable rules contained elsewhere in Sections 42 and 44.

The distances specified in Tables 441-1, 441-2, 441-3, or 441-4 shall be maintained from all grounded objects and from other conductors, lines, and equipment having a potential different from that to which conductive equipment and devices are bonded in order to maintain the equipotentially energized work environment in an isolated state.

A. Training

Employees shall be trained in live work practices, which include rubber glove, hot stick, or bare-hand method, before being permitted to use these techniques on energized lines.

B. Equipment
   1. Insulated aerial devices, ladders, and other support equipment used in live work shall be evaluated for performance at the voltages involved. Tests shall be conducted to ensure the equipment's integrity. Insulated aerial devices used in bare-hand work shall be tested before the work is started to ensure the integrity of the insulation. See applicable references in Section 3, specifically IEEE Std 516-1987 and ANSI/SIA A92.2-1992.
   2. Insulated aerial devices and other equipment used in this work shall be maintained in a clean condition.
   3. Tools and equipment shall not be used in a manner that will reduce the overall insulating strength of the insulated aerial device.

C. When working on insulators under live-line procedures, the clear insulation distance shall be not less than the distances required by Tables 441-1, 441-2, 441-3, and 441-4.

D. Bonding and Shielding for Bare-Hand Method
   1. A conductive bucket liner or other suitable conducting device shall be provided for bonding the insulated aerial device to the energized line or equipment.
   2. The employee shall be bonded to the insulated aerial device by use of conducting shoes, leg clips, or other suitable means.
   3. Adequate electrostatic shielding in the form of protective clothing that has been evaluated for electrical performance shall be provided and used where necessary.

      NOTE: Electrostatic Shielding. Evaluation of protective clothing designed for this purpose is covered in IEEE Std 516-1987.
   4. Before the employee contacts the energized part to be worked on, the aerial device shall be bonded to the energized conductor by means of a positive connection.

447. Protection Against Arcing and Other Damage While Installating and Maintaining Insulators and Conductors

In installing and maintaining insulators and conductors, precautions shall be taken to limit the opportunity for, as far as is practical, any damage that might render the conductors or insulators liable to fall. Precautions shall also be taken to prevent, as far as is practical, any arc from forming and to prevent any arc that might be formed from injuring or burning any parts of the supporting structures, insulators, or conductors.

Appendix A

(This Appendix is not part of Accredited Standards Committee C2-2002, National Electrical Safety Code, 2002 Edition, and is included for information only.)

Uniform System of Clearances Adopted in the 1990 Edition

Rules 232, 233, and 234

Introduction

The original format or system for stating NESC requirements was developed before 1920 and recognized the practical constraints of that time: Clearances were specified for a set of basic conditions. Some *basic clearances* included conductor movement; adders were used for nonbasic conditions. Although easy to use, it was unduly conservative in many cases and did not adequately recognize new materials and construction in others. Various additional clearance requirements were added over the years.

For the 1990 edition, an intensive study by the NESC Clearances Subcommittee identified:

- Modern utility practices and capabilities that remove the previous clearance measurement constraints,
- Apparent inconsistencies in certain clearance treatments, and
- The need to develop a uniform clearance system independent of materials used for conductors and cables, stringing tensions, operating temperatures, and similar constraints.

The new uniform clearance system contained in the 1990 and subsequent editions reflects the dimensions of expected activities in each area (reference component), as well as the relative potential problem caused by each type of facility (mechanical and electrical component).

Conductor clearance in the 1990 and subsequent editions is stated in terms of the "closest approach," i.e., the clear distance that must be maintained under specified conditions.

- Vertical clearances are required during maximum sag conditions; they provide for expected activity beneath a line.
- Horizontal clearances are required when the conductor is at rest; they provide for expected activity alongside a line. In addition, displacement of conductors by wind is considered under certain conditions.

Under the new system, users consider the actual characteristics of the materials and construction, rather than the reference characteristics built into the early code requirements.

While some clearance values in the new system may appear to be larger and some smaller, the net effective clearances for conductors and cables are, for most of the clearance values, essentially unchanged. Some few values required minor adjustments of the effective clearances to make them uniform with the other values, thus illustrating one of the needs for these changes.

The 1990 and subsequent editions provide the following user benefits:

- Simple code language in performance-standard format (as opposed to the prior design-manual style),
- Readily understandable intent,
- Uniform clearance values,
- Integration of prior rules for long-span construction and/or high-conductor temperature operation, and
- Reduced number of footnotes required to cover exceptions.

The 1990 Changes

Rules 232, 233, and 234 were revised based on a coordinated, uniform system of clearances developed under a building-block approach. Three components were considered to determine the total clearance required:

- *A reference component* to cover activity in the area to be cleared by the overhead supply and/or communication lines. For example, truck height for over-the-road transport is limited to 4.3 m (14 ft) by state regulation. Thus the reference component for roads in Table 232-3 is 4.3 m (14 ft). Reference components included in the required clearances are shown in Table A-2.

- *A mechanical component* appropriate for the supply or communication line item. The mechanical component for open supply conductors is 610 mm (2 ft) (Table A-1).
- *An electrical component* appropriate for the voltage involved. The electrical component for open supply conductors, over 750 V to 22 kV, is 760 mm (2.5 ft) (Table A-1).

The required clearance is the sum of the three components: thus, 5.6 m (18.5 ft) is required for open supply conductors, over 750 V to 22 kV, over roads (Table 232-1). For purposes of illustration, the mechanical and electrical components are combined in Table A-1, and items with the same total mechanical and electrical components are grouped into similar clearance categories. Six groups are thus created.

Table A-1

Group	M&E (mm/m)	M&E (ft)	Category	R/NR	GI/O	M	E
I	305 mm	1.0	Support Arms	1.0/—	0.0/—	1.0	0.0
			Effectively Grounded Equipment Cases	1.0/—	0.0/—	1.0	0.0
II	455 mm	1.5	Insulated Communication Conductors and Cables	—/1.5	0.0/—	1.5	0.0
			Messengers	—/1.5	0.0/—	1.5	0.0
			Surge Protection Wires	—/1.5	0.0/—	1.5	0.0
			Grounded Guys	—/1.5	0.0/—	1.5	0.0
			230E1	—/1.5	0.0/—	1.5	0.0
			230C1	—/1.5	0.0/—	1.5	0.0
III	610 mm	2.0	URLP, 0 V to 750 V	1.0/—	—/0.5	1.5	0.5
			Noninsulated Communication Conductors	—/1.5	—/0.5	2.0	0.0
			230C2, 0 V to 750 V	—/1.5	0.0/—	1.5	0.5
			230C3, 0 V to 750 V	—/1.5	0.0/—	1.5	0.5
			Ungrounded Cases of Equipment at 0 to 750 V	1.0/—	—/0.5	1.5	0.5
IV	760 mm	2.5	230C2, Greater than 750 V	—/1.5	0.0/—	1.5	1.0*
			230C3, Greater than 750 V	—/1.5	0.0/—	1.5	1.0*
			Open Supply Conductors, 0 to 750 V	—/1.5	0.0/—	2.0	0.5*
V	1.2 m	4.0	URLP, Greater than 750 V to 22 kV	1.0/—	—/0.5	1.5	2.5
			Ungrounded Cases of Equipment at Greater than 750 V to 22 kV	1.0/—	—/0.5	1.5	2.5
VI	1.37 m	4.5	Open Supply Conductors, Greater than 750 V to 22 kV	—/1.5	—/0.5	2.0	2.5

LEGEND:

URLP — Unguarded rigid live parts
R — Rigid = 305 mm (1.0 ft)
NR — Nonrigid = 155 mm (1.5 ft)
GI — Grounded or insulated = 0.0 m (0.0 ft)
O — Bare, ungrounded, or open conductor or part = 152 mm (0.5 ft)
M — Mechanical component = R/NR plus GI/O
E — Electrical Component

- Grounded & Communication Conductor = 0.0 m (0.0 ft)
- Supply Line 0 V to 750 V = 152 mm (0.5 ft)
- Supply Line Greater than 750 V to 22 kV =760 mm (2.5 ft)

M&E — Sum of M and E values

NOTES: (1) Ungrounded guys and ungrounded portions of guys between guy insulators have clearances based on the highest voltage to which they are exposed.
(2) An asterisk (*) beside a value indicates an exception to the legend.

m

Table A-2a
Reference Components of Rule 232

	Table 232-1		Table 232-2	
	Item	Ref (m)	Item	Ref (m)
Track rails	1	6.7	—	—
Roads, streets, alleys, etc.	2	4.3	1a	4.3
Residential driveways, etc.	3	4.3	1b	4.3
Other land traversed by vehicles	4	4.3	1c	4.3
Spaces and ways—pedestrians	5	2.45/3.0	1d	3.0
Water areas—no sailboating	6	3.8	—	—
Water areas—sailboating	7		—	
(a) Less than 0.08 km^2		4.9		—
(b) Over 0.08 to 0.8 km^2		9.0		—
(c) Over 0.8 to 8 km^2		9.0		—
(d) Over 8 km^2		11.0		—
Areas posted for rigging or launching sailboats	8	See 7	—	
Within or along:				
Roads, streets, or alleys	9	4.3	2a	4.3
Rural districts, vehicles unlikely	10	3.65	2b	3.65

ft

Table A-2a
Reference Components of Rule 232

	Table 232-1		Table 232-2	
	Item	Ref (ft)	Item	Ref (ft)
Track rails	1	22.0	—	—
Roads, streets, alleys, etc.	2	14.0	1a	14.0
Residential driveways, etc.	3	14.0	1b	14.0
Other land traversed by vehicles	4	14.0	1c	14.0
Spaces and ways—pedestrians	5	8.0/10.0	1d	10.0
Water areas—no sailboating	6	12.5	—	—
Water areas—sailboating	7		—	
(a) Less than 20 acres		16.0		—
(b) Over 20 to 200 acres		30.0		—
(c) Over 200 to 2000 acres		30.0		—
(d) Over 2000 acres		36.0		—
Areas posted for rigging or launching sailboats	8	See 7	—	
Within or along:				
Roads, streets, or alleys	9	14.0	2a	14.0
Rural districts, vehicles unlikely	10	12.0	2b	12.0

Table A-2b
Reference Components of Rule 234

Table	Item	Ref (mm/m)	Ref (ft)
234-1	1. Buildings		
	a. Horizontal		
	(1) Walls, projections, and guarded windows	915 mm	3.0
	(2) Unguarded windows	915 mm	3.0
	(3) Balconies and areas accessible to pedestrians	915 mm	3.0
	b. Vertical		
	(1) Roofs/projections not accessible to pedestrians	2.44 m	8.0
	(2) Balconies and roofs accessible to pedestrians	2.74 m	9.0
	(3) Roofs—vehicles not over 2.4 m (8 ft)	2.74 m	9.0
	(4) Roofs—vehicles over 2.4 m (8 ft)	4.3 m	14.0
	2. Signs, chimneys, billboards, antennas, tanks, etc.		
	a. Horizontal	9.15 mm	3.0
	b. Vertical over or under	1.07 m	3.5
234-2	1. Over bridges		
	a. Attached	305 mm	1.0
	b. Not attached	2.44 m	8.0
	2. Beside, under, or within bridge structure		
	a. Accessible		
	(1) Attached	305 mm	1.0
	(2) Not attached	2.44 m	3.0
	b. Inaccessible		
	(1) Attached	305 mm	1.0
	(2) Not attached	610 mm	2.0
234-3	A. From water level, edge of pool, etc.	6.25 m	20.5
	B. From diving platform or tower	3.8 m	12.5

Application rules were revised in 1990 to coordinate with clearances developed under the component or building-block approach.

Vertical clearances now apply at the maximum conductor sag condition, such as outlined in Rule 232A, rather than at a 15 °C (60 °F) conductor temperature condition as used in the 1987 Edition. This is illustrated in Fig A-1: 5.6 m (18.5 ft) is required for open supply conductors, over 750 V to 22 kV, over roads, for any sag condition or span length.

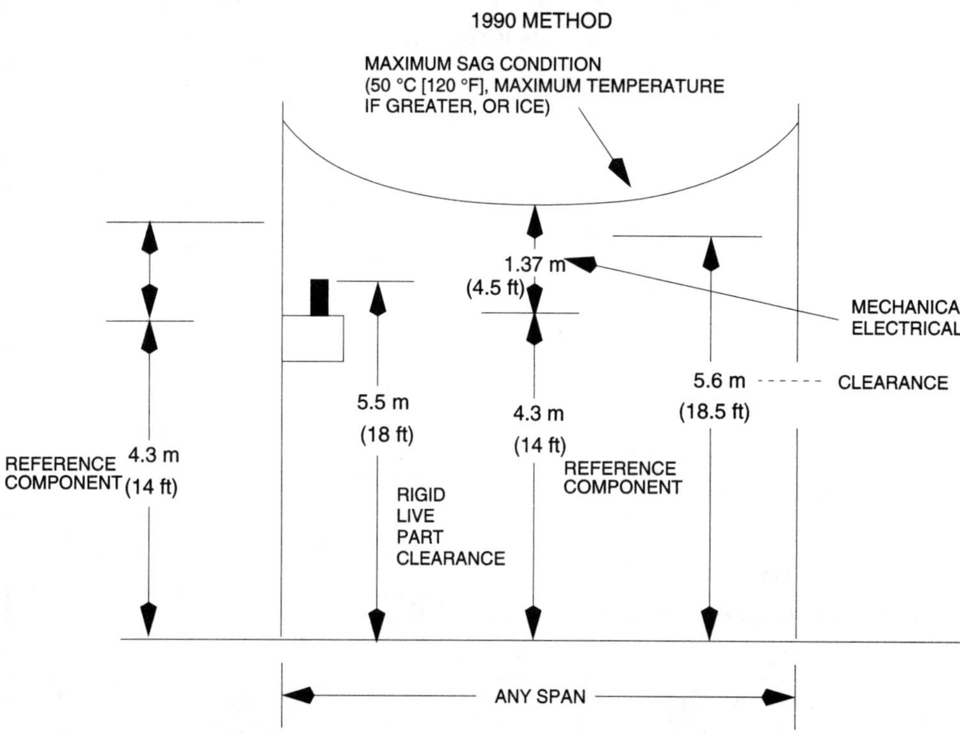

CONDITION—CLEARANCE OF OPEN SUPPLY CONDUCTORS, 750 V–22 kV, ABOVE ROADS
CLEARANCE (5.6 m [18.5 ft])= REFERENCE COMPONENT (4.3 m [14 ft])
+ MECHANICAL AND ELECTRICAL COMPONENT (1.37 m [4.5 ft])

Fig A-1
Clearance at Maximum Sag

Not only is the intended vertical clearance shown, *rule simplification* is also achieved. Rules for long-span construction and/or high-temperature operation were removed because they are no longer necessary.

Horizontal clearances to buildings and other installations now apply with the conductor at rest (no wind displacement) as outlined in Rule 234A, rather than at a wind displacement condition. The horizontal clearance for open supply conductors from buildings (over 750 V to 22 kV) at rest is now 2.3 m (7.5 ft) (Table 234-1). This clearance is essentially the same as the 2.44 m (8 ft) required by early code editions.

Wind displacement need be considered only for energized open supply conductors and 230C2–230C3 cables energized at more than 750 V; see Rule 234C1. In the above example, a minimum clearance of 1.37 m (4.5 ft) is required under wind displacement conditions. While less than the 1.5 or 1.8 m (5 or 6 ft) required under prior codes, people will not be working on outside building walls during high wind conditions.

Because application rules have been revised, *it must be understood that clearance values cannot be directly compared between the 1987 and 1990, 1993, or 1997 Editions.* Vertical clearance values *appear* smaller because sag changes formerly included in clearance values are now addressed in the application rules. Horizontal clearance values *appear* larger because wind displacement is now applicable to energized conductors and certain supply cables only; clearances for all wires, conductors, and cables are shown in the tables under at-rest conditions.

The following changes were also made to consolidate requirements and simplify application:

- Voltages in the tables are limited to 0 V to 750 V and over 750 V to 22 kV, normal secondary and primary distribution ranges respectively. Voltages in the 22 kV to 50 kV range are covered by a 10 mm-per-kV (0.4-in-per-kV) adder; see Rules 232C1a, 232C2a, and 234G1. Exceptions at 22 kV to 50 kV are noted where they apply.
- Rules for voltages above 22 kV and the alternate clearances for voltages above 98 kV are consolidated.
- Clearances for equipment cases are relocated from Rules 286E and 286F to Rules 232B3 and 234J.

Clearance Rules and Tables Prior to 1990

In editions prior to 1990, clearances shown in the tables were *basic clearances,* applied under specified conditions of conductor temperature and sag, span length, and voltage range. For example, vertical clearances in Table 232-1 of the 1987 Edition applied at a conductor temperature of 15 °C (60 °F), no wind, final unloaded sag. Span lengths were limited by loading district. Voltages up to 50 kV were covered in the table. Conditions outside these basic conditions required additional clearances.

Actual clearances vary from the values required at 15 °C (60 °F) as conductor sag changes due to conductor movement under loading. Table 232-1 allowed 610 mm (24 in) for ice loading, higher conductor temperatures [to 50 °C (120 °F)], and structure flexure. The actual allowance was 460 mm (18 in) for ice loading or higher conductor temperature plus 152 mm (6 in) for miscellaneous causes. Thus, the true clearance requirement was 610 mm (24 in) less than the value shown in the table (i.e., equivalent to the clearance required for rigid parts).

Table 232-2, vertical clearance of *rigid* live parts, was consistent with the 610 mm (24 in) allowance in Table 232-1. A rigid live part energized at 750 V to 22 kV over a road (item 1a, middle column) required a 5.5 m (18 ft) clearance. A conductor energized at the same voltage over a road required a 6.1 m (20 ft) clearance (Table 232-1, item 2). The additional 610 mm (24 in) was the allowance for conductor movement, and the true clearance that may be experienced is 5.5 m (18 ft)—the same as specified for rigid live parts.

Additional clearances above the basic values shown in Table 232-1 were required when the limiting conditions were exceeded. Rule 232B2c covered long-span construction and Rule 232B2d covered high-temperature operation. Both rules recognized and allowed for additional conductor movement. Finally, Rule 232B1 required additional clearance for voltages exceeding 50 kV. Note that this was an electrical requirement apart from conductor movement.

Application of basic and additional clearances is illustrated in the drawings that follow.

Fig A-2 shows the basic clearance applied at a 15 °C (60 °F) conductor temperature, at the limiting span length. When conductor movement is considered, the actual clearance at maximum sag is less than the basic clearance.

Fig A-3 shows application of an additional clearance for conductor temperature over 50 °C (120 °F), at the limiting span. Note that the additional clearance is added to the basic clearance to determine the required clearance. Again, the actual clearance at maximum sag is essentially the same as in Fig A1.

Fig A-4 shows application of a long-span additional clearance, for a conductor operating at or within the 50 °C (120 °F) limit. As before, the additional clearance is added to the basic clearance, and the actual conductor clearance at maximum sag is essentially the same as in Fig A-2.

To repeat, *both the basic clearances shown in the tables and the additional clearances required by the rules apply only when the conductor temperature is 15 °C (60 °F). Actual clearances are expected and intended to be less due to conductor movement.* Only the voltage adder (for voltages above those shown in the tables) is a true clearance requirement.

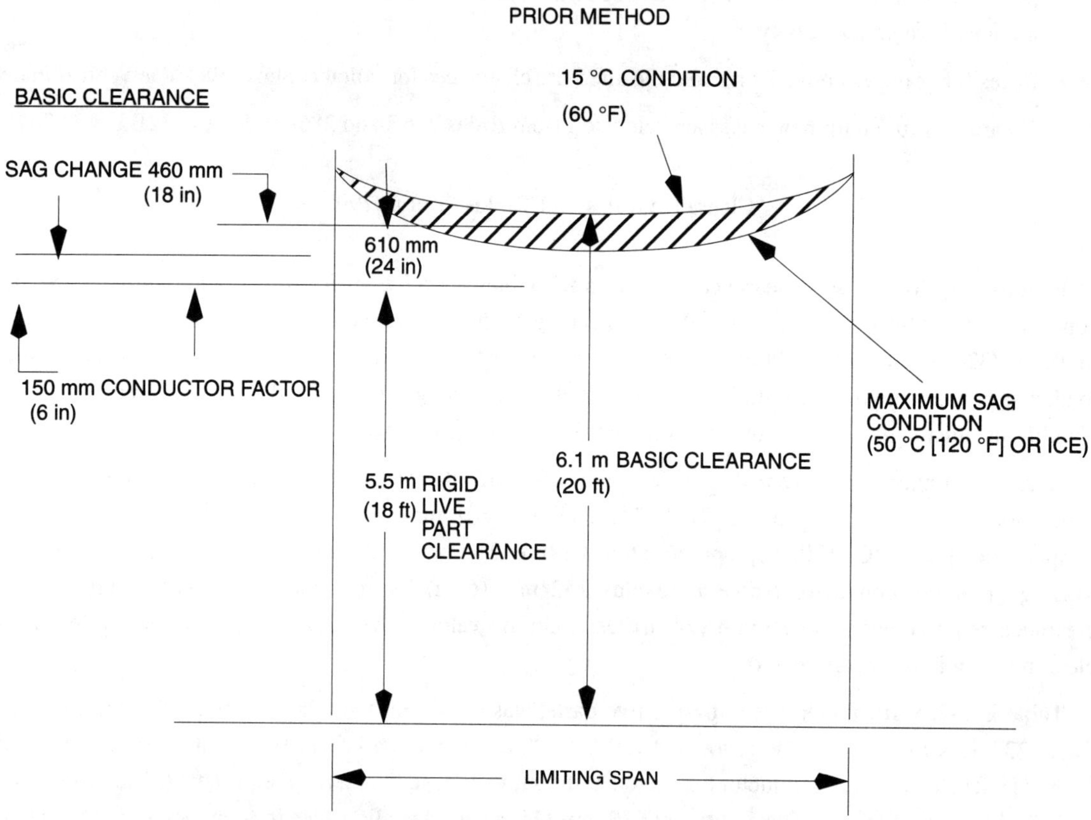

CONDITION—CLEARANCE OF OPEN SUPPLY CONDUCTORS, 750 V–22 kV, ABOVE ROADS

LIMITING SPAN—53.3 m (175 ft) MAXIMUM IN HEAVY LOADING DISTRICT

/ / / / / / / —CONDUCTOR MOVEMENT

BASIC CLEARANCE (6.1 m [20 ft]) = ACTUAL CLEARANCE (5.5 m [18 ft]) + CONDUCTOR MOVEMENT 0.6 m (2 ft)

Fig A-2
Basic Clearance

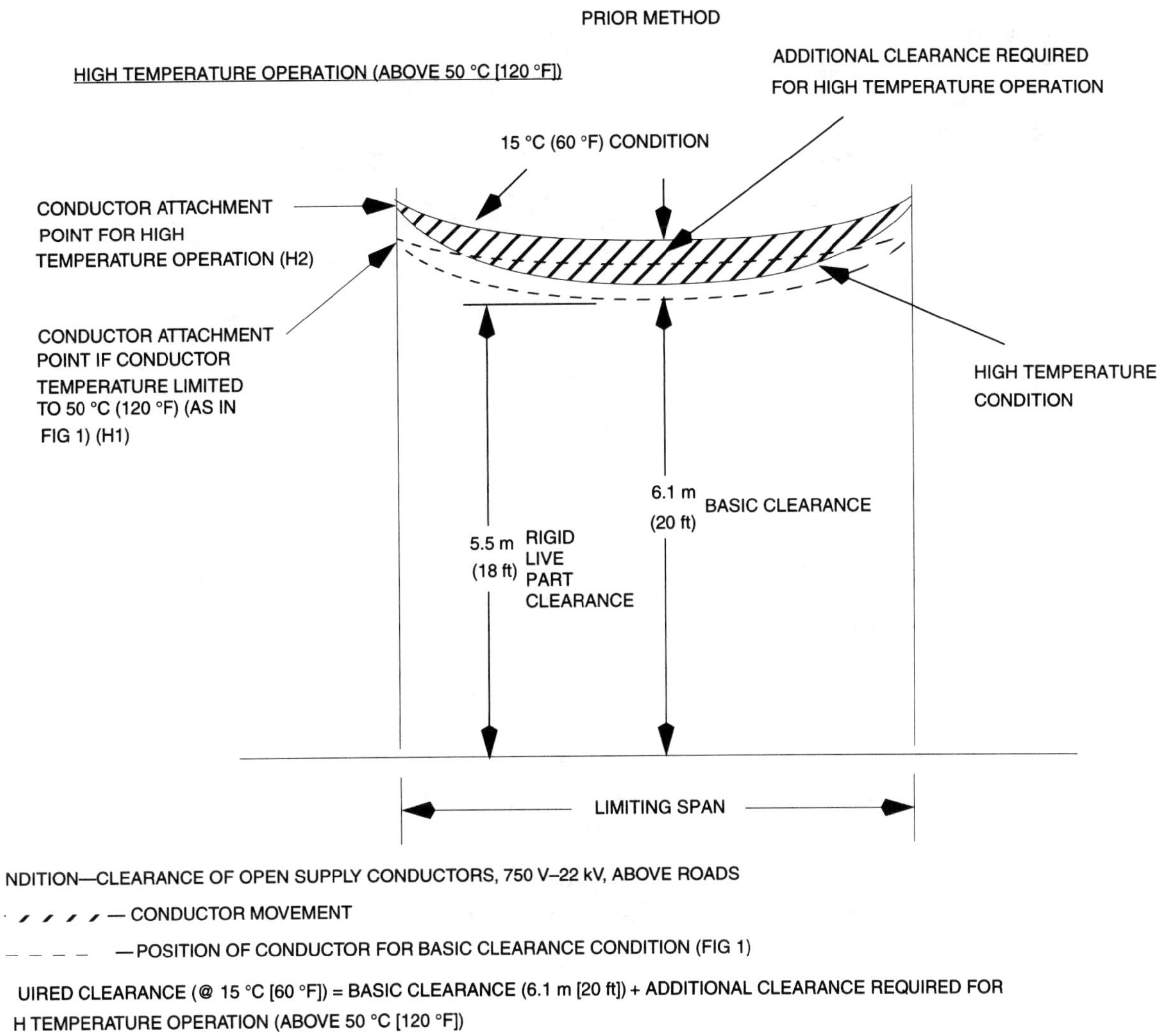

Fig A-3
Additional Clearance Required for High-Temperature Operation

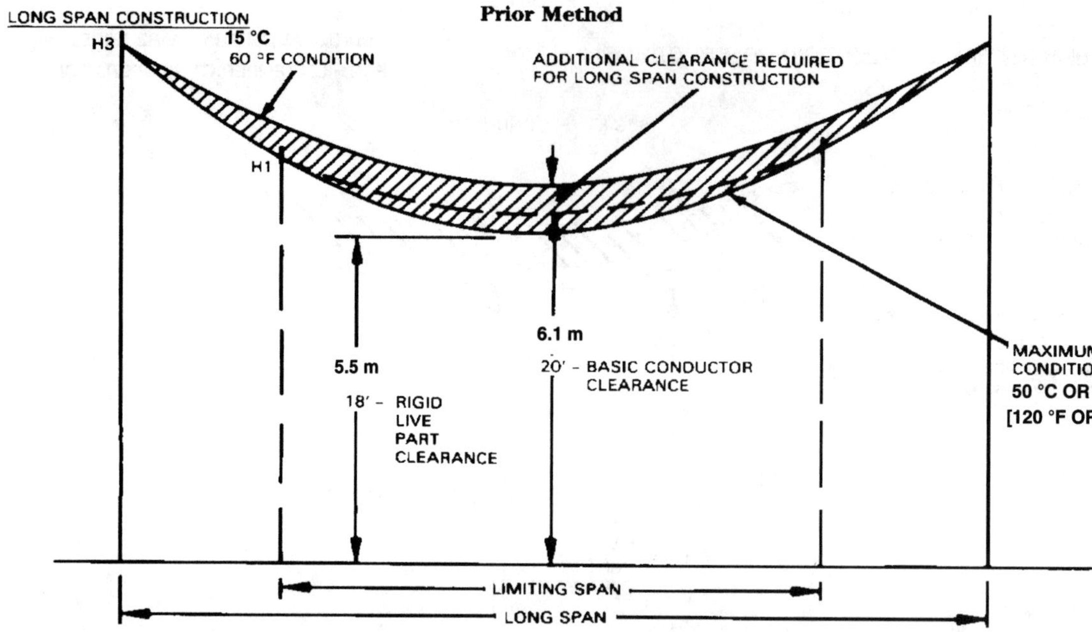

CONDITION—CLEARANCE OF OPEN SUPPLY CONDUCTORS, 750 V–22 kV, ABOVE ROADS

/ / / / / .— CONDUCTOR MOVEMENT

- – – – – – —POSITION OF CONDUCTOR FOR BASIC CLEARANCE CONDITION (FIG 1)

REQUIRED CLEARANCE (@ 15 °C [60 °F]) = BASIC CLEARANCE (6.1 m [20 ft]) + ADDITIONAL CLEARANCE REQUIRED FOR LONG SPAN CONSTRUCTION (ASSUMING MAXIMUM CONDUCTOR TEMPERATURE LIMITED TO 50 °C [120 °F])

Fig A-4
Additional Clearance Required for Long-Span Construction

Clearance Values Prior to 1990

Clearance values were based on experience. They were developed over time, at different times, for different reasons.

While those clearance values worked well, partly because several of them proved to be overly conservative, they did not follow a uniform system. For example, Table 234-1 showed vertical clearances from buildings. In line 1b(1) of the 1987 Edition, vertical clearance of open supply conductors over roofs not accessible to pedestrians was the same for all voltages from 0 to 22 kV. However, in line 1b(3), the vertical clearance of open supply conductors over roofs accessible to vehicles but not subject to truck traffic varied with voltage: 3.6 m (12 ft) for conductors energized at 0 V to 300 V, 4.6 m (15 ft) for 300 V to 750 V, and 6.1 m (20 ft) for 750 V to 22 kV.

Summary—Prior Editions

The clearance section was complex, with requirements stated in design manual format. Clearance was a mixture of basic clearance, clearance to cover conductor movement, and voltage clearance. Required clearance applied only at a 15 °C (60 °F) conductor temperature; actual clearance was not shown. Finally, clearance values were empirical; they were not systematized.

Clearances Subcommittee Activities

Subcommittee 4 discussed problems with the clearance section of the code during the 1987 revision cycle. Because there was insufficient time to develop a comprehensive proposal, Subcommittee 4 recommended formation of a special working group. The NESC Committee approved this recommendation and established Working Group 4.2 to:

- Review overhead line clearances, primarily Rules 232 and 234 and Section 28, and
- Investigate feasibility of a uniform method of determining clearances under all conditions of conductor movement.

The working group concluded that:

- A uniform system for determining clearances could be developed utilizing a building-block approach.
- Vertical clearance values could be stated for maximum sag conditions to cover conductor movement.
- Horizontal clearances could be stated under at-rest conditions, with special requirements to cover displacement of energized conductors during wind conditions.
- The revisions proposed in the working group report were explicit, readily understood, and performance oriented.

Subcommittee 4 reviewed and approved the Working Group 4.2 report with minor modifications. Further enhancements were made in response to public comment.

Summary

The 1990 changes constitute a comprehensive revision of Section 23, which incorporates related provisions of Section 28 and provides significant user benefits, as detailed in the introduction at the beginning of this discussion. While some clearance values may appear to be larger and some smaller, the net effective clearances for energized conductors and cables are essentially unchanged.

Bibliography

[B1] ANSI A14.1-1994, American National Standard Safety Requirements for Portable Wood Ladders. [Rule 323F REC.]

[B2] ANSI A14.2-1990, American National Standard Safety Requirements for Portable Metal Ladders [and supplement ANSI 14.2a (1985)]. [Rule 323F REC.]

[B3] ANSI A14.3-1992, American National Standard Safety Requirements for Fixed Ladders. [Rule 323F REC.]

[B4] ANSI A14.5-1992, American National Standard Safety Requirements for Portable Reinforced Plastic Ladders [and supplement ANSI A14.5a (1985)]. [Rule 323F REC.]

[B5] ANSI A1264.1-1995, American National Safety Requirements for Workplace Floor and Wall Openings. [Rule 112 D NOTE]

[B6] ANSI C2-1973, National Electrical Safety Code. [Rule 402 NOTE]

[B7] ANSI C29.9-1983, American National Standard for Apparatus, Post Type Wet-Process Porcelain Insulators. [Rule 277 NOTE 2]

[B8] ANSI C29.11-1989, American National Standard Tests to Composite Suspension Insulation for Overhead Transmission Lines. [Rule 277 NOTE 1b]

[B9] API RP 500, Recommended Practice for Classification of Locations for Electrical Installations at Petroleum Facilities, 1 June 1991. [Rule 127L NOTE]

[B10] ASTM F 478-92, Standard Specification for In-Service Care of Insulating Line Hose and Covers.

[B11] ASTM F 479-95, Standard Specification for In-Service Care of Insulating Blankets.

[B12] ASTM F 496-96, Standard Specification for In-Service Care of Insulating Gloves and Sleeves.

[B13] ASTM F 696-91, Standard Specification for Leather Protectors for Rubber Insulating Gloves and Mittens.

[B14] ASTM F 711-89, Standard Specification for Fiberglass-Reinforced Plastic (FRP) Rod and Tube Used in Live Line Tools.

[B15] ASTM F 855-96, Standard Specification for Temporary Grounding Systems to Be Used on De-energized Electric Power Lines and Equipment.

[B16] ASTM F 887-91a, Standard Specification for Personal Climbing Equipment.

[B17] ASTM F 914-91, Standard Test Method for Acoustic Emission for Insulated Aerial Personnel Devices.

[B18] ASTM F 968-93, Standard Specification for Electrically Insulating Plastic Guard Equipment for Protection of Workers.

[B19] ASTM F 1116-88 (1995), Standard Test Method for Determining Dielectric Strength of Overshoe Footwear.

[B20] ASTM F 1236-96, Standard Guide for Visual Inspection of Electrical Protective Rubber Products.

[B21] IEEE/ASTM SI 10-1997, Standard for the Use of the International System of Units (SI). [Rule 17A, Footnote 2]

[B22] IEEE Std 80-2000, IEEE Guide for Safety in AC Substation Grounding. [Rule 96A NOTE and Rule 123B NOTE]

[B23] IEEE Std 268-1992, IEEE Standard for Metric Practice (ANSI) (DoD adopted).

[B24] IEEE Std 524-1992, IEEE Guide to the Installation of Overhead Transmission Line Conductors (ANSI).

[B25] IEEE Std 738-1993, IEEE Standard for Calculating the Current-Temperature of Bare Overhead Conductors (ANSI).

[B26] IEEE Std 751-1990, IEEE Trial-Use Design Guide for Wood Transmission Structures (ANSI).

[B27] IEEE Std 935-1989 (R1995), IEEE Guide on Terminology for Tools and Equipment to Be Used in Live Line Working (ANSI).

[B28] IEEE Std 951-1996, IEEE Guide to the Assembly and Erection of Metal Transmission Structures (ANSI).

[B29] IEEE Std 957-1995, IEEE Guide for Cleaning Insulators (ANSI).

[B30] IEEE Std 977-1991, IEEE Guide to Installation of Foundations for Transmission Line Structures (ANSI).

[B31] IEEE Std 978-1984 (R1990), IEEE Guide for In-Service Maintenance and Electrical Testing of Live-Line Tools (ANSI).

[B32] IEEE Std 987-1985, IEEE Guide for Application of Composite Insulators (ANSI).[12]

[B33] IEEE Std 1024-1988, IEEE Recommended Practice for Specifying Distribution Composite Insulators (Suspension Type) (ANSI).[13]

[B34] IEEE Std 1048-1990, IEEE Guide for Protective Grounding of Power Lines (ANSI).

[B35] IEEE Std 1067-1996, IEEE Guide for In-Service Use, Care, Maintenance, and Testing of Conductive Clothing for Use on Voltages up to 765 kV AC (ANSI).

[B36] IEEE Std 1070-1988, IEEE Guide for the Design and Testing of Transmission Modular Restoration Structure Components (ANSI).

[B37] IEEE Std 1119-1988 (R1993), IEEE Guide for Fence Safety Clearances in Electric-Supply Stations (ANSI).

[B38] IEEE Std C62.1-1989 (R1994), IEEE Standard for Gapped Silicon-Carbide Surge Arresters for AC Power Circuits (ANSI). [Rule 190 NOTE]

[B39] IEEE Std C62.11-1999, IEEE Standard for Metal-Oxide Surge Arresters for Alternating Current Power Circuits (ANSI). [Rule 190 NOTE]

[B40] IEEE Working Group on Switching Surges, "Switching Surges, pt. IV—Control on AC Transmission Lines," *IEEE Transactions on Power Apparatus and Systems,* vol. PAS-101, No. 8, pp. 2694–2702, Aug. 1982.

[B41] NFPA 77-1993, Recommended Practice on Static Electricity. [Rule 127D3 NOTE]

---

[12]IEEE Std 987-1985 has been withdrawn; however, copies can be obtained from Global Engineering, 15 Inverness Way East, Englewood, CO 80112-5704, USA, tel. (303) 792-2181.

[13]IEEE Std 1024-1988 has been withdrawn; however, copies can be obtained from Global Engineering.

[B42] NFPA 497M-1991, Classification of Gases, Vapors, and Dusts for Electrical Equipment in Hazardous (Classified) Locations. [Rule 127L NOTE]

[B43] OSHA 29 CFR 1926 Subpart V—Power Transmission and Distribution. [Rule 402 NOTE]

[B44] OSHA Federal Register [01/31/89] 54: 4974–5024—Electric Power Generation, Transmission, and Distribution; Electrical Protective Equipment; Proposed Rule (1910.269). [Rule 402 NOTE]

[B45] OSHA Federal Register [01/31/94] 59: 4320–4476—Electric Power Generation, Transmission, and Distribution; Electrical Protective Equipment (1910.269). [Rule 402 NOTE]

Index

Italic type is used for C2 *rule*, *section*, *part*, *figure*, or *table* identification. The corresponding page numbers follow in upright type.

A

Adjustable-speed motors, *130E;* 49

Administrative authority, definition of, *Sec. 2;* 4

Aerial cable systems, insulation requirements, *278;* 186

Aerial services, *239G4;* 151

Altitude corrections (for approach distance to live parts), *431, 441A5, Table 441-5;* 222, 225, 232

American Wire Gage (AWG), *Sec. 2;* 13

Ampacity
Definition of, *Sec. 2;* 4
Grounding conductor, *93C;* 19

Anchorage, definition of, *Sec. 2;* 4

Anchors, *253, 261B;* 170, 175

Annunciators, *180E;* 56

Antennas, conductor clearance from, *234C, Table 234-1, Table 234-5, 235I;* 99, 102, 121, 141

Apparent sag at any point in a span, definition of, *Sec. 2;* 10

Apparent sag of a span, definition, *Sec. 2;* 10

Application of National Electrical Safety Code, *013;* 1

Approach distance
Live parts, *441A;* 224
Minimum, definition of, *Sec. 2;* 8

Arcing, *420F, 447;* 217, 240

Armless construction, *243C, 261G;* 160, 177

Augers (grain bins), clearance considerations, *234F, Fig 234-4;* 114, 115

Authority, administrative, definition of, *Sec. 2;* 4

Automatic, definition of, *Sec. 2;* 4

Automatic overspeed trip device for prime movers, *130A;* 49

AWG (American Wire Gage), *Sec. 2;* 13

B

Backfill, *321B;* 193
Definition of, *Sec. 2;* 4

Ballast section (railroads), definition of, *Sec. 2;* 4

Basic Impulse Insulation Level (BIL), *Table 124-1;* 37

Batteries. *See* Storage batteries

Beaches, conductor clearance from, *234E2;* 109

Belt (line-worker's body), *Sec. 2, 420K;* 6, 201

B grade construction. *See* Grades of construction

BIL (Basic Impulse Insulation Level), *Table 124-1;* 37

Billboards, conductor clearance from, *234C, Table 234-1;* 99, 102

Birmingham Wire Gage, *Sec. 2;* 13

Boilers, *127F, 180E1;* 47, 56

Bonding, definition of, *Sec. 2;* 4

Bonding jumpers, buried, *92E, 93E6b;* 18, 21

Boring, *352C;* 204

Braces for line supports, *261D;* 176

Breaker, circuit. *See* Circuit breakers

Bridges
Clearance from, *234;* 96
Conduit location, *320A4;* 192
Trolley guards under, *225E;* 68

Brown & Sharpe (B&S), *Sec. 2;* 13

Buckarm construction, *236F, 237D, Fig 237-1;* 144, 146

Buildings, clearance from, *234, Fig 234-1, Table 234-1, Table 234-4;* 96, 97, 101, 120
Buried cable under, *351C2;* 203

Bundled conductors, definition of, *Sec. 2;* 5

Buried lines. *See* Underground communication lines; Underground electric supply lines

Burner safety devices, *180E;* 56

Bus, metal-enclosed, *181;* 56
Isolated-phase, *181B;* 56

C

Cable jacket, definition of, *Sec. 2;* 4

Cables
Aerial, insulation requirements, *278;* 186
Communication, *261K;* 178
Definition of, *Sec. 2;* 4
Fiber-optic—communication
Clearances, *230F;* 70
Definition of, *Sec. 2;* 6
Fiber-optic—supply
Clearances, *230F;* 70
Definition of, *Sec. 2;* 6
Fireproofing, definition of, *Sec. 2;* 6
Insulation, definition of, *Sec. 2;* 7
Spacer, definition of, *Sec. 2;* 11
Supply, *230C, 241A;* 69,155

Cable sheath, definition of, *Sec. 2;* 4

Cable terminal, definition of, *Sec. 2;* 4

Capacitors, *443I;* 236

C grade construction. *See* Grades of construction

Chimneys, conductor clearance from, *234C, Table 234-1, Table 234-5;* 99, 102, 121

Circuit
Definition of, *Sec. 2;* 4
Voltage in, definitions of, *Sec. 2;* 12

Circuit breakers
Application, *171;* 54
Arrangement, *170;* 54
Capacity, *173A;* 54
Definition of, *Sec. 2;* 4
Oil-containing, *172;* 54
Provisions for disconnecting, *173B;* 54
Visible break switch, *173C;* 54

Classified locations, electrical installations in
Boilers, *127F, 180E1;* 47, 56
Coal-handling areas, *127A;* 46
Flammable and combustible liquids, *127B;* 46
Loading and unloading facilities, *127D;* 47
Storage areas, *127C;* 47
Gaseous hydrogen systems for supply equipment, *127G;* 47
Gasoline-dispensing stations, *127E;* 47
Liquid hydrogen systems, *127H;* 48
Liquid petroleum gas (LPG), *127K;* 48
Loading and unloading facilities, *127D;* 47
Natural gas (methane), *127L;* 48
Sulfur, *127I;* 48

Clearance. *See* Overhead communication lines, clearances; Overhead electric supply lines, clearances

Climbable, readily, definition of, *Sec. 2;* 9

Climber, qualified, *Sec. 2, 420K2;* 9, 217

Climbing, definition of, *Sec. 2;* 4

Climbing space, *230A;* 69
Buckarm construction on, *236F;* 144
Conductors bounding, *236E, Table 236-1;* 142
Dimensions, *236A;* 141
Equipment location in relation to, *236D;* 142
Inhibiting climbing of supporting structures, *217A2;* 62
Lateral conductors, past, *239B;* 149
Location, *236A;* 141
Longitudinal runs not on support arms, past, *236G;* 144
Portions of supporting structures in, *236B;* 141
Ridge-pin conductors near, *236I;* 144
Steps and standoff brackets for supporting structures, *217A2;* 62
Support arm location relative to, *236C;* 141
Vertical conductors, past, *236H, 239B;* 144, 149

Clothing, protective, *420I;* 217

Coal-handling areas, *127A;* 46

Combustible liquids, *127B;* 46
- Loading and unloading facilities, *127D;* 47
- Storage areas, *127C;* 47

Common use, definition of, *Sec. 2;* 4

Communication lines. *See* Lines, communication

Communication worker safety zone, *235C4, 238E;* 133, 147

Computers, *180E1;* 56

Concentric neutral cable as grounding electrode, *94B5;* 23

Concrete-encased electrodes as ground electrodes, *94B6, 95A;* 23

Concrete structures. *See* Prestressed-concrete structures; Reinforced concrete structures

Conductors, *Sec. 16;* 53
- Application, *160;* 53
- Bundled, definition of, *Sec. 2;* 5
- Clearance envelope, *233A2, Fig 233-1, Fig 233-3, Fig 234-4;* 85, 86, 89, 115
- Climbing space bounding, *236E;* 142
- Communication. *See also* Lines, communication
  - Open-wire, *261J;* 178
  - Paired, *261L;* 178
- Constant-current, *242A;* 156
- Covered, *Sec. 2, 230D;* 5, 70
- Definition of, *Sec. 2;* 5
- Electrical protection, *161;* 53
  - Grounded, *161B;* 53
  - Insulated power cables, *161C;* 53
  - Overcurrent protection required, *161A;* 53
- Fastenings, *243C, 252, 261F;* 160, 169, 177
- Fiber-optic, definition of, *Sec. 2;* 6
- Fire-alarm circuits, *242D;* 156
- Grades of construction
  - Communication conductors, *242C, Table 241-1, Table 242-2;* 156, 157, 159
  - Constant-current circuit conductors, *242A;* 156
  - Fire-alarm circuit conductors, *242D;* 156
  - Neutral conductors of supply circuits, *242E;* 156
  - Railway feeder conductors, *242B;* 156
  - Supply, *241A, Table 242-1;* 155, 157
  - Surge-protection wires, *242F;* 156
  - Trolley-contact circuit conductors, *242B;* 156
- Grounded, *161B;* 53
  - Definition of, *Sec. 2;* 5
- Grounding. *See* Grounding, conductors
- Installing, *447;* 240
- Insulated, *161C;* 53
  - Definition of, *Sec. 2;* 5
- Isolation, *163;* 53
- Lateral, definition of, *Sec. 2;* 5
- Line. *See* Overhead communication lines; Overhead electric supply lines
- Loading, *251;* 168
- Maintaining, *447;* 240
- Mechanical protection and support, *162;* 53
- Movement envelope, *233A1, Fig 233-1, Fig 233-2;* 85, 86, 87
- Neutral, *Sec. 2, 230E, 242E;* 8, 70, 156
- Open, *Sec. 2, 261H;* 5, 177
- Overcurrent protection required, *161A;* 53
- Railway feeder, *242B;* 156
- Ridge pin, *236I;* 144
- Sag, definition of, *Sec. 2;* 9
- Shielded, *92B2b;* 16
- Shielding, definition of, *Sec. 2;* 5
- Spacing, *235G;* 140
- Strength requirements. *See* Strength requirements
- Supply. *See* Overhead communication lines; Overhead electric supply lines
- Surge-protection, *242F;* 156
- Terminations
  - Insulation, *164A;* 53
  - Metal-sheathed or shielded cable, *164B;* 53
- Trolley-contact circuit, *225A, 234A3, 234D2, 239G, 241C3b, 242B, 243, 261H3;* 67, 96, 109, 150, 155, 156, 160, 178
- Voltage to ground, definition of, *Sec. 2;* 13

Conduit
- Definition of, *Sec. 2, Sec. 32;* 5, 192
- Multiple duct, *Sec. 2;* 5
- Single duct, *Sec. 2;* 5

Conduit systems, definition of, *Sec. 2;* 5

Conflict
- Between lines, avoidance of, *221;* 65
- Structure, definition of, *Sec. 2;* 11

Constant-current circuit, voltage of (definition), *Sec. 2;* 12

Control, motor, *130, 131, 180D;* 49, 56

Control switchboards, *180E;* 56

Conveyors (grain bins), clearance considerations, *234F, Fig 234-4;* 114, 115

Cover, manhole, definition of, *Sec. 2;* 8

Covered conductors, *230D;* 70
Definition of, *Sec. 2;* 5

Crossarms for line supports. *See also* Supporting structures
Grades of construction, *243B, 263C;* 160, 181
Loads upon, *252, 253;* 169, 170
Strength, determining, *261D;* 176

Crossings, *225D, 241C;* 67, 155

Current-carrying part, definition of, *Sec. 2;* 5

Current transformers, *150;* 51. *See also* Transformers

Cylindrical structures and components, *252B2a;* 169

D

DC. *See* Direct current systems

Deadends, *252C3, 261F1c;* 170, 177

Dead-front power switchboards, *180C;* 56

Decorative lighting, *217A5;* 63

De-energized, definition of, *Sec. 2;* 5

De-energizing lines and equipment, *444;* 237

Definitions of National Electrical Safety Code terms, *Sec. 2;* 4

Designated person, definition of, *Sec. 2;* 5

Diagonal clearance, *235D;* 133

Direct current systems
Clearances, *230G;* 70
Grounding methods, *123D;* 35
Grounding point, *92A;* 16
Disconnecting switch, *Sec. 17;* 54
Definition of, *Sec. 2;* 5

Districts
Rural, definition of, *Sec. 2;* 9
Urban, definition of, *Sec. 2;* 12

Drop, service, definition of, *Sec. 2;* 10

Ducts, *322;* 193
Definition of, *Sec. 2, Sec. 32;* 5, 192
Installation, *322B;* 194

E

Effective date of National Electrical Safety Code, *016;* 3

Effectively grounded, definition of, *Sec. 2;* 6

Effectively grounded circuits, voltage of (definition), *Sec. 2;* 12

Electrical code. *See* National Electrical Code; National Electrical Safety Code

Electric railway construction, *225;* 67. *See also* Railroads
Guards under bridges, *225E;* 68
High-voltage contact conductors, *225B;* 67
Prevention of contact loss at railroad crossings at grade, *225D;* 67
Third rails, *225C;* 67
Trolley-contact conductors, *225A, 234A3, 234D2, 239G, 241C3b, 242B, 243, 261H3;* 67, 96, 109, 150, 155, 156, 160, 178

Electric supply equipment, definition of, *Sec. 2;* 6

Electric supply lines. *See* Lines, electric supply

Electric supply stations. *See* Supply stations, electric

Electric supply systems, operation of. *See* Operation of electric supply systems, employee rules; Operation of electric supply systems, employer rules

Electrodes, grounding. *See* Grounding, electrodes

Electron-tube-containing devices, *180E;* 56

Elevation
Correction for approach distance to live parts, *441A5, Table 441-5;* 225, 232
Isolated by, definition of, *Sec. 2;* 7

Elevators, grain, clearance considerations, *234F, Fig 234-4;* 114, 115

Emergency installations, *014A;* 2
Clearances for, *230A2;* 69
Inspections, *121C;* 34

Emergency lighting, supply station, *111B;* 30

Emergency procedures, *410B;* 214

Employee rules. *See* Operation of electric supply systems, employee rules

Employer rules. *See* Operation of electric supply systems, employer rules

Enclosed, definition of, *Sec. 2;* 6

Enclosure of equipment, *110A, 124C2;* 28, 35

Energized
- Definition of, *Sec. 2;* 6
- Equipment
  - Guarding, *237E;* 146
  - Working on. *See* Operation of electric supply systems, employee rules, energized lines and equipment
- Parts, equipment for work on, *126, 446B;* 46, 239

Envelope, clearance, *233A2, Fig 233-1, Fig 233-3, Fig 234-4;* 85, 86, 89, 115

Equipment
- Cases, *234J2;* 118
- Clearance, *234J;* 118. *See also* Overhead communication lines, clearance; Overhead electric supply lines, clearance
- Climbing space, location relative to, *236D;* 142
- Definition of, *Sec. 2;* 6
- Electric supply. *See* Supply stations, electric
- Grounding conductors for, *93C5;* 19
- Installation. *See* Installation and maintenance of equipment
- Maintenance. *See* Installation and maintenance of equipment
- Remotely operable, definition of, *Sec. 2;* 9
- Rotating. *See* Rotating equipment
- Underground, *Sec. 38;* 210
  - Design, *381;* 210
  - Grounding, *384;* 211
  - Identification, *385;* 211
  - Installation, *383;* 211
  - Location, *382;* 211
- Utilization, definition of, *Sec. 2;* 12

Excavation, *321A, 423D;* 193, 221

Existing installations, *013B;* 2

Exits, supply station, *113;* 33

Exposed, definition of, *Sec. 2;* 6

Extension of installations, *013A;* 1

F

Fall arrest system, definition of, *Sec. 2;* 6

Fall prevention system, definition of, *Sec. 2;* 6

Fall protection, *411F, 420K;* 215, 217

Fall protection program, definition of, *Sec. 2;* 6

Fall protection system (hardware), *Sec. 2, 420K;* 6, 217

Fastenings, conductor, *243C, 252, 261F;* 160, 169, 177

Fences
- Grounding of, *93E, 93C6;* 18, 19
- Safety clearance zone, *110A2, Fig 110-1;* 28, 29
- Types of, *110A1;* 28

Fiber-optic cable—communication
- Clearances, *230F;* 70
- Definition of, *Sec. 2;* 6

Fiber-optic cable—supply
- Clearances, *230F;* 70
- Definition of, *Sec. 2;* 6

Fiber-optic conductor, definition of, *Sec. 2;* 5

Fiber stress, wood, *261A2;* 174

Final sag, definition of, *Sec. 2;* 10

Final unloaded sag, definition of, *Sec. 2;* 10

Final unloaded tension, definition of, *Sec. 2;* 12

Fire-alarm circuit conductors, *242D;* 156

Fire-extinguishing equipment, *114, 420L;* 33, 218

Fire hydrants, clearances from, *231A;* 71

Fireproofing (of cables), definition of, *Sec. 2;* 6

First aid rules, *410B;* 214

Flammable liquids, *127B;* 46
- Loading and unloading facilities, *127D;* 47
- Storage areas, *127C;* 47

Flashover voltage, insulation, *272;* 185

Flat-surfaced structures and components, *252B2b;* 169

Floors, supply station, *112A;* 32

Foundations, loads upon, *252, 253, 261B;* 169, 170, 175

Fuel lines, separation from underground conduit, *320B5;* 193

Fuses, *Sec. 17, 420N;* 54, 218
Application, *171;* 54
Arrangement, *170;* 54
Capacity, *173A;* 54
Disconnecting, *173B, 174;* 54
Visible-break, *173C;* 54

G

Gages, wire, definition of, *Sec. 2;* 13

Gaseous hydrogen systems for supply equipment, *127G;* 47

Gas-insulated equipment, *443J;* 236

Gas lines, separation from underground conduit, *320B5;* 193

Gasoline-dispensing stations, *127E;* 47

Gates, *92E;* 18

Generating station, definition of, *Sec. 2;* 6

Generators, *Sec. 13;* 49
Motor control, *131;* 49
Short-circuit protection, *133;* 49
Speed control and stopping devices, *130;* 49

Grades of construction

Application, *241;* 155
B and C
Loading. *See* Loading, line
Strength requirements. *See* Strength requirements
Conductors
Communication conductors, *242C, Table 242-1, Table 242-2;* 156, 157, 159
Constant-current circuit conductors, *242A;* 156
Fire-alarm circuit conductors, *242D;* 156
Neutral conductors of supply circuits, *242E;* 156
Railway feeder conductors, *242E;* 156
Supply, *241A, Table 242-1;* 155, 157
Surge-protection wires, *242F;* 156
Trolley-contact circuit conductors, *242B;* 156
Crossings, *241C;* 155
Line supports, *243;* 160
Armless construction brackets, *243C;* 160
Conductor construction grade, change in, *252C1;* 169
Conductor fastenings, *234C;* 160
Crossarms, *243B;* 160
Insulators, *243C;* 160
Pins, *243C;* 160
Structures, *243A;* 160
Support arms, *243B;* 160
N, strength requirements. *See* Strength requirements
Order of grades, *241B;* 155

Grain bins, conductor clearance from, *234F, Fig 234-4;* 114, 115

Grating, manhole, definition of, *Sec. 2;* 8

Ground, clearances above, *231B, 232, Table 232-1, Table 232-2;* 71, 74, 81. *See also* Overhead communication lines, clearances; Overhead electric supply lines, clearances

Grounded, definition of, *Sec. 2;* 7

Grounded, effectively, definition of, *Sec. 2;* 6

Grounded circuit, voltage of (definitions), *Sec. 2;* 12

Grounded conductor, *161B;* 53
Definition of, *Sec. 2;* 5

Grounded system, definition of, *Sec. 2;* 7

Grounding
Circuits and equipment, underground, *314;* 190
Communication lines, *99, 315;* 26, 191
Conductors, *215B;* 61
Ampacity, *93C;* 19
Common, for circuits and equipment, *93F;* 21
Composition of, *93A;* 18
Connection methods, *93, 95;* 18, 23
Connection points, *92, 95B;* 16, 24
Current in, *92D;* 17
Definition of, *Sec. 2;* 5
Guarding, *93D;* 20
Protection, *93D;* 20
Separate, *92B3;* 17
Separation of, *97;* 25
Strength, *93C;* 19
Underground, *93E;* 20

Direct current systems, *123D;* 35
Electrodes, *94, 99;* 21, 26
Existing, *94A;* 21
Made, *94B;* 21
Equipment, *93C5, 123;* 19, 34
During maintenance, *123C;* 35
Guys, *92C, 93C5, 215C2;* 17, 19
Messenger wires, *92C, 93C5;* 17, 19
Purpose of code, *90;* 16
Resistance requirements, *96;* 24
Multi-grounded systems, *96C;* 24
Single-grounded systems, *96B;* 24
Supply stations, *96A;* 24
Scope of code, *91;* 24
Supporting structures, *215C1;* 61

Ground wire, definition of, *Sec. 2;* 10

Guarded, definition of, *Sec. 2;* 7

Guarding
Definition of, *Sec. 2;* 7
Grounding conductors, *93D, 239D;* 20, 149
Live parts, *124, 237E;* 35, 146
Strength of guards, *124B;* 35
Transmission machinery and suddenly moving parts, *122;* 34
Types of guards, *124C;* 35
Where required, *124A;* 35

Guys
Anchors, *253, 261B;* 170, 175
Grounding, *92C, 93C5, 215C2;* 17, 19, 61
Insulators, *261C, 279A;* 175, 187
Overload factors, *253;* 170
Side guys, *261A3;* 175
Strength of, *261C, 263B;* 175, 183

H

Handhole. *See* Manholes, handholes, and vaults

Harness, *420K;* 217
Definition of, *Sec. 2;* 7

Highways. *See also* Roadways

Limited access, definition of, *Sec. 2;* 8

Horizontal clearance, *233B, 234A2, 234C1b, 234D1;* 89, 96, 99, 107
Between line conductors, *235B, Table 235-1, Table 235-2, Table 235-3;* 121, 123, 124, 125
From live parts, *Fig 124-1, Table 124-1;* 36, 37

Horizontal loading component, *251B2;* 168

Hydrogen systems
Gaseous, for supply equipment, *127G;* 47
Liquid, *127H;* 48
Mobile, *129;* 48

I

Ice. *See* Loading, line

Identification, *128, 411E, 423E;* 48, 215, 221

Idle equipment, inspection of, *121B;* 34

Illumination
Attachment plugs and receptacles, *111D;* 30
Battery areas, *145;* 50
Decorative, *217A5;* 63
Emergency lighting, *111B;* 30
Fixtures, *111C;* 30
Levels, *Table 111-1;* 31
Receptacles in damp or wet locations, *111E;* 30
Street and area, *263H, 420P;* 182, 219
Supply stations, *111;* 30
Under normal conditions, *111A;* 30

Inch-foot-pound system, *017A;* 3

Indoor installations
Power transformers and regulators, *152B;* 51
Surge arresters, *191;* 58

Initial unloaded sag, definition of, *Sec. 2;* 9

Initial unloaded tension, definition of, *Sec. 2;* 12

In service, definition of, *Sec. 2;* 7

In-service equipment, inspection of, *121A, 214A;* 34, 60

Inspections, *121, 214;* 34, 60
Emergency equipment, *121C;* 34
Idle equipment, *121B;* 34
In-service lines and equipment, *121A, 214A;* 34, 60
lines, *214;* 60
New equipment, *121D;* 34
Out-of-service lines and equipment, *214B;* 60
Protective devices, *411C;* 215
Underground lines and equipment, *313;* 190
When in service, *313A;* 190
When out of service, *313B;* 190

Installation and maintenance of equipment, *Sec. 12;* 34
Classified locations, *127;* 46
Boilers, *127F;* 47
Coal-handling areas, *127A;* 46
Flammable and combustible liquids, *127B;* 46
Flammable liquid storage area, *127C;* 47
Gaseous hydrogen systems for supply equipment, *127G;* 47
Gasoline-dispensing stations, *127E;* 47
Liquefied petroleum gas (LPG), *127K;* 48
Liquid hydrogen systems, *127H;* 48
Loading and unloading facilities, *127D;* 47
Natural gas (methane), *127L;* 48
Sulfur, *127I;* 48
Emergency equipment, *121C;* 34
Energized parts, equipment for work on, *126;* 46
General requirements, *120;* 34
Guarding live parts, *124, 237E;* 35, 146
Strength of guards, *124B;* 35
Types of guards, *124C;* 35
Where required, *124A;* 35
Guarding shaft ends, pulleys, belts, and suddenly moving parts, *122;* 34
Identification, *128;* 48
Idle equipment, *121B;* 34
In-service equipment, *121A;* 34
Inspections, *121;* 34
Mobile hydrogen equipment, *129;* 48
New equipment, *121D;* 34
Protective grounding, *123;* 34
During maintenance, *123C;* 35
Working space about electric equipment, *125;* 45

Installation and maintenance of overhead communication lines. *See* Overhead communication lines

Installation and maintenance of overhead electric supply lines. *See* Overhead electric supply lines

Installation and maintenance of underground communication lines. *See* Underground communication lines

Installation and maintenance of underground electric supply lines. *See* Underground electric supply lines

Instrument transformers. *See also* Transformers

Grounding of, *151;* 51
Grounding conductor ampacity, *93C3;* 19

Insulated
Conductor, *161C;* 53
Definition of, *Sec. 2;* 5
Definition of, *Sec. 2;* 7

Insulation

Aerial cable systems, *278;* 186
Definition of, *Sec. 2;* 7
Factory tests, *274;* 185
Guy insulators, *279A;* 187
Level, *273;* 185
Material and marking, *271;* 185
Mechanical strength, *277;* 186
Open-conductor supply lines. *See* Overhead electric supply lines, insulation
Ratio of flashover to puncture voltage, *272;* 185
Shielding, definition of, *Sec. 2;* 7
Span-wire insulators, *279B;* 187
Special applications, *275;* 185

Insulator
Definition of, *Sec. 2;* 7
Grade of construction, *243C, 263I;* 160, 182
Installing, *447;* 240
Loads upon, *252;* 169
Maintaining, *447;* 240
Span, *279B;* 187
Suspension, *235B2, 235E2;* 126, 138

Intent of National Electrical Safety Code, *015;* 2

Interactive system, utility, definition of, *Sec.2;* 12

International System of Units (SI), *017A;* 3

Introduction to National Electrical Safety Code, *Sec. 1;* 1

Inverters, *180E1;* 56

Iron, pulling, definition of, *Sec. 2;* 9

Isolated, definition of, *Sec. 2;* 7

Isolated by elevation, definition of, *Sec. 2;* 7

Isolated-phase bus, *181B;* 56

Isolating switch, definition of, *Sec. 2;* 5

Isolation of conductors, *163B;* 53

Isolator. *See* Disconnecting switch; Isolating switch

J

Jacket, cable, definition of, *Sec. 2;* 4

Jacket, definition of, *Sec. 2;* 7

Joints, *322;* 193

Joint use, *222;* 65
  Definition of, *Sec. 2;* 7

L

Ladders, *420J;* 217

Lanyards, *441B, 411C;* 215
  Definition of, *Sec. 2;* 7

Lateral conductor, definition of, *Sec. 2;* 5

Lattice structures, *252B2c;* 169

Length, span, definition of, *Sec. 2;* 11

Lighting. *See* Illumination

Lightning circuits, *92B1, 97A;* 16, 25

Lightning protection equipment, *93D4;* 20

Limited access highway, definition of, *Sec. 2;* 8

Line conductor, definition of, *Sec. 2;* 5

Lines. *See* Lines, communication; Lines, electric supply

Lines, communication
  Definition of, *Sec. 2;* 8
  Fiber-optic cable, *Sec. 2, 230F;* 6, 70
  Grounding, *97, 99, 315;* 26, 191
  Installation and maintenance
    Overhead. *See* Overhead communication lines
    Underground. *See* Underground communication lines
  Operation. *See* Operation of communication systems, employee rules; Operation of communication systems, employer rules

Lines, electric supply
  Definition of, *Sec. 2;* 8
  Fiber-optic cable, *Sec. 2, 230F;* 6, 70
  Installation and maintenance
    Overhead. *See* Overhead electric supply lines
    Underground. *See* Underground electric supply lines
  Operation. *See* Operation of electric supply systems, employee rules; Operation of electric supply systems, employer rules

Line supports. *See* Supporting structures

Line-worker's body belt, *Sec. 2, 420K;* 8, 217
  Definition of, *Sec. 2;* 4

Liquefied petroleum gas (LPG), *127K;* 48

Liquid-cell batteries, *420G;* 217

Liquid hydrogen systems, *127H;* 48

Liquids, flammable. *See* Flammable liquids

Live lines and equipment, work on
  Approach distance, *441A, 441B;* 224, 232
  Bare-hand method, bonding and shielding for, *446D;* 239
  Capacitors, *443I;* 236
  Clear live-line tool length, *441C;* 233
  Connections, making, *443F;* 236
  Current transformer secondaries, *443H;* 236
  De-energizing to protect employees, *444;* 237
    Employee's protective grounds, *444D;* 237
    Employee's request, *444B;* 237
    Operating switches, disconnectors, and tagging, *444C;* 237
    Proceeding with work, *444E;* 238
    Re-energizing, sequence of, *444H;* 238
    Reporting clear, *444F1;* 238
    Tags, removal of, *444G;* 238
    Transferring responsibility, *444F2;* 238
  Delta circuits, unintentional grounds on, *443L;* 237
  Employee-assisting requirements, *443B;* 236
  Equipment, *446B;* 239
    Gas-insulated, *443J;* 236
  General requirements, *443A;* 235
  Protective grounds, *445;* 238
    Installing, *445A;* 238
    Removing, *445B;* 239
  Switching control procedures, *442, 443C, 443E, 443G;* 233, 236
    Designated person, *442A;* 233
    Operation at stations, *442C;* 234
    Oral messages, repeating, *442G;* 235
    Re-energizing after work, *442D;* 234
    Restoration of service after automatic trip, *442F;* 235
    Specific work, *442B;* 234
    Tagging circuits associated with work activities, *442E;* 234
  Training, *446A;* 239
  Working position, *443D;* 236

Live parts
  Approach distance to, *441A, 441B;* 224, 232
  Clearance from, *232C, 234C, 234D;* 72, 99, 107
  Guarding, *124, 234C2;* 35, 99

Load control devices, *180E1;* 56

Loading, line, *Sec. 25;* 161
  Combined ice and wind loading, *250B, 253, Table 253-1, Table 253-2;* 161, 170, 171, 172
  Components, *251B;* 168
  Conductors, *251;* 168
  Extreme wind loading, *250C, 253, Table 253-1, Table 253-2, 260B2;* 161, 170, 171, 172, 173
  Line supports, *252;* 169
    Longitudinal, assumed, *252C;* 169
    Simultaneous, application of loads, *252D;* 170
    Transverse, assumed, *252B;* 269
    Vertical, assumed, *252A;* 169
  Maps, *Fig 250-1, Fig 250-2;* 162, 166

Loading facilities, flammable and combustible liquids, *127D;* 47

Logic devices, *180E1;* 56

Longitudinal loading, assumed, *252C;* 169
  Changes in construction grade, *252C1;* 169
  Communication conductors on unguyed supports, *252C7;* 170
  Deadends, *252C3;* 170
  Longitudinal capability, *252C6;* 170
  Poles, jointly used, *252C2;* 170
  Stringing loads, *252C5;* 170
  Unequal spans, *252C4;* 170

Longitudinal strength requirements, *261A4;* 175

LPG (liquefied petroleum gas), *127K;* 48

Luminaires, *238C, 238D, Table 238-2, 239E2b, 239G3;* 147, 148, 150, 151

M

Maintenance. *See* Installation and maintenance

Manholes, handholes, and vaults, *323;* 194
  Access, *323C, 423A, 443K;* 196, 220, 237
  Covers, *323D;* 196
    Definition of, *Sec. 2;* 8
  Definitions of, *Sec. 2;* 7, 8, 12
  Dimensions, *323B;* 196
  Drainage, *323G;* 197
  Gratings, definition of, *Sec. 2;* 8
  Identification, *323J;* 197
  Ladder requirements, *323F;* 197
  Mechanical protection, *323I;* 197
  Routing, *320A;* 192
    Bridges, *320A4;* 192
    Highways and streets, *320A3;* 192
    Natural hazards, *320A2;* 192
    Railroad tracks, *320A5;* 192
    Submarine crossings, *320A6;* 192
    Tunnels, *320A4, Sec. 39;* 192, 212
  Separation from other underground installations, *320B;* 192
    Fuel lines, *320B5;* 193
    Sewers, sanitary and storm, *320B3;* 193
    Steam lines, *320B6;* 193
    Supply conduit systems, *320B2;* 193
    Water lines, *320B4;* 193
  Strength, *323A;* 194
  Testing for gas in, *423B;* 220
  Vault and utility tunnel access, *323E;* 196
    Ventilation, *323H;* 197

Manual, definition of, *Sec. 2;* 8

Manual stopping devices, *130B;* 49

Maps, overhead line loading, *Fig 250-1, Fig 250-2;* 162, 166

Mats, *124C4;* 35

Maximum total sag, definition of, *Sec. 2;* 10

Measure, units of, *017;* 3

Mechanical protection of conductors, *162, 239D;* 53, 149

Mechanical transmission machinery, guarding, *122A;* 34

Messenger wires, grounding, *92C, 93C5, 215C3;* 17, 19, 61

Metal supporting structures, *261A, 261C1;* 173, 175

Methane (natural gas), *127L;* 48

Metric system, *017A;* 3

Microwave radios, totalizing, *180E1;* 56

Minimum approach distance, definition of, *Sec. 2;* 8

Mobile hydrogen systems, *129;* 48

Motor generators. *See* Generators

Motors
Adjustable-speed, *130E;* 49
Control, *130, 131, 180D;* 49, 56
Short-circuit protection, *133;* 49
Speed limit, *130C;* 49

Movement envelope, conductor, *233A1, Fig 233-1, Fig 233-2;* 85, 86, 87

Multigrounded/multiple grounded systems, *96C, 97D2;* 24, 25
Definition of, *Sec. 2;* 8
Multiple-duct conduit, definition of, *Sec. 2;* 5

N

National Electrical Code (NEC), *011;* 1

National Electrical Safety Code (NESC)
Application, *013;* 1
Definitions of special terms, *Sec. 2;* 4
Effective date, *016;* 3
Intent, *015;* 2
Introduction to, *Sec. 1;* 1
Purpose, *010;* 1
Rules, *012;* 1
Scope, *011;* 1
Units of measure, *017;* 3
Waiver, *014;* 2

Natural gas (methane), *127L;* 48

NEC. *See* National Electrical Code

NESC. *See* National Electrical Safety Code

Neutral conductors, *230E, 242E;* 70, 156
Definition of, *Sec. 2;* 8

New equipment, inspection of, *121D;* 34

New installations, *013A;* 1

N grade construction. *See* Grades of construction

Nonshielded conductors, *92B2a;* 16

O

Oil-containing circuit-interrupting devices, *172;* 54

Open conductors, *261H;* 177
Definition of, *Sec. 2;* 5

Operation of communication systems, employee rules, *Sec. 42, Sec. 43;* 216, 222
General operating routines, *421, 430;* 219, 222
Area protection, *421B;* 219
Duties of first-level supervisor or person in charge, *421A;* 219
Escort, *421C;* 219
Overhead line operating procedures, *422;* 220
Checking structures before climbing, *422B;* 220
Installing and removing wires or cables, *422C;* 220
Joint-use structures, *432;* 222
Setting, moving, or removing poles in or near energized electric supply lines, *422A;* 220
Personal general precautions, *420;* 216
Arcing conditions, *420F;* 217
Cable reels, *420O;* 218
Clothing, *420I;* 217
Energized or unknown conditions, *420D, 431, Table 431-1;* 216, 222, 223
Fall protection, *420K;* 217
Fire extinguishers, *420L;* 218
Fuses, *420N;* 218
Ladders and supports, *420J;* 217
Lighting, street and area, *420P;* 219
Liquid-cell batteries, *420G;* 217
Machines or moving parts, *420M;* 218
Purpose of code, *400;* 213
Qualifications of employees, *420B;* 216
Rules and emergency methods, *420A;* 216
Safeguarding oneself and others, *420C;* 216
Scope of code, *401;* 213
Tools and protective equipment, *420H;* 217
Ungrounded metal parts, *420E;* 216
Purpose of code, *400;* 213
Scope of code, *401;* 213
Underground line operating procedures, *423;* 220
Excavation, *423D;* 221
Flames, *423C;* 221
Guarding manhole and street openings, *423A, 433;* 220, 222
Identification, *423E;* 221
Operation of power-driven equipment, *423F;* 221
Sheath continuity, *434;* 222
Testing for gas in manholes and unventilated vaults, *423B;* 220

Operation of communication systems, employer rules, *Sec. 41;* 214
Emergency procedures and first-aid rules, *410B;* 214
General requirements, *420;* 214
Protective methods and devices, *411;* 214
Devices and equipment, *411B;* 215
Fall protection, *411F;* 215
Identification and location, *411E;* 215

Inspection and testing of protective devices, *411C;* 215
Methods, *411A;* 214
Warning signs, *411D;* 215
Purpose of code, *400;* 213
Scope of code, *401;* 213

Operation of electric supply systems, employee rules, *Sec. 42, Sec. 44;* 216, 224
Energized lines and equipment. *See also* Live lines and equipment, work on
Approach distance, *441A, 441B;* 224, 232
Clear live-line tool length, *441C;* 233
De-energizing to protect employees, *444;* 237
Work on, *443, 446;* 235, 239
General operating routines, *421, 440;* 219, 224
Area protection, *421B;* 219
Duties of first-level supervisor or person in charge, *421A;* 219
Escort, *421C;* 219
Overhead line operating procedures, *422;* 220
Checking structures before climbing, *422B;* 220
Installing and removing wires or cables, *422C;* 220
Setting, moving, or removing poles in or near energized electric supply lines, *422A;* 220
Personal general precautions, *420;* 216
Arcing conditions, *420F;* 217
Cable reels, *420O;* 218
Clothing, *420I;* 217
Energized or unknown conditions, *420D, 441;* 216, 224
Fall protection, *420K;* 217
Fire extinguishers, *420L;* 218
Fuses, *420N;* 218
Ladders and supports, *420J;* 217
Lighting, street and area, *420P;* 219
Liquid-cell batteries, *420G;* 217
Machines or moving parts, *420M;* 218
Purpose of code, *400;* 213
Qualifications of employees, *420B;* 216
Rules and emergency methods, *420A;* 216
Safeguarding oneself and others, *420C;* 216
Scope of code, *401;* 213
Tools and protective equipment, *420H;* 217
Ungrounded metal parts, *420E;* 216
Protective grounds, *445;* 238
Installing, *445A;* 238
Removing, *445B;* 239
Purpose of code, *400;* 213
Scope of code, *401;* 213
Switching control procedures, *442;* 233
Designated person, *442A;* 233
Operation at stations, *442C;* 234
Oral messages, repeating, *442G;* 235
Re-energizing after work, *442D;* 234
Restoration of service after automatic trip, *442F;* 235
Specific work, *442B;* 234
Tagging circuits associated with work activities, *442E;* 234
Underground line operating procedures, *423;* 220
Excavation, *423D;* 221
Flames, *423C;* 221
Guarding manhole and street openings, *423A, 443K;* 220, 237
Identification, *423E;* 221
Operation of power-driven equipment, *423F;* 221
Testing for gas in manholes and unventilated vaults, *423B;* 220

Operation of electric supply systems, employer rules, *Sec. 41;* 214
Emergency procedures and first-aid rules, *410B;* 214
General requirements, *410;* 214
Protective methods and devices, *411;* 214
Devices and equipment, *411B;* 215
Fall protection, *411F;* 215
Identification and location, *411E;* 215
Inspection and testing of protective devices, *411C;* 215
Methods, *411A;* 214
Warning signs, *411D;* 215
Purpose of code, *400;* 213
Scope of code, *401;* 213

Outdoor installations, power transformers and regulators, *152A;* 51

Out of service, definition of, *Sec. 2;* 8

Overhead communication lines
Accessibility, *213, 216;* 60, 62
Application of rules, *202;* 59
Clearances, *Sec. 23;* 69
Antennas, *234, Table 234-1, Table 234-5, 235I;* 96, 101, 121, 141
Application of code, *230A;* 69
Beaches, *234E2;* 109
Billboards, *234C, Table 234-1;* 99, 101
Bridges, *234, Table 234-1;* 96, 101
Buildings, *234, Fig 234-1, 234C, Table 234-1, Table 234-4;* 96, 97, 99, 101, 120
Chimneys, *234C, Table 234-1, Table 234-5;* 99, 101, 121
Climbing space. *See* Climbing space
Covered conductors, *230D;* 70
DC circuits, *230G;* 70
Definition of, *Sec. 2;* 4
Diagonal, *235D;* 133
Envelope, *233A2, Fig 233-1, Fig 233-3, Fig 234-4;* 85, 86, 89, 115
Equipment, *234J;* 118
Fiber-optic cable, *230F;* 70
Grain bins, *234F;* 114

Ground, *231B, 232, Table 232-1, Table 232-2;* 71, 74, 81
Horizontal, *233B, 234A, Fig 234-1, 234C1, 234D1, 235B, Table 235-1, Table 235-2, Table 235-3;* 89, 96, 97, 99, 107, 121, 123, 124, 125
Measurement, *230B;* 69
Movement envelope, *233A1, Fig 233-1, Fig 233-2;* 85, 86, 87
Neutral conductors, *230E;* 70
Rail cars, *234I, Fig 234-5;* 118, 119
Roadways, *231B, 232, Table 232-1, Table 232-2;* 71, 74, 81
Signs, *Table 234-5;* 121
Spacing between, *235H;* 141
Supporting structures, *231, 233, 234B, 234J;* 71, 85, 96, 118
Swimming pools, *234, Fig 234-3, Table 234-3;* 96, 109, 113
Tanks, *234C, Table 234-1, Table 234-5;* 99, 101, 121
Uniform system of (adopted in 1990 NESC edition), *Appendix A;* 241
Vertical, *232, 233C, 234, Fig 234-1, 235C, 238;* 71, 89, 96, 97, 126, 147
Water surfaces. *See* Water, clearance from
Wind displacement. *See* Loading, line
Working space. *See* Working space
Climbing space. *See* Climbing space
Communication circuits located within the supply space, *224A;* 66
Communication and supply facilities located on same structure, *238;* 147
Conductors. *See* Conductors
Conflict between lines, avoiding, *221;* 65
Electric railway construction, *225;* 67
Guards under bridges, *225E;* 68
High-voltage contact conductors, *225B;* 67
Prevention of contact loss at railroad crossings at grade, *225D;* 67
Third rails, *225C;* 67
Trolley-contact conductor fastenings, *225A, 234A3, 234D2, 241C3b, 242B, 243, 261H3;* 67, 96, 109, 155, 156, 160, 178
General requirements, *Sec. 21;* 60
Grades of construction. *See* Grades of construction
Grounding, *215;* 61
Circuits, *215B;* 61
Guys, *92C, 93C5, 215C2;* 17, 19, 61
Messengers, *92C, 93C5, 215C3;* 17, 19, 61
Supporting structures, *215C1;* 61
Identification, *220D;* 65
Inspection
When in service, *214A;* 60
When out of service, *214B;* 60
Joint use of structures, *222;* 65
Loading. *See* Loading, line
Protective requirements, *223;* 66
Purpose of code, *200;* 59
Relative levels, *220;* 64
Scope of code, *201;* 59
Span wires, *238C, Table 238-2;* 147, 148
Strength requirements. *See* Strength requirements
Supply circuits located within the communication space, *224B, 239F, 239G;* 67, 150
Supporting structures. *See* Supporting structures
Switches. *See* Switches
Tests
When in service, *214A, 313A;* 60, 190
When out of service, *214B, 313B;* 60, 190
Tree trimming, *218;* 63

Overhead electric supply lines

Accessibility, *213, 216;* 60, 61
Application of rules, *202;* 59
Clearances, *Sec. 23;* 69
Antennas, *234, Table 234-1, Table 234-5, 235I;* 96, 102, 121, 141
Application of code, *230A;* 69
Beaches, *234E2;* 109
Billboards, *234C, Table 234-1;* 99, 102
Bridges, *234, Table 234-4;* 96, 120
Buildings, *234, Fig 234-1, 234-C, Table 234-1, Table 234-4;* 96, 98, 99, 101, 120
Chimneys, *234C, Table 234-1, Table 234-5;* 99, 101, 121
Climbing space. *See* Climbing space
Covered conductors, *230D;* 70
DC circuits, *230G;* 70
Definition of, *Sec. 2;* 4
Diagonal, *235D;* 133
Envelope, *233A2, Fig 233-1, Fig 233-3;* 85, 86, 89
Equipment, *234J;* 118
Fiber-optic cable, *230F;* 70
Grain bins, *234F;* 114
Horizontal, *233B, 234A, Fig 234-1, 234C1, 234D1, 235B, Table 235-1, Table 235-2, Table 235-3;* 89, 96, 98, 99, 107, 121, 123, 124, 125
Measurement, *230B;* 69
Movement envelope, *233A1, Fig 233-1, Fig 233-2;* 85, 86, 87
Neutral conductors, *230E;* 70
Rail cars, *234I, Fig 234-5;* 118, 119
Roadways, *231B, 232, Table 232-1, Table 232-2;* 71, 74, 81
Signs, *Table 234-5;* 121
Supply cables, *230C;* 69
Supporting structures, *231, 233, 234B, 234J;* 71, 85, 96, 118
Swimming pools, *234, Fig 234-3, Table 234-3;* 96, 109, 113
Tanks, *234C, Table 234-1, Table 234-5;* 99, 102, 121
Uniform system of (adopted in 1990 NESC edition), *Appendix A;* 241
Vertical, *232, 233C, 234, Fig 234-1, 235C, 238;* 71, 89, 96, 98, 126, 147

Water surfaces. *See* Water, clearance from
Wind displacement. *See* Loading, line
Climbing space. *See* Climbing space
Communication and supply facilities located on same structure, *238;* 147
Communication circuits located within the supply space, *224A;* 66
Conductors. *See* Conductors
Conflict between lines, avoiding, *221;* 65
Electric railway construction, *225;* 67
Guards under bridges, *225E;* 68
High-voltage contact conductors, *225B;* 67
Prevention of contact loss at railroad crossings at grade, *225D;* 67
Third rails, *225C;* 67
Trolley-contact conductor fastenings, *225A, 234A3, 234D2, 241C3b, 242B, 243, 261H3;* 67, 96, 109, 155, 156, 160, 178
General requirements, *Sec. 21;* 60
Grades of construction. *See* Grades of construction
Grounding, *215B;* 61
Circuits, *215B;* 61
Guys, *92C, 93C5, 215C2;* 17, 19, 61
Messengers, *92C, 93C5, 215C3;* 17, 19, 61
Supporting structures, *215C1;* 61
Identification, *220D;* 65
Inspection
When in service, *214A;* 60
When out of service, *214B;* 60
Insulation, *Sec. 27;* 185
Aerial cable systems, *278;* 186
Definition of, *Sec. 2;* 7
Factory tests, *274;* 185
Guy insulators, *279A;* 187
Level, *273;* 185
Material and marking, *271;* 185
Mechanical strength, *277;* 186
Open-conductor supply line. *See* Overhead electric supply lines, insulation
Ratio of flashover to puncture voltage, *272;* 185
Shielding, definition of, *Sec. 2;* 7
Span-wire insulators, *279B;* 187
Special applications, *275;* 169
Joint use of structures, *222;* 65
Loading. *See* Loading, line
Purpose of code, *200;* 59
Relative levels, *220;* 64
Scope of code, *201;* 59
Strength requirements. *See* Strength requirements
Supply circuits located within the communication space, *224B, 239F, 239G;* 67, 150
Supporting structures. *See* Supporting structures
Switches. *See* Switches
Tests
When in service, *214A, 313A;* 60, 190
When out of service, *214B, 313B;* 60, 190
Tree trimming, *218;* 63

Overhead shield wires, *261H;* 177

Overspeed trip device for prime movers, automatic, *130A;* 49

P

Pad-mounted equipment, definition of, *Sec. 2;* 8

Part, current-carrying, definition of, *Sec. 2;* 5

Passageways, supply station, *112B;* 32

Person, designated, definition of, *Sec. 2;* 5

Pins (line support), *234C, 252, 261F;* 160, 169, 177

Piping systems as grounding electrodes, *94A, 95B;* 21, 24

Plates, buried, as grounding electrodes, *94B3c;* 22

Plowing, *352B;* 204

Poles
See also Supporting structures
Butt plates as grounding electrodes, *94B4;* 22
Loads on, *252;* 169
Strength requirements, *Sec. 26;* 173

Positioning device system, definition of, *Sec. 2;* 9

Positioning straps, *411B, 420K;* 215, 217
Definition of, *Sec. 2;* 9

Power transformers, *152;* 51. *See also* Transformers
Indoor installations, *152B;* 51
Outdoor installations, *152A;* 51
Short-circuit protection, *153;* 51

Precipitator logic devices, *180E1;* 56

Pressure, side-wall, definition of, *Sec. 2;* 11

Prestressed-concrete structures, *261A1, Table 261-1, 261C1;* 173, 175, 180
Definition of, *Sec. 2;* 9

Protective grounding. *See* Grounding

Protective methods and devices. *See* Operation of electric supply systems, employee rules; Operation of electric supply systems, employer rules

Pulling iron, definition of, *Sec. 2;* 9

Pulling tension, definition of, *Sec. 2;* 9

Puncture voltage, insulation, *272;* 185

Purpose of National Electrical Safety Code, *010;* 1

Q

Qualified, definition of, *Sec. 2;* 9

Qualified climber, *Sec. 2, 420K2;* 9, 201

R

Raceway
- Definition of, *Sec. 2;* 9
- Grounding conductors for, *93C5;* 19

Racks, battery, *143;* 50

Radios, microwave, totalizing, *180E1;* 56

Railings, *112, Fig 124-2;* 32, 44

Railroads. *See also* Electric railway construction
- Ballast section, definition of, *Sec. 2;* 4
- Clearances, *231C, 232, 234;* 71, 96
  - Underground, *320A5, 351C3;* 192,203
- Grades of construction, *241C3b, 242B, 243;* 155, 156, 160

Random separation
- Definition of, *Sec. 2;* 9
- Direct buried cable, *354;* 205

Readily climbable supporting structure, *Sec. 2, 217A2;* 9, 62

Reclosers, *Sec. 17;* 54
- Application, *171;* 54
- Arrangement, *170;* 54
- Capacity, *173A;* 54
- Oil-containing, *172;* 54
- Provisions for disconnecting, *173B;* 54
- Visible break switch, *173C;* 54

Re-energizing after work, *442D, 444H;* 234, 238

Regulators
- Indoor installations, *152B;* 51
- Outdoor installations, *152A;* 51

Reinforced concrete structures, *Table 253-2, 261A1, 261C2, Table 261-1;* 172, 173, 175, 180

Reinforcing bars as grounding electrodes, *94A3, 95A3;* 21, 23

Relay logic devices, *180E1;* 56

Remotely operable (as applied to equipment), definition of, *Sec. 2;* 9

Resistance requirements, grounding systems, *96;* 24
- Multi-grounded systems, *96C;* 24
- Single-grounded systems, *96B;* 24
- Supply stations, *96A;* 24

Ridge-pin conductors, *236I;* 144

Risers, *Sec. 36;* 208
- Installation, *361;* 208
- Pad-mounted installations, *363;* 208
- Pole risers, *362;* 208

Roadways
- Clearances from
  - Overhead, *231B, 232, Table 232-1, Table 232-2;* 71, 74, 81
  - Underground, *320A3, 351C4;* 192, 204
- Definition of, *Sec. 2;* 9
- Shoulder, definition of, *Sec. 2;* 11
- Traveled way, definition of, *Sec. 2;* 12

Rods as grounding electrodes, *94B2;* 22

Rotating equipment, *Sec. 13;* 49
- Motor control, *131;* 49
- Short-circuit protection, *133;* 49
- Speed control and stopping devices, *130;* 49

Rules, National Electrical Safety Code, *012;* 1

Rural districts, definition of, *Sec. 2;* 9

S

Safety, employee. *See* Operation of electric supply systems, employee rules

Safety, employer. *See* Operation of electric supply systems, employer rules

Safety clearance zone, electric supply stations, *110A2, Fig 110-1;* 28, 29

Safety zone, communication worker, *235C4, 238E;* 133, 147

Sag
Apparent, at any point in a span (definition), *Sec. 2;* 10
Apparent, of a span (definition), *Sec. 2;* 10
Clearance considerations
Conductors of different sag on same support, *235C2b;* 127
Horizontal clearance, *235B1b, Table 235-2, Table 235-3;* 122, 124, 125
Conductor, definition of, *Sec. 2;* 10
Definition of, *Sec. 2;* 9
Final, definition of, *Sec. 2;* 10
Final unloaded, definition of, *Sec. 2;* 10
Initial unloaded, definition of, *Sec. 2;* 9
Total, definition of, *Sec. 2;* 10

Scope of National Electrical Safety Code, *011;* 1

Separation, definition of, *Sec. 2;* 10

Separation, random
Definition of, *Sec. 2;* 9
Direct buried cable, *354;* 205

Separation of grounding conductors, *97;* 25

Service. *See* In service; Out of service

Service drop, definition of, *Sec. 2;* 10

Service point, definition of, *Sec. 2;* 10

Settings, strength of, *261B;* 175

Sewers, separation from underground lines, *320B3, 354E;* 193, 207

Sheath, cable, definition of, *Sec. 2;* 4

Sheath transposition connections (cross-bonding), *93E6;* 21

Shielded conductors, *92B2b;* 16

Shielding, conductor, definition of, *Sec. 2;* 5

Shielding, insulation, definition of, *Sec. 2;* 7

Shields, *124C2;* 35

Shield wire/conductor, definition of, *Sec. 2;* 10

Short-circuit protection
Motors, *133;* 49
Power transformers, *153;* 51

Short-time ampacity, grounding conductor, *93C;* 19

Shoulder, definition of, *Sec. 2;* 11

SI (International System of Units), *17A;* 3

Side-wall pressure, definition of, *Sec. 2;* 11

Signs, *217A4, Table 234-5, 411D;* 63, 121, 215

Single-duct conduit, definition of, *Sec. 2;* 5

Single-grounded system, definition of, *Sec. 2;* 11

Snaphooks, positioning strap, *420K;* 217

Solid-state logic devices, *180E1;* 56

Soot blower control devices, *180E1;* 56

Spacer cable, definition of, *Sec. 2;* 11

Span
Length, definition of, *Sec. 2;* 11
Sag, definitions of, *Sec. 2;* 10
Wire
Clearances from communication lines, *238C, Table 238-2;* 147, 148
Definition of, *Sec. 2;* 11
Insulators, *279B;* 187

Speed control and stopping devices, *130;* 49
Adjustable-speed motors, *130E;* 49
Automatic overspeed trip device for prime movers, *130A;* 49
Manual stopping devices, *130B;* 49
Protection of control circuits, *130F;* 49
Speed limit for motors, *130C;* 49

Stairs and steps
Supply stations, *112;* 32
Supporting structures, *217A2, 239B;* 62, 149

Standoff brackets, *217A2c, 239J;* 63, 152

Static wire, definition of, *Sec. 2;* 10

Station, electric supply, definition of, *Sec. 2;* 6

Station, generating, definition of, *Sec. 2;* 6

Steam lines, separation from underground conduit, *320B6;* 193

Steel Wire Gage (Stl WG), *Sec. 2;* 13

Steps. *See* Stairs and steps

Stopping devices. *See* Speed control and stopping devices

Storage batteries, *Sec. 14;* 50
- Employee precautions, *420G;* 217
- Floors in battery areas, *144;* 50
- Illumination for battery areas, *145;* 50
- Location, *141;* 50
- Racks, *143;* 50
- Service facilities, *146;* 50
- Ventilation, *142;* 50

Strap (positioning), *411B, 420K;* 215, 217
- Definition of, *Sec. 2;* 9

Streets, clearances from, *231B;* 71
- Underground, *320A3, 351C4;* 192, 204

Strength requirements, *Sec. 26;* 173
- Grades B and C construction, *261;* 173
  - Armless, *261G;* 177
  - Braces, *261D;* 176
  - Communication cables, *261K;* 178
  - Communication conductors, open-wire, *261J;* 178
  - Communication conductors, paired, *261L;* 178
  - Crossarms, *261D;* 176
  - Fastenings, *261F;* 177
  - Foundations, *261B;* 175
  - Guy anchors, *261B;* 175
  - Guys and guy insulators, *261C;* 175
  - Open supply conductors, *261H;* 177
  - Overhead shield wires, *261H;* 177
  - Pins, *261F;* 177
  - Settings, *261B;* 175
  - Supply cable messengers, *261I;* 178
  - Support hardware, *261M;* 179
  - Supporting structures, *261A;* 173
- Grade N construction, *263;* 181
  - Communication conductors, *263G;* 182
  - Crossarm strength, *263C;* 181
  - Guys, *263B;* 181
  - Insulators, *263I;* 182
  - Poles, *263A;* 181
  - Service drops, *263E, Table 263-2;* 181, 183
  - Street and area lighting equipment, *263H;* 182
  - Supply-line conductors, *263D, Table 263-1;* 181, 182
  - Trolley-contact conductors, *263F;* 182
- Guying and bracing, *264;* 183
  - Anchor rods, *264G;* 184
  - Electrolysis, *264F;* 184
  - Fastenings, *264D;* 183
  - Markers and Protection, *264E;* 184
  - Point of attachment, *264C;* 183
  - Strength, *264B;* 183
  - Where used, *264A;* 183

Strips, buried, as grounding electrodes, *94B3b;* 22

Structure conflict, definition of, *Sec. 2;* 11

Structures, supporting. *See* Supporting structures

Submarine crossings, *320A6, 351C5;* 192, 204

Substation, definition of, *Sec. 2;* 6

Suddenly moving parts, guarding, *122B;* 34

Sulfur dust, *127I;* 48

Supply equipment, electric, definition of, *Sec. 2;* 6

Supply lines, electric. *See* Lines, electric supply

Supply stations, electric
- Definition of, *Sec. 2;* 6
- Generating station, definition of, *Sec. 2;* 6
- Ground resistance requirements, *96A;* 24
- Protective arrangements, *Sec. 11;* 28
  - Electric equipment, *110C;* 30
  - Enclosures, *110A;* 28
  - Exits, *113;* 33
  - Fire-extinguishing equipment, *114;* 33
  - Floors, *112A;* 32
  - Illumination, *111;* 30
  - Passageways, *112B;* 32
  - Railings, *112C;* 33
  - Rooms and spaces, *110B;* 28
  - Stair guards, *112D;* 33
  - Top rails, *112E;* 33
- Purpose of code, *100;* 27
- Scope of code, *101;* 27
- Substation, definition of, *Sec. 2;* 6

Support arms, *232B, 243B;* 72, 160

Support hardware, *253, 261M;* 170, 179

Supporting structures
- Armless construction brackets, *243C;* 160
- Clearances from other objects, *231;* 71
  - Fire hydrants, *231A;* 71
  - Railroad tracks, *231C;* 71

Streets, roads, highways, *231B;* 71
Climbing. *See* Climbing space
Conductor fastenings, *243C, 252, 261F;* 160, 169, 177
Crossarms, *243B, 253, Table 253-1;* 160, 170, 171
Decorative lighting, *217A5;* 63
Definition of, *Sec. 2;* 11
Different, clearances between conductors carried on, *233, Table 233-1;* 85, 91
Grades of construction, *243A;* 160
Grounding of, *215C1;* 61
Identification, *217A3, 220E;* 63, 65
Insulators, *243C;* 160
Loads upon, *252;* 169
Longitudinal, assumed, *252C;* 169
Simultaneous application of, *252D;* 170
Transverse, assumed, *252B;* 169
Vertical, assumed, *252A;* 169
Metal, *261A, 261C1, Table 261-1;* 173, 175, 180
Obstructions, *217A4;* 63
Pins, *243C;* 160
Prestressed concrete, *Sec. 2, 261A1, 261C1, Table 261-1;* 9, 173, 175, 180
Protection of, *217A1;* 62
Readily climbable, *Sec. 2, 217A2;* 9, 62
Reinforced concrete, *Table 253-2, 261A1, 261C2, Table 261-1;* 172, 173, 175, 180
Standoff brackets, *217A2c;* 63
Steps, *217A2;* 62
Strength requirements. *See* Strength requirements
Support arms, *243B;* 160
Support hardware, *253, 261M;* 170, 179
Unusual, *217B;* 63
Wood, *261A2, 261C2, Table 261-2;* 174, 175, 181
Working space. *See* Working space

Surge arresters, *Sec. 19;* 58
Grounding conductors, *192;* 58
Ampacity, *93C4;* 19
Indoor locations, *191;* 58
Installation, *193;* 58

Susceptiveness, definition of, *Sec. 2;* 11

Suspension insulators, *235B2, 235E2;* 126, 138

Swimming pools, *234, Fig 234-3, Table 234-3, 351C1;* 96, 109, 113, 203

Switchboards
Control, *180E;* 56
Dead-front, *180C;* 56
Definition of, *Sec. 2;* 11

Switches
Application, *171;* 54
Arrangement, *170, 216;* 54, 62
Capacity, *173A;* 54
Definition of, *Sec. 2;* 11
Disconnecting, *Sec. 17;* 54
Definition of, *Sec. 2;* 5
Isolating, definition of, *Sec. 2;* 5
Oil-containing, *172;* 54
Protecting employees by, *443E;* 236
Visible-break, *173C;* 54

Switchgear assemblies
Control switchboards, *180E;* 56
Dead-front power switchboards, *180D;* 56
De-energizing, *443G;* 236
General requirements, *180A;* 55
Metal-enclosed power switchgear, *180B;* 55
Motor control centers, *180D;* 56

System, conduit, definition of, *Sec. 2;* 5

System, grounded, definition of, *Sec. 2;* 7

Systeme International d'Units (SI), *017A;* 3

T

Tag, definition of, *Sec. 2;* 11

Tanks, conductor clearance from, *234C, Table 234-1, Table 234-5;* 99, 102, 121

Telemetering devices, *180E1;* 56

Temporary overhead installations, *014B;* 2

Tension, pulling, definition of, *Sec. 2;* 9

Tension, unloaded
Final, definition of, *Sec. 2;* 12
Initial, definition of, *Sec. 2;* 12

Terminal, cable, definition of, *Sec. 2;* 4

Terminations, conductor, *164;* 53

Terminations, underground supply cable, *Sec. 37;* 209
Clearances in enclosures or vaults, *373;* 209
Grounding, *374;* 209
Identification, *372;* 209
Support, *371;* 209

Terms, definitions of, *Sec. 2;* 4

Tests
For gas in manholes and unventilated vaults, *423B;* 220
Insulation, *274;* 185
Lines and equipment
When in service, *214A, 313A;* 60, 190
When out of service, *214B, 313B;* 60, 190
Protective devices, *411C;* 215

Third rails, *225C;* 67

Totalizing microwave radios, *180E1;* 56

Total sag, definition of, *Sec. 2;* 10

Total sag, maximum, definition of, *Sec. 2;* 10

Towers. *See* Supporting structures

Traffic signals, *238, Table 238-2, 239G;* 147, 148, 150

Training, *446A;* 239

Transferring (fall protection), definition of, *Sec. 2;* 11

Transformers, *Sec. 15;* 51
Current-transformer secondary circuits protection when exceeding 600 V, *150;* 51
Grounding secondary circuits of instrument transformers, *151;* 51
Instrument, *93C3, 151;* 19, 51
Location and arrangement of power transformers and regulators, *152;* 51
Indoor installations, *152B;* 51
Outdoor installations, *152A;* 51
Short-circuit protection, *153;* 51

Transformer vault, definition of, *Sec. 2;* 11

Transitioning (fall protection), definition of, *Sec. 2;* 11

Transverse loading, assumed, *252B;* 169
At angles, *252B3;* 169
Conductors, *252B1;* 169
Messengers, *252B1;* 169
Span lengths, *252B4;* 169
Structure, *252B2, 261A3;* 169, 175

Traveled way, definition of, *Sec. 2;* 12

Tree trimming, *218;* 63

Trenching, *352A;* 204

Trolley-contact conductors, *225A, Table 233-1, 234A3, 234D2, 238C, 239G, 241C3b, 242B, 243, 261H3;* 64, 91, 96, 109, 147, 150, 155, 156, 160, 178

Tunnels, *320A4, Sec. 39;* 192, 212

U

Underground communication lines, *Part 3;* 189
Accessibility, *312;* 190
Application of rules, *302;* 189
Direct buried cable, *Sec. 35;* 202
Deliberate separation, *353;* 205
Installation, *352;* 204
Random separation, *354;* 205
Routing, *351;* 202
Symbols for identification of, *Fig 350-1;* 203
Equipment, *Sec. 39;* 210
Design, *381;* 210
Grounding, *384;* 211
Identification, *385;* 211
Installation, *383;* 211
Location in underground structures, *382;* 211
General requirements, *Sec. 31;* 190
Grounding, *314;* 190
Circuits, *314C;* 191
Conductive parts to be grounded, *314B;* 190
Methods, *314A;* 190
Induced voltage, *316;* 191
Inspection
When in service, *313A;* 190
When out of service, *313B;* 190
Installation, *311;* 190
Maintenance, *311;* 190
Operating procedures, *423;* 220
Protective requirements, *315;* 191
Purpose of code, *300;* 189
Risers, *Sec. 36;* 208
Installation, *361;* 208
Pad-mounted, *363;* 208
Pole, *362;* 208
Scope of code, *301;* 189
Supply cable
Cable accessories and joints, *333;* 198
Sheaths and jackets, *331;* 198
Shielding, *332;* 198
Terminations, *Sec. 37;* 209
Tunnels, installation in, *Sec. 39;* 212
Environment, *391;* 212
Underground structures, cable in, *Sec. 34;* 199
Bonding, *342;* 200
Communication cables containing special supply circuits, *344;* 200
Fireproofing, *343;* 200
Grounding, *342;* 200
Installation, *341;* 199
In manholes and vaults, *341B;* 199

Underground electric supply lines, *Part 3;* 189
Accessibility, *312;* 190
Application of rules, *302;* 189
Direct buried cable, *Sec. 35;* 202
Deliberate separation, *353;* 205
Installation, *352;* 204
Random separation, *354;* 205
Routing, *351;* 202
Symbols for identification of, *Fig 350-1;* 203
Equipment, *Sec. 38;* 210
Design, *381;* 210
Grounding, *384;* 211
Identification, *385;* 211
Installation, *383;* 211
Location in underground structures, *382;* 211
General requirements, *Sec. 31;* 190
Grounding, *314;* 190
Circuits, *314C;* 190
Conductive parts to be grounded, *314B;* 190
Methods, *314A;* 190
Induced voltage, *316;* 191
Inspection
When in service, *313A;* 190
When out of service, *313B;* 190
Installation, *311;* 190
Maintenance, *311;* 190
Operating procedures, *423;* 220
Purpose of code, *300;* 189
Risers, *Sec. 36;* 208
Installation, *361;* 208
Pad-mounted, *363;* 208
Pole, *362;* 208
Scope of code, *301;* 189
Supply cable
Cable accessories and joints, *333;* 198
Sheaths and jackets, *331;* 198
Shielding, *332;* 198
Terminations, *Sec. 37;* 209
Tunnels, installation in, *Sec. 39;* 212
Environment, *391;* 212
Underground structures, cable in, *Sec. 34;* 199
Bonding, *342;* 200
Communication cables containing special supply circuits, *344;* 200
Fireproofing, *343;* 200
Grounding, *342;* 200
Installation, *341;* 199
in manholes and vaults, *341B;* 199

Underground grounding conductors, *93E;* 20

Ungrounded circuits, voltage of (definitions), *Sec. 2;* 13

Ungrounded system, definition of, *Sec. 2;* 12

Uniform system of clearances (adopted in 1990 NESC edition), *Appendix A;* 241

Unigrounded system, definition of, *Sec. 2;* 11

Units of measure, *017;* 3

Unloaded sag
Final, definition of, *Sec. 2;* 10
Initial, definition of, *Sec. 2;* 9

Unloaded tension
Final, definition of, *Sec. 2;* 12
Initial, definition of, *Sec. 2;* 12

Unloading facilities, flammable and combustible liquids, *127D;* 47

Urban districts, definition of, *Sec. 2;* 12

Use, common, definition of, *Sec. 2;* 4

Utility, definition of, *Sec. 2;* 12

Utility interactive system, definition of, *Sec. 2;* 12

Utilization equipment, definition of, *Sec. 2;* 12

V

Vaults. *See* Manholes, handholes, and vaults

Ventilation, battery area, *142;* 50

Vertical clearance, *232, 233C, 234, 235C, 238;* 71, 89, 96, 126, 147
From live parts, *Fig 124-1, Table 124-1;* 36, 37

Vertical loading component, *251B1, 252A;* 168, 169

Vertical racks, *235G;* 140

Visible break switches, *173C;* 54

Voltage
Of circuit not effectively grounded, definition of, *Sec. 2;* 12
Of constant-current circuit, definition of, *Sec. 2;* 12
Definition of, *Sec. 2;* 12
Of effectively grounded circuit, definition of, *Sec. 2;* 12
To ground of conductor of grounded circuit, definition of, *Sec. 2;* 13
To ground of conductor of ungrounded circuit, definition of, *Sec. 2;* 13
To ground of grounded circuit, definition of, *Sec. 2;* 13

To ground of ungrounded circuit, definition of, *Sec. 2;* 13

W

Waiver of National Electrical Safety Code rules, *014;* 2

Water, clearance from, *232, Table 232-1, Table 232-2;* 71, 74, 81
Beaches, *234E2;* 109
Swimming pools, *234, Fig 234-3, Table 234-3, 351C1;* 96, 109, 113, 203
Water lines, separation from underground lines, *320B4, 354E;* 193, 207

Water piping systems as grounding electrodes, *94A, 95B;* 21, 24

Way, traveled, definition of, *Sec. 2;* 12

Weather loading. *See* Loading, line

Wind. *See* Loading, line

Wire, buried, as grounding electrode, *94B3a;* 22

Wire, span, definition of, *Sec. 2;* 11

Wire gages, definition of, *Sec. 2;* 13

Wire wraps as grounding electrodes, *94B4;* 22

Wood structures, *Table 253-2, 261A2, Table 261-1, Table 261-2;* 172, 174, 180, 181

Wording of NESC rules, *015;* 2

Working space
Buckarms, location relative to, *237D, Fig 237-1;* 146
Dimensions, *237B;* 145
Energized equipment
Guarding of, *237E;* 146
Working clearances from, *237F;* 146
Equipment over 600 V, *125B,* 45
Equipment 600 V or less, *125A;* 45
Location, *237A;* 145
Vertical and lateral conductors, location relative to, *237C, 239B;* 146, 149

Work rules. *See* Operation of electric supply systems, employee rules; Operation of electric supply systems, employer rules

Worksite (fall protection), definition of, *Sec. 2;* 13

THIS PAGE IS INTENTIONALLY BLANK

Procedure for Revising the National Electrical Safety Code

1. Preparation of Proposals for Amendment

NOTE: The procedures for the collection of Change Proposals are subject to change as the revision of the 2007 Edition of the NESC approaches. Logon to the NESC Zone at http://standards.ieee.org/nesc/ for updates.

1.1 A proposal may be prepared by any
- Substantially interested person
- Interested organization
- NESC Subcommittee
- Member of the NESC Committee or its subcommittees

1.2 Proposals shall be submitted to: Secretary, National Electrical Safety Code Committee (at the address listed on the following Change Proposal Form on p. 261). Copies must be suitable for reproduction and shall be on the forms for new proposals available on p. 261. A diskette containing the change proposal text in any electronic form should be submitted in addition to the hard copy.

1.3 Each separate topic shall begin on a separate form, printed or typed on one side only. If a proposal references documents not readily available to all subcommittee members, sufficient copies of the referenced documents to supply the subcommittee must be furnished.

1.4 The proposal shall consist of

a. a statement, in NESC rule form, of the exact change, rewording, or new material proposed. Words to be deleted shall be stricken through and words to be added shall be underlined. (See example on p. 260.)

b. the name of the submitter (organization or individual as applicable).

c. supporting comment, giving the reasons why the NESC should be so revised.

2. The NESC Secretary will

(a) Acknowledge receipt of proposals for revision. (If the submitter does not receive an acknowledgment within 30 days of mailing his or her proposal, the Secretary should be contacted.)

(b) Distribute to each member of the appropriate NESC Subcommittee all of the proposals received, arranged in a coordinated sequence.

3. Subcommittee Recommendation

The NESC Subcommittee responsible will consider each proposal and take one or more of the following steps:

(a) Endorse the proposal as received.

(b) Prepare a proposed revision or addition for the NESC (this may be a coordination of several comments, or a committee consensus on a modification of a proposal).

(c) Refer the proposal to a technical working group for detailed consideration.

(d) Request coordination with other NESC Subcommittees.

(e) Recommend rejection of the proposal, for stated reasons.

For each item, the responsible subcommittee shall prepare a voting statement, accompanied by all members' statements concerning their votes (cogent reasons are required for negative votes). Steps (c) and (d) are intended to result, eventually, in a proposal of category (b).

Action under steps (c) or (d) shall be completed and reported to the subcommittee before the end of the public review period if the item is to be included in the upcoming revision.

4. Preprint of Proposals

The NESC Secretary shall organize and publish a preprint of the proposed revisions including

(a) The original proposal as received from the submitter.

(b) The recommendation of the subcommittee with respect to the proposal (including a voting statement and subcommittee members' statements).

(c) Copies of submittal form for comments.

The preprint shall be distributed to all members of NESC Subcommittees and representatives of organizations comprising the NESC Committee. Copies shall be available for sale to other interested parties. Notice of availability of the preprint shall be submitted to ANSI for publication in *ANSI Standards Action*. The preprint shall carry information on how to submit comments on the proposals and the final date for such submissions.

5. Final Processing of Proposed Revisions and Comments

5.1 Following the public review period, the Secretary shall organize and distribute for subcommittee consideration all comments received.

5.2 The preprint and the comments received shall be reconsidered by the subcommittees. No new change proposals may be considered.

(a) The subcommittee may recommend adoption or rejection of the proposal by majority vote.

(b) When extended technical consideration or resolution of differing or conflicting points of view is necessary, the subcommittee shall refer the problem to a working group of the subcommittee for proposed resolution. If expeditious resolution is not possible, the subject shall be held on the docket.

Each working group shall provide, to its parent subcommittee, recommendations on matters considered as a result of subcommittee referrals under items 3(c) and 5.2(b).

Each subcommittee shall prepare a report showing its proposed revisions and all items held on the docket together with a plan for their disposition.

5.3 The Secretary shall provide commentors with copies of actions taken on the rules affected by their comments, and shall make all such reports available for examination upon request.

6. Final Approval

6.1 Based upon the subcommittee reports, the Secretary shall prepare a draft of the revision of the NESC and distribute copies to

(a) The NESC Committee for approval by a six-week letter ballot.

(b) The American National Standards Institute Board of Standards Review for concurrent 60-day public review.

6.2 Comments received in response to the letter ballot and public review shall be referred to the Executive Subcommittee for resolution or referral to the appropriate subcommittee. Those items on which consensus cannot be reached shall be referred to the appropriate subcommittee for consideration during the next revision cycle. Unless a consensus for revision is established, the requirements of the current edition shall carry over to the proposed edition.

Request for Change
National Electrical Safety Code

Rule 233A1b(5), revise to read:

(5) Where crossing is not at midspan of the upper conductor ~~and under conditions where the upper span exceeds those specified in the Rule 233A1b(2)~~, the additional sag determined in accordance with Rules 233A1b(3) and 233A1b(4) may be reduced by multiplying ~~the additional sag determined by Rules 233A1b(3) and 233A1b(4)~~ by the following factors.

Supporting Comment: It is believed that the suggested revision clarifies the intent of the Rule. Additionally, the revision properly addresses Rule 233A1b(4), which applies to upper conductors designed to operate at a conductor temperature above 120 °F (50 °C), *regardless of span length* [based on companion proposal for Rule 233A1b(4)].

FORM FOR CHANGE PROPOSALS ON THE NATIONAL ELECTRICAL SAFETY CODE

(A separate form must be used for each Change Proposal)

Name ______________________________ Date ____________

Address ______________________________

Organization Represented______________________________

1. Rule: ____________

2. Proposal: Include proposed exact wording, or identification of wording to be deleted. (Underline added words, strike through deleted words)

3. Statement of Problem and Supporting Comments

Signature ____________________

Mail to: Secretary-NESC Committee, IEEE Standards Department
445 Hoes Lane, P.O. Box 1331, Piscataway, NJ 08855-1331

This form may be duplicated. A diskette containing the change proposal text in any electronic form should be submitted, in addition to the hard copy.

Time Schedule for the Next Revision of the National Electrical Safety Code

17 July 2003	Final date for receipt of proposals from the public for revision of the 2002 Edition of the National Electrical Safety Code, preparatory to the publication of a 2007 Edition. Proposals should be forwarded in the prescribed form to: Secretary National Electrical Safety Code IEEE Standards Department 445 Hoes Lane P.O. Box 1331 Piscataway, NJ 08855-1331
5–24 October 2003	NESC Subcommittees consider proposals for changes to the NESC and prepare their recommendations.
1 September 2004	Preprint of Proposed Amendments for incorporation into the 2007 Edition of the NESC published for distribution to the NESC Committee and other interested parties.
1 May 2005	Period for study of proposed amendments and submittal by interested parties of recommendations concerning the proposed amendments. Submit recommendations to the Secretary, NESC Committee, at the above address.
2–20 October 2005	Period for NESC Subcommittee Working Groups and NESC Subcommittees to reconsider all recommendations concerning the proposed amendments and prepare final report.
15 January 2006	Proposed revision of the NESC, Accredited Standards Committee C2, submitted to NESC Committee for letter ballot and to the American National Standards Institute for concurrent public review.
15 May 2006	NESC Committee approved revisions of the NESC submitted to the American National Standards Institute for recognition as an ANSI standard.
1 August 2006	Publication of the 2007 Edition of the National Electrical Safety Code.

Working Group Assignments

In response to comments received on the *NESC Preprint 2002 Proposals*, and proposals for change to the 1997 edition of the Code, members of the technical subcommittees recommended new assignments for existing or new working groups. All of these recommendations and subsequent actions are summarized below.

These working groups are preparing recommendations for the next edition of the NESC. NESC Subcommittee membership is not required for working group participation. For more information or to participate on any working group(s), please contact the NESC Secretary.

Subcommittee 1, Coordination (SC1)

Working Group 1.10, Definition and Use of Grounding Terms. A Working Group (WG) of Subcommittee 1 was formed to clarify what is meant by the terms "grounded," "effectively grounded," etc., as relating to neutral conductors. See CP 2182 on page 63 of the NESC Preprint 2002 Proposals.
Chair: Allen L. Clapp

Subcommittee 2, Grounding Methods (SC2)

Working Group 2.2, Steel Poles as Made Electrodes. A WG was formed to investigate the feasibility of accepting steel poles as a man-made electrode in response to CP 2284, Comment 3247 (Rule 94B).
Chair: John Dagenhart

Subcommittee 3, Electric Supply Stations (SC3)

None.

Subcommittee 4, Clearances-Overhead Lines (SC4)

Working Group 4.8, Aerial Facilities Clearances. Subject existing aerial clearance rules between communications and supply facilities to a comprehensive review that will mirror the significant changes in business and technology over the past 1/2 century, with the objective of relieving pole space congestion. The WG will concentrate on definitions and Sections 21, 22, and 23, but no work will be done on clearances over terrain or from structures. Section 9, Grounding, will be addressed only if required by changes made elsewhere. Parts 1, 3, and 4 will not be examined. During the proposal portion of the 2002 Code cycle, SC4 voted to extend the scope of the WG to include the responsibility of evaluating all new communication technology for any impact that it may have on the responsibilities of SC4.
Chair: Chuck Amrhyn

Working Group 4.9, Wire, Conductor and Cable Terminology. Subcommittee 4 created a working group to address the use of terms line wires, conductors, and cables. This working group was formed as a result of CP 2147.
Chair: Stephen Olinick

Working Group 4.10, New Ice Loads and Clearances. Subcommittee 4 created a working group to deal with pending changes in the ice loading map that will affect rules contained in Section 23. See CP 2209 and CP 2372.
Chair: Larry Slavin

Subcommittee 5, Overhead Lines-Strength and Loadings (SC5)

Working Group 5.1, Continuity of Sections 24, 25, and 26. This working group is examining the appropriate placement of requirements within Sections 24, 25, and 26 of the present code system and developing a basis for coordinating the relative values of present and future requirements for loading factors, material strength factors, importance factors, and similar requirements that will assure appropriate results from the present or from a possible future calculation system.
Chair: L. Kempner
This working group also contains the following Task Forces:
TF 5.1.1, Load Factors. Chair: R. Kluge
TF 5.1.4, Old/Alternate vs Recommended Method. Chair: L. Kempner
TF 5.1.7, Fiber Reinforced Composite Structures. Chair: N. Bingel
TF 5.1.8, Grade B vs Grade C Reliability. Chair: N. Bingel

Working Group 5.2, Complete Revisions of Sections 25 and 26. The scope of this work is to continue to review CP 2372, which deals with a major revision of Sections 24–27 (the "new method").
Chair: Robert C. Peters
This working group also contains the following task forces:
TF 5.2.1, Ice Loading. Chair: Clayton Clem

Working Group 5.6, Line Insulation. Continuing review of overhead line insulation component performance, analysis of insulator problems due to component age and other factors, along with the recommendations for appropriate recognition of newly developing synthetic insulating components and the appropriate methods to test for normal and extreme loading conditions.
Chair: Thomas A. Pinkham.
This working group contains the following task forces:
TF 5.6.1, Coordinate changes and improvements. Chair: T. A. Pinkham
TF 5.6.2, Test Methods/Extreme Loading. T. A. Pinkham

Working Group 5.7, Seminars and Presentations. A working group was set up to develop a Panel Session to explain the changes in the NESC 2002 Edition to industry and the public. Panel Sessions are currently scheduled at the Summer Power Engineering Society meeting in July, 2001, and at the Winter Transmission and Distribution Expo in October 2001. Other public presentations are provided, as appropriate.
Chair: Lawrence Slavin
TF 5.7.1, Technical Presentations, 2002 Edition. Chair: Lawrence Slavin
TF 5.7.2, Field Surveys on Relative Impact of Proposed New and Old System, (Sections 23 and 24–27). Chair: Lawrence Slavin

Working Group 5.8, Application of Extreme Wind to All Structures. A working group was set up to develop recommendations for appropriate differences in wind treatment or other limitations for lower profile structures and lines vs higher structures and lines for the 2007 Edition. See CP 2151 regarding evaluation of the 60-ft exclusion in the application of high wind loading to overhead lines.
Chair: Donald Heald

Subcommittee 7, Underground Lines (SC7)

Working Group 7.5, Manholes and Vaults. SC7 created a working group to review definitions and usage of the terms "manholes" and "vaults."
Chair: Les Shankland

Subcommittee 8, Work Rules (SC8)

Working Group 8.10, Worker Clothing. WG 8.10 was formed to address CPs 2122, 2357, 2358, and 2343. WG will develop a report and change proposal(s) for the 2007 Edition.
Chair: Jim Tomaseski

Tentative Interim Amendment 2002-1 to the National Electrical Safety Code ANSI C2-2002

9 May 2001

In accordance with Section 13 of its Procedures, the National Electrical Safety Code Committee has issued the following Tentative Interim Amendment (TIA) to ANSI C2, National Electrical Safety Code, 2002 Edition. The TIA was issued by the Secretariat on 9 May 2001, as a result of a proposal submitted by a member of the NESC Main Committee.

A Tentative Interim Amendment is tentative because it has not been processed through the entire standards-making procedure. It is interim because it is effective only between editions of the code. A TIA automatically becomes a Proposal of the proponent for the next edition of the code; as such, it is then subject to all the procedures of the standards-making process.

Rule 123C: Delete the last sentence and add the following underlined sentence.

c. Provisions for Grounding Equipment During Maintenance.
Electric equipment or conductors normally operating at more than 600 V between conductors, on or about which work is occasionally done while isolated from a source of electric energy by disconnecting or isolating switches only, shall be provided with some means for grounding, such as switches, connectors, or a readily accessible means for connecting a portable grounding conductor. ~~When necessary, grounding may be omitted on conductors normally operating at 25 kV or less and not influenced by higher voltage conductors, where visible openings in the source of supply are available, and are properly tagged in the open position.~~ See Part 4 of this code.

Rule 173C: Delete entire rule.

Tentative Interim Amendment 2002-2
to the
National Electrical Safety Code
ANSI C2-2002

23 April 2001

In accordance with Section 13 of its Procedures, the National Electrical Safety Code Committee has issued the following Tentative Interim Amendment (TIA) to ANSI C2, National Electrical Safety Code, 2002 Edition. The TIA was issued by the Secretariat on 23 April 2001, as a result of a proposal submitted by a member of the NESC Main Committee.

A Tentative Interim Amendment is tentative because it has not been processed through the entire standards-making procedure. It is interim because it is effective only between editions of the code. A TIA automatically becomes a Proposal of the proponent for the next edition of the code; as such, it is then subject to all the procedures of the standards-making process.

Table 232-2 m and ft are revised as follows:

Replace footnote 1 in row 1b, column 2 with unused footnote 6 in both the meter and feet Tables 232-2 thusly:

[6]~~This footnote not used in this edition.~~ This clearance may be reduced to the following values for driveways, parking lots, and alleys not subject to truck traffic:

		meters	feet
(a)	Insulated live parts limited to 300 V to ground	3.6	12
(b)	Insulated live parts limited to 150 V to ground	3.0	10

Also replace footnote 1(b) in row 1d, column 2 with footnote 1 and revise footnote 1 thusly:

[1]~~This clearance may be reduced to the following values:~~

		~~meters~~	~~feet~~
~~(a)~~	~~Insulated live parts limited to 300 V to ground~~	~~3.6~~	~~12~~
~~(b)~~	~~Insulated live parts limited to 150 V to ground~~	~~3.0~~	~~10~~

[1]For insulated live parts limited to 150 V to ground, this value may be reduced to 3.0 m (10 ft).

Since the footnote only references 1(b) making this revision eliminates the need for the present footnote 1(a) and 1(b).